SUPERMASSIVE BLACK HOLES

Written by an international leader in the field, this is a coherent and accessible account of the concepts that are now vital for understanding cutting-edge work on supermassive black holes. These include accretion disc misalignment, disc breaking and tearing, chaotic accretion, the merging of binary supermassive holes, the demographics of supermassive black holes, and the defining effects of feedback on their host galaxies. The treatment is largely analytic and gives in-depth discussions of the underlying physics, including gas dynamics, ideal and non-ideal magnetohydrodynamics, force-free electrodynamics, accretion disc physics, and the properties of the Kerr metric. It stresses aspects where conventional assumptions may be inappropriate and encourages the reader to think critically about current models. This volume will be useful for graduate or master's courses in astrophysics, and as a handbook for active researchers in the field. eBook formats include colour figures while print formats are greyscale only.

ANDREW KING is a professor at the University of Leicester and a visiting professor at Leiden Observatory, and is Long-Term Visitor at the University of Amsterdam. His academic awards include a PPARC Senior Fellowship; Gauss Professor, Göttingen; Professeur Invité, Université Paris VII; a Royal Society Wolfson Merit Award; and the Eddington Medal of the Royal Astronomical Society. He is a co-author of *Accretion Power in Astrophysics* (1985, 1992, 2002) and *Astrophysical Flows* (2007), and author of *Stars: A Very Short Introduction* (2012).

SUPERMASSIVE BLACK HOLES

ANDREW KING

University of Leicester

CAMBRIDGE
UNIVERSITY PRESS

Shaftesbury Road, Cambridge CB2 8EA, United Kingdom

One Liberty Plaza, 20th Floor, New York, NY 10006, USA

477 Williamstown Road, Port Melbourne, VIC 3207, Australia

314–321, 3rd Floor, Plot 3, Splendor Forum, Jasola District Centre, New Delhi – 110025, India

103 Penang Road, #05–06/07, Visioncrest Commercial, Singapore 238467

Cambridge University Press is part of Cambridge University Press & Assessment, a department of the University of Cambridge.

We share the University's mission to contribute to society through the pursuit of education, learning and research at the highest international levels of excellence.

www.cambridge.org
Information on this title: www.cambridge.org/9781108488051
DOI: 10.1017/9781108768849

First published 2023

A catalogue record for this publication is available from the British Library.

ISBN 978-1-108-48805-1 Hardback

Faced with the choice between changing one's
mind and proving there is no need to do so,
almost everyone gets busy on the proof.

J. K. Galbraith

Contents

Preface

More than a century after Einstein formulated general relativity (GR), black holes are firmly established as one of its most striking and inescapable consequences. The perceived complexity of the theory meant that this realization itself arrived only after half a century and several missed opportunities. But the mathematics leaves no room for doubt – GR describes all non-quantum properties of black holes in the form of the Kerr (1963) family of exact solutions. General relativity itself has survived unscathed a very large number of observational tests with exquisite precision, notably involving stellar-mass pulsar binary systems, the detailed orbital dynamics of stars orbiting the black hole at the Galactic Centre, and LIGO–Virgo observations of gravitational waves from black hole mergers.

A vast body of observations now shows that black holes are not simply a theoretical possibility, but have central importance in the real Universe. Gas infall – accretion – on to a black hole is the most efficient way of getting energy from ordinary matter. Only matter–antimatter annihilation is more efficient, but probably impossible to realize on scales larger than subnuclear. It follows that black hole accretion must power the most luminous astrophysical sources at every mass scale. The discovery of the huge intrinsic luminosities of quasars led Salpeter (1964) and Zeldovich (1964) to suggest independently that they contained supermassive black holes (SMBH) with masses $\gtrsim 10^8 M_\odot$, accreting at suitably high rates from some unknown source. The idea that accretion on to the hole took place through a gas disc, allowing mass and angular momentum to diffuse in opposite directions, quickly followed (Lynden-Bell, 1969), as well as the suggestion that the hole might in some way channel energy outwards in collimated jets (Rees, 1971). These ideas remain the paradigm for understanding active galactic nuclei (AGN).

Not long afterwards, astronomers realized that a scaled-down version of the same process, with black holes of stellar masses, was a likely driver of some stellar-mass X-ray binary systems. Here the source of the accreting matter was easier to understand in terms of mass transfer from a companion star, driven by stellar evolutionary expansion or systemic loss of orbital angular momentum, or capture of some of the companion's stellar wind. This picture related accretion in these systems closely to their stellar evolution, and together with their conveniently observable timescales explains why most progress in understanding accretion has until recently come largely from studying accreting stellar-mass binaries. The three editions of the book *Accretion Power in Astrophysics* (Frank, King & Raine, 1985, 1992, 2002) (denoted as APIA1, APIA2, APIA3 – collectively APIA – in this book) illustrate this clearly. The wealth of data on stellar-mass systems has driven continued theoretical progress in areas such as accretion disc stability, although the need for a directly applicable first-principles treatment of the fundamental mechanism transporting angular momentum outwards in accretion discs remains. Theory has otherwise progressed relatively steadily, with probably only one major surprise (ultraluminous X-ray sources – see Section 6.3 of this book) since APIA3 in 2002. In contrast, although the idea of SMBH accretion remains unchallenged, and indeed strengthened by recent discoveries, progress in understanding it was until recently slower, because of the far longer timescales needed for meaningful observations, and the continuing difficulty in understanding how galaxies supply large amounts of mass to the SMBH.

The major change came with the discovery of scaling relations between SMBH and their host galaxies, particularly the M–σ relation (Ferrarese & Merritt, 2000; Gebhardt et al., 2000). The idea that some of the SMBH's huge binding energy is communicated to its host galaxy through feedback from accretion implies a set of constraints as tight as those relating stellar evolution and accretion in X-ray binaries. These insights have already largely overthrown the older view that SMBH disc accretion is essentially a scaled-up version of the same process in close binaries.

We can now see that these two accretion regimes differ in very significant ways. In stellar-mass binary systems the binary orbit generally defines a stable plane for at least the outer accretion disc, mass supply to the disc is often effectively steady on timescales of interest, and the disc mass is almost always low enough that self-gravity is negligible.

It is likely that none of these restrictions hold for SMBH accretion. Mass is supplied sporadically, so the disc is effectively always globally time-dependent, probably undergoing repeated episodes of evolution from fixed initial masses rather than transmitting a constant supply of mass from outside. Each feeding event like this probably has completely random orientation compared with earlier ones, and in most cases the outer parts of the disc spread to take up the angular momentum lost by the inner parts, and so eventually reach the self-gravity radius. These features have the surprising consequence that in some ways understanding aspects of SMBH accretion physics may have more in common with following the evolution of discs around protostars than those in close stellar-mass binaries. One can add more complexities – where binary stellar evolution offers a stable framework for understanding close binaries, for SMBH we have to connect with the far more uncertain details of how galaxies form and evolve at high redshift. But despite the lack of a full understanding of angular momentum transport in accretion discs, there have been significant advances in understanding at both stellar-mass and supermassive scales.

The scaling relations show that supermassive black hole growth is organically connected with galaxy evolution, and so ultimately with how the contents of the Universe as a whole came to be as we observe them. The aim of this book is to introduce the main ideas at play in current SMBH research. Some of these, such as the basics of disc accretion, are already widely known from the close binary context (e.g. in APIA), and I have tried to highlight how things differ in the SMBH context.

The plan of the book is straightforward. After a general introduction (Chapter 1), I detail some of the basic physics needed, starting with the description of black holes in GR (Chapter 2), and astrophysical gas dynamics, magnetohydrodynamics (MHD), and force-free electrodynamics (FFE) (Chapter 3). The next two chapters discuss accretion on to SMBH, gradually moving outwards from disc accretion (Chapter 4), followed by Chapter 5 on what we know of how the SMBH environment feeds gas into these discs. Chapters 6 and 7 discuss how SMBH have definitive effects on their host galaxies, and Chapter 8 outlines some ideas of how the demographics of SMBH in the Universe may have come about.

The treatments of GR and MHD/FFE are perhaps more detailed than usual in books of this kind, but for good reasons. The classical (non-quantum) aspects of black hole physics were all largely established in a golden decade, roughly from the mid-1960s to the mid-1970s. Since then, frontier GR research has largely moved on to less directly applicable areas, and in parallel, university courses on advanced classical GR at a level suitable for astrophysical applications have become rarer in many institutions. Some important insights from GR are now not always fully appreciated, for example that coordinates in general have no physical meaning, so that approximations and expansions are of doubtful value unless proved otherwise; that properties such as the existence of event horizons are global, and do not require extreme local physics; and that numerical treatments can have the unwanted side effect of subverting light-travel arguments even when applied to properly GR-invariant equations. Because almost all the GR literature uses geometrized units, it is often not appreciated just how weak gravity is in comparison to all other forces, particularly electromagnetism. This means, for example, that black holes can in principle (like stars) have electric charge that is completely negligible in terms of the spacetime metric, but nevertheless has significant effects on the motions of charged particles near them.

Magnetohydrodynamic treatments are now strongly prevalent in theoretical SMBH research. Although MHD is a largely classical theory, the peculiarly powerful – and therefore restrictive – nature of the ideal MHD approximation is often underestimated. In particular, by assuming complete charge neutrality everywhere, it effectively removes currents and charges from the physics, relegating them to the status of balancing items in equations. As a result, thinking about currents is often positively counterproductive in ideal MHD. Further, numerical treatments inevitably introduce spurious non-ideal MHD terms. This has long been recognized as a significant barrier to theoretical progress in the solar physics and MHD literature. But because astrophysics rarely has practical consequences, errors and misconceptions in theoretical treatments can survive for a long time.

Astrophysics is a dynamical and developing field, and observations of astrophysical phenomena often initially appear to defy current theoretical understanding. Many results like this have been important in deepening understanding of fundamental parts of the subject, but there are others whose significance is unclear. In the SMBH context a good example is the mass of results on the optical and UV emission line properties of AGN. For stellar-mass binaries a similar status applies to the large body of observations showing that their orbital periods can change over time far more rapidly (and often in the opposite direction!) compared with expectations from stellar evolutionary processes. In both cases it is not obvious that understanding the problem would really constrain the fundamental picture of the system. Although AGN emission lines are important in allowing estimates of SMBH masses and lengthscales, particularly through reverberation mapping, these lines do not carry a large fraction of the total luminosity of accreting SMBH, and their interpretation as a physical diagnostic is complex and indirect. Similarly, observed orbital period changes in stellar-mass binaries are often dominated by short-term effects that drown out the far slower systematic trends resulting from binary evolution.

The attitude I take in this book is to discuss in depth only phenomena where there is a promising route for physical understanding, or better still, a real observational challenge in detail to what might otherwise seem a temptingly reasonable theoretical picture.[1] I have not tried to aim at completeness, but rather to stimulate readers to think about the subject. As well as APIA, I refer to the book *Astrophysical Flows* (Pringle & King, 2007) (denoted as AF) for some fluid-dynamical results.

Most astrophysical publications use the cgs (centimetre, gram, second) system of units, and I have followed this practice. For electromagnetic quantities the equations differ in form between the cgs and SI (mks, i.e. metre, kilogram, second) systems. I have followed the convention of APIA that equations are given in the

---

[1] I recall once at a meeting remarking rather unnecessarily that science progresses by trial and error, and getting the tart response that I must have been making a lot of progress recently.

cgs system, but with multiplicative factors in square brackets giving the conversion to SI units (so a cgs reader should mentally set these square-bracket quantities to unity). These factors always involve the quantities ϵ_0, μ_0, or c, and there should be no confusion with other uses of square brackets in algebraic formulae.

I have added some problems at the end of the book. Some of these refer to published papers, and many of them are aimed at encouraging the reader to see for themselves what is involved in various topics, sometimes asking them to derive results given without proof in the relevant chapter. I have made no attempt to devise problem sets of equal difficulty.

Acknowledgements

Some of the material of this book formed the basis of a master's-level lecture course I taught at the University of Leicester for several years. I thank the students of that course for their patience, and for their many questions. These materially helped in removing obscurities in the presentation. I apologize in advance for those that remain.

I thank Jaime Tung for sharing her publishing expertise, which was particularly helpful at an early stage in the planning of this book. The staff of Cambridge University Press were – as ever – a pleasure to work with, from my initial and continuing contact with the senior commissioning editor Vince Higgs, and with Rowan Groat, and especially Sarah Amstrong, who in particular guided me through the process of obtaining permissions to reproduce figures, and ensuring that the results were of suitable quality. Suresh Kumar gave instant and accurate help with the intricacies of producing a book in LaTeX. My copy-editor Judy Napper was superb in getting the draft into its final form, spotting errors, omissions, duplications, and clumsy wording. I thank them all for their encouragement, help, and sound advice.

I have benefitted enormously from interactions over the years with many people working in the general areas covered in this book, among them Marek Abramowicz, Richard Alexander, Hossam Aly, Ski Antonucci, Steve Balbus, Mitch Begelman, Brandon Carter, Walter Dehnen, Suzan Doğan, George Ellis, Martin Elvis, Pepi Fabbiano, Claude-André Faucher-Giguère, Rob Fender, Juhan Frank, Reinhard Genzel, Andrea Ghez, Gary Gibbons, John Hawley, Luis Ho, Knud Jahnke, Sri Kulkarni, Jean-Pierre Lasota, Doug Lin, Matthew Liska, Andrew Lobban, Giuseppe Lodato, Steve Lubow, Roberto Maiolino, Alessandro Marconi, Sera Markoff, Rebecca Martin, Dean McLaughlin, David Merritt, Selma de Mink, Felix Mirabel, Ramesh Narayan, Priya Natarajan, Sergei Nayakshin, Rebecca Nealon, Chris Nixon, Gordon Ogilvie, Ken Ohsuga, Jerry Ostriker, Roger Penrose, Ken Pounds, Chris Power, Dan Price, Jim Pringle, James Reeves, Elena Rossi, Stephan Rosswog, Joop Schaye, Henk Spruit, Francesco Tombesi, Phil Uttley, Sylvain

Veilleux, Jian-Min Wang, Ralph Wijers, Mark Wilkinson, and Kastytis Zubovas. I thank all the authors (cited in the respective captions) who gave permission to reproduce figures, and who in many cases gave generous help in adapting them. I remain, of course, solely responsible for the errors which are inevitable in any book of this kind.

I wrote most of this book during the Covid-19 pandemic, in Amsterdam and in rural Leicestershire. I thank my many friends in both places, and all my online friends, colleagues, and students everywhere for their cheerful support and kindness in the dark months when both the Concertgebouw and the Cock Inn were closed. I will single out Martin Kenworthy and Oded Regev, who from Leicester and Manhattan, respectively, supplied streams of jokes, ribald and admirably scurrilous comments, and shrewd judgements on football and other sports. And in particular I thank my family, who despite the long separations delighted me so often with their news and their online presence.

Above all I thank Nicole, for all that we have shared. *Jij bent geweldig.*

1

Black Holes and Galaxies

1.1 Basic Properties

A black hole is an object so dense that light cannot escape its gravity. The proper description of black holes requires general relativity (GR), and we discuss this in Chapter 2. But simple Newtonian ideas already give us some insight into their properties if we bear in mind the restrictions to velocities below the speed of light and energies smaller than the rest-mass value.

The escape velocity from the surface of a star of mass M and radius R is $v = (2GM/R)^{1/2}$, where G is the gravitational constant. This velocity reaches the speed of light, c, for a radius

$$R = \frac{2GM}{c^2}. \tag{1.1}$$

We see that for $M = 1\mathrm{M}_\odot$ (the solar mass), the 'star' must have a tiny radius $R \lesssim 3\,\mathrm{km}$. The characteristic property of a black hole is that it is small for a given mass, making the gravitational field very strong in its immediate vicinity. But it is important to remember that at large distances from a black hole, the gravitational field strength is the same as for any gravitating object of the same mass. Black holes have their distinctive properties only because they are small enough to allow matter to get very close, so that orbital speeds approach that of light. The characteristic size of a black hole is its gravitational radius

$$R_g = \frac{GM}{c^2} \simeq 1.5 \times 10^{14} M_8\ \mathrm{cm}, \tag{1.2}$$

where we have parametrized the black hole mass $M = 10^8 M_8 \mathrm{M}_\odot$, as this is a typical SMBH mass.

Matter falling radially towards an object like this acquires very high speeds because of the large gravitational potential energy available near the black hole. If nothing intervenes to stop it getting very close to the hole the matter eventually gains gravitational energy approaching $\sim 0.5c^2$ times its rest mass. If the matter is gaseous, it is very likely that the process is sufficiently untidy that much of this energy is dissipated as radiation. The luminosity this releases is far greater than would happen if the same mass was consumed in nuclear burning. Transmuting hydrogen to helium or heavier elements, which powers most stars, releases energy only $0.007c^2$ times the rest mass.

In reality two things complicate this comparison a little. First, in reality matter is almost certain to orbit the black hole with some angular momentum, and fall towards it more slowly, as it gradually loses this angular momentum to matter further out. This kind of configuration is called an *accretion disc*, and will appear everywhere in this book. Accretion through a disc gives up the gravitational binding energy of orbits close to the black hole, which is somewhat less than the radial infall kinetic energy (a factor of two for circular Newtonian orbits).

Second, GR changes these binding energies slightly from Newtonian values. A full GR calculation (see Section 4.1) refines the estimate of the infall energy $\sim 0.5c^2$ to a value ~ 0.1–$0.4c^2$, but this slightly reduced 'accretion yield' is still far higher than the nuclear yield $0.007c^2$. Matter–antimatter annihilation releases the full restmass energy, but is very unlikely on any scale larger than an atomic nucleus, so that accretion on to a black hole is the most efficient way of getting energy from normal matter. We conclude that

> *accretion on to black holes must power the most luminous objects in the Universe.*

The obvious candidates here are quasars and active galactic nuclei, collectively called AGN, which harbour the most massive black holes. Their luminosities can reach 10^{46}–10^{48} erg s^{-1} or even more.[1] At typical distances of Mpc, the angular sizes of their gravitational radii R_g are extremely small. But radio interferometry with extremely long baselines is now able to resolve some of the nearer AGN, giving spectacular images (e.g. Figure 1.1).

At a smaller mass scale, the same argument tells us that binary systems where a stellar-mass black hole accretes gas at a high rate from a companion star are good candidates for explaining some of the stellar-mass X-ray binaries, with luminosities up to 10^{38}–10^{39} erg s^{-1}. Of course there are no AGN analogues of neutron stars, which cannot have masses larger than about $3M_\odot$.[2]

---

[1] Gamma-ray bursts can for extremely short times exceed this luminosity, and reach $\sim 10^{53}$ erg s^{-1}. This approaches the total luminosity L_* of all the stars in the observable Universe, which contains $\sim 10^{11}$ galaxies, each with $\sim 10^{11}$ stars emitting roughly solar luminosities $\sim L_\odot \sim 10^{33}$ erg s^{-1}. This gives $L_* \sim 10^{55}$ erg s^{-1}. See Problem 1.1 at the end of the book which investigates if this is a coincidence.

Figure 1.1 The immediate surroundings of the supermassive black hole in the galaxy M87 as imaged in the radio by the Event Horizon Telescope (Event Horizon Telescope Collaboration, 2019). The scale of the solar system is shown for comparison. Credit: Randall Munro (2019).

Observations clearly distinguish between stellar-mass accretors and SMBH. X-ray binaries are in spatially extended populations in their host galaxies, while AGN are point sources close to the centres of their hosts, and are generally intrinsically far brighter than X-ray binaries. They cannot simply be unresolved collections of X-ray binaries because they are often observed to vary by factors of $\gtrsim 2$.

1.2 The Eddington Limit

The accretion luminosity L of a black hole must be related to its mass accretion rate $\dot{M}$ by

$$L = \eta \dot{M} c^2, \tag{1.3}$$

---

[2] The minor complication here is that neutron stars are almost as compact as black holes – they have radii ~10 km for masses ~1–3M$_\odot$. Both black hole and neutron star X-ray binaries are fairly common, although there are probably far more neutron star systems in total.

where we see from the discussions of the last previous section that the *accretion efficiency* η is a dimensionless quantity of order at most a few times 0.1. But there is a limit to the luminosity that any gravitating object, accreting or otherwise, can emit, since radiation produces a pressure force which tends to disperse the matter producing the luminosity. This force acts on electrons because they scatter electromagnetic radiation, which carries momentum $(1/c)$ times its energy flux $L/4\pi r^2$. Protons have little effect on radiation, but make up most of the mass of the gas. For simplicity we consider a spherically symmetric situation, so that the radiation pressure force acts radially outwards. Its magnitude at radius r from the centre is

$$F_{\text{rad}} = \frac{L\sigma_T}{4\pi cr^2},\tag{1.4}$$

where $\sigma_T \simeq 6.65 \times 10^{-25}$ cm^2 is the Thomson cross-section, the effective blocking area of an electron in a beam of radiation. The electron is not free to move in response to this outward force, since charge neutrality means that it is strongly bound by Coulomb attraction to a mass of gas carrying one proton charge. Most astrophysical gases are largely hydrogen, so this mass is of order the proton mass m_p. Then the gravity force resisting the radiation pressure is

$$F_{\text{grav}} \simeq \frac{G(m_p + m_e)}{r^2} \simeq \frac{GMm_p}{r^2},\tag{1.5}$$

since the electron mass m_e is much smaller than m_p. Both F_{rad} and F_{grav} vary as r^{-2}, so we see that if one of them exceeds the other at any one radius, it does so at all radii. Then in spherical symmetry, accretion must be at least inhibited once L is large enough to make $F_{\text{rad}} = F_{\text{grav}}$. This defines the *Eddington luminosity* or *Eddington limit* as

$$L_{\text{Edd}} = \frac{4\pi GMc}{\kappa},\tag{1.6}$$

where we have used the electron opacity $\kappa \simeq \sigma_T/m_p \simeq 0.34$ cm^2 g^{-1} for an astrophysical gas of typical composition, giving

$$L_{\text{Edd}} \simeq 1.3 \times 10^{46} M_8 \text{ erg s}^{-1}.\tag{1.7}$$

(The opacity κ is roughly halved, so L_{Edd} doubled, if the accreting gas is hydrogen-poor.) This result, and the huge difference in the mass that is accreting gas, explains why AGN can be far more intrinsically luminous than X-ray binaries.

Although we have derived it here for spherically symmetric systems, the limit (1.6) holds to factors of order unity for almost any other geometry. In particular, this is true even for matter falling in through a sequence of orbits of decreasing angular momentum – that is, in an *accretion disc*. The appropriate form of the Eddington limit applies to any luminous object, whatever powers its luminosity.

The importance of the Eddington luminosity was first realized a century ago in the context of stellar structure, where the radiation comes from nuclear burning. Massive hot stars radiate luminosities close to the limit (1.7). A nuclear luminosity even slightly above L_{Edd} would make them expand a little, lowering the density ρ in the central nuclear-burning core. The nuclear luminosity L_{nuc} varies as ρ^2 and so drops below L_{Edd}. This self-limiting property means that hot stars can remain stably in equilibrium very close to L_{Edd}. The source of the luminosity reacts sensitively – and negatively – to the luminosity itself, rather like a thermostat.

But this kind of self-limiting behaviour does not apply to accretion-powered objects. The mass supply rate driving accretion is in general given by some process totally unaffected by changes in the accretion luminosity, and so is unlikely to adjust to respect the Eddington limit. For example, in close binary systems there is no reason why the evolution of the donor star, and so the resulting mass transfer rate $\dot{M}_{supp}$ it supplies to a companion black hole, should know or care about the possibility that $\dot{M}_{supp}$ might exceed the rate

$$\dot{M}_{Edd} = \frac{L_{Edd}}{\eta c^2} = \frac{4\pi GM}{\eta \kappa c} \qquad (1.8)$$

that would make the accretion luminosity $L_{acc} = \eta c^2 \dot{M}_{acc} = \eta c^2 \dot{M}_{trans}$ greater than L_{Edd}. This possibility was already recognized in the very first papers discussing realistic accretion processes: it is entirely possible for a black hole (or any other accreting object) to be supplied with mass at rates $\dot{M} > \dot{M}_{Edd}$, and for this situation to persist over significant timescales. We will discuss in detail what happens in such cases in Section 4.6, but it already is clear that there are only two routes to dealing with the mismatch – either

(a) preventing much of the matter getting too close to the hole, where it would gain and then radiate the full accretion energy, or
(b) ensuring that the matter close to the hole has unusually low accretion efficiency and so does not radiate a strongly super-Eddington luminosity.

The mild complexity of these two possibilities has generated a cloud of confused and confusing language in the astrophysical literature. The phrase 'super-Eddington accretion' is deeply ambiguous if not carefully qualified, as it is often used to denote either one of the outcomes (a, b) (or in the worst cases, both simultaneously!). To avoid this ambiguity, this book uses the description super-Eddington *mass supply* (or *feeding*) to describe cases where $\dot{M}_{supp} > \dot{M}_{Edd}$. In treating these, it is vital to distinguish between the outcomes (a) and (b), which differ markedly.[3]

---

[3] This muddle is maximal in discussions of the (stellar-mass) ultraluminous X-ray sources (ULXs). ULXs have very anisotropic radiation patterns (see Section 4.10), and when viewed from tightly defined directions appear

In case (a), the black hole mass cannot grow faster than the rate $\dot{M}_{\mathrm{Edd}}$. Then the shortest e-folding timescale for mass growth is the Salpeter timescale

$$t_{\mathrm{Sal}} = \frac{M}{\dot{M}_{\mathrm{Edd}}} \simeq 5 \times 10^7 \eta_{0.1} \, \mathrm{yr} \tag{1.9}$$

(Salpeter, 1964), where $\eta_{0.1}$ is the efficiency of conversion of rest-mass energy to radiation in units of $0.1c^2$. We see that high radiative efficiency implies slower mass growth, as the limiting luminosity L_{Edd} is produced by a smaller accretion rate.

If instead (b) holds, the accretor mass grows faster than the Eddington rate, that is, on a timescale $< t_{\mathrm{Sal}}$.

In both cases (a) and (b) it is difficult for the total luminosity (correctly evaluated over all directions in the case of anisotropy – see footnote 3) of an accreting object to exceed the Eddington luminosity L_{Edd} by large amounts, except in impulsive or explosive situations such as supernovae or gamma-ray bursts. If we are confident that a given object is not of this type, and its luminosity is not markedly anisotropic, its luminosity gives us a lower limit to its mass through (1.7).

1.3 SMBH Accretion

By the argument detailed previous section, observation places tight constraints on the total mass in SMBHs in the local (low-redshift) Universe. AGN spectra (cf. Figure 1.2) typically peak in the soft X-ray–far UV region, with almost always a significant component in the medium-energy X-rays. This latter component is relatively easy to observe, as it is often fairly immune to interstellar absorption or scattering. In addition, very few astronomical objects other than those that accrete produce substantial X-ray emission, so it is usually safe to assume that all the detected emission comes from the AGN itself.

Medium-energy cosmic X-ray detectors find non-zero fluxes even when not observing specific point sources – this is called the X-ray background. Since the emission from AGN in this spectral band is far more powerful than from anything else, such as X-ray binaries or supernovae, we can identify this background flux as the result of the collective emission from AGN – that is, growing SMBH – in the local Universe. From the typical X-ray spectrum (cf. Figure 1.2) this gives us the total SMBH growth in this region, and so a lower limit on the total SMBH mass there. This is in outline the *Soltan argument* (Soltan, 1982). It tells us that the

to have luminosities $\gg L_{\mathrm{Edd}}$, if the observed flux is (wrongly) assumed to be isotropic. Confusingly, the indirect cause of the anisotropic emission, and so of the apparent super-Eddington luminosity, is super-Eddington *feeding*, since radiation pressure blows matter away from the accretion disc (see (a)) except near the rotational axes of the accretion flow, so collimating the escaping radiation tightly. The accretors in ULXs do not gain mass at significantly super-Eddington rates, even though the mass transfer rates from their companions are super-Eddington. Their luminosities are *apparently* super-Eddington, but in reality are not.

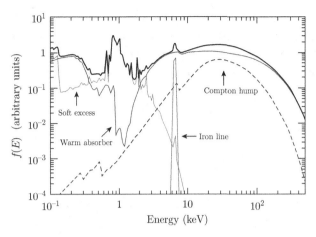

Figure 1.2 Average total spectrum (thick line) and main components (thin lines) in the X-ray spectrum of a type I AGN. The main primary continuum component is a power law with a high energy cut-off at $E \sim 100$–300 keV, absorbed at soft energies by warm gas with $N_H \sim 10^{21}$–10^{23} cm^{-2}. A cold reflection component is also shown. The most relevant narrow feature is the iron Kα emission line at 6.4 keV. Finally, a 'soft excess' is shown, resulting from thermal emission of a Compton thin plasma with temperature $kT \sim 0.1$–1 keV. Credit: Risaliti and Elvis (2004).

average ration of SMBH mass is $\gtrsim 10^8 M_\odot$ per medium-size galaxy. This mass scale agrees with the values we can deduce from the Eddington limit (1.7), and we shall see that it is similar to the masses found by various kinds of direct observation, so it cannot be concentrated in just a minority of galaxies. We conclude that

> *the centre of almost every medium to large-mass galaxy in the local Universe must host a supermassive black hole, whose mass grew via luminous accretion of gas.*

Despite this, observations show that only a minority (less than 1%) of low-redshift galaxies have active nuclei, where the SMBH are currently growing their masses. But since we have just concluded that almost every galaxy has an SMBH, this must mean that accretion and SMBH growth occur only in short-lived phases – AGN must be strongly variable, not simply on the timescales we observe directly, but on longer ones also.

1.4 SMBH Locations

The reasoning of Section 1.3 implies that almost all low-redshift galaxies must host SMBH. Observations of the active minority where the SMBH is caught in the act of accreting always find the AGN close to the dynamical centre (hence the 'nucleus' part of 'AGN'). This is no accident, but results because the gravity of an SMBH moving through a galaxy makes it pull large numbers of stars along behind it, in a kind of gravitational wake (see Figure 1.3). This process is called 'dynamical friction', and is directly analogous to the way the Coulomb attraction of a charged particle slows its motion through a plasma. The result (Chandrasekhar, 1943) is a drag force – the speed v of an SMBH of mass M moving through a galaxy with stellar mass density ρ obeys

$$\frac{dv}{dt} = -\frac{4\pi CG^2 M \rho}{v^2}. \tag{1.10}$$

Here $C \simeq 10$ is a constant (the Coulomb logarithm, measuring the cumulative effect of the weak drag forces of many distant stars ('small-angle scattering') compared with the individually stronger drag forces ('large-angle scattering') of a few nearby stars). If ρ is constant this equation integrates as

$$v^3 = v_0^3 \left(1 - \frac{t}{t_{\text{fric}}}\right) \tag{1.11}$$

where

$$t_{\text{fric}} = \frac{v_0^3}{12\pi CG^2 M \rho}, \tag{1.12}$$

with v_0 the initial velocity. In a time $\sim t_{\text{fric}}$ the SMBH is reduced to rest, which is only possible if it has spiralled in to the dynamical centre of the galaxy. Crudely modelling the central region of a galaxy as a uniform sphere of stars with total mass $M_* = 10^{11} M_\odot$ and radius $R_* =$ a few kpc, a $10^8 M_\odot$ SMBH with speed $v_0 \sim (2GM_*/R_*)^{1/2}$ has $t_{\text{fric}} \lesssim 10^8$ yr, much smaller than the age $\sim 10^{10}$ yr of a low-redshift galaxy.

So unless a galaxy has recently been disturbed, which is often obvious because it has an irregular shape, its SMBH is very likely to be at its dynamical centre. If it somehow has more than one SMBH, dynamical friction makes them collect rapidly in its centre, and then orbit under their mutual gravitational attraction. Here they lose energy and spiral inwards as they emit gravitational radiation, eventually merging. The merger may eject the lightest hole(s) if there are more than two, since these must be moving fastest if the holes have similar orbital energies, as is likely. So in most galaxies we expect to find just one SMBH, in its centre. (In very small galaxies the shallow gravitational potential may be unable to retain a merging pair of black holes as they recoil under anisotropic gravitational wave emission, so some may have no central black hole). By the same reasoning, a merger of two galaxies is likely to produce a single larger galaxy with a merged SMBH in its centre (see Figure 1.4).

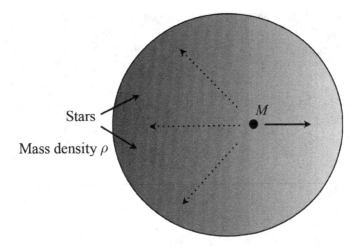

Figure 1.3 dynamical friction: a massive object moving through a collection of stars raises a gravitational 'wake' and slows.

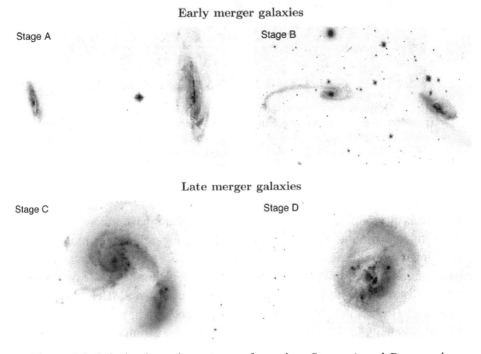

Figure 1.4 Galaxies in various stages of merging. Stages A and B are early mergers and stages C and D are late mergers. Credit: Ricci et al. (2017).

In summary, we expect almost every galaxy, except possibly a few dwarfs, to have a single SMBH at its centre. Observations appear to agree with this very simple one-to-one relation. Our own Galaxy's central region is well studied. Infrared

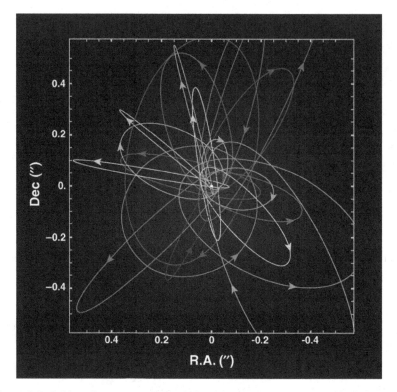

Figure 1.5 Orbits of stars around the Galactic Centre supermassive black hole Sgr A* (position given by the white dot in the centre of the figure). All of these orbits, observed with great precision over decades, require a unique black hole mass $4 \times 10^6 M_\odot$. The closest orbits now show evidence for the Einstein precession (advance of the pericentre, as predicted by general relativity). Credit: S. Gillessen.

observations by groups in Germany and the United States have followed the proper motions of stars around it in exquisite detail (see Figure 1.5) for more than 25 years.[4] Interpreting these motions as Kepler orbits shows that the moving stars orbit a central mass of order $4.5 \times 10^6 M_\odot$. Constraints on its size leave no room for doubt that this object (Sgr A*) must be the Galaxy's own SMBH.

This SMBH is remarkably inactive at the current epoch, but there is indirect evidence of activity in the past. In particular, two large gamma-ray-emitting lobes (the 'Fermi bubbles' – see Figures 1.6 and 1.7) are symmetrically placed each side of the Galactic plane with Sgr A* at the centre of symmetry. These are probably the result of an energetic outflow event from the SMBH about 6 Myr ago – the event was probably roughly isotropic, but the greater density of the Galactic plane means that propagation only occurs along the axis of the Galaxy.

---

[4] Reinhard Genzel and Andrea Ghez shared half the 2020 Nobel Prize in Physics for leading this work.

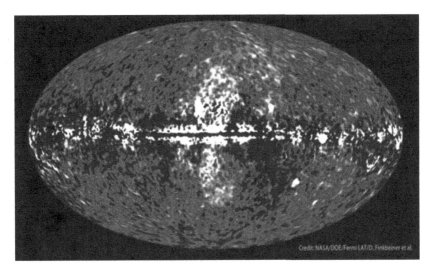

Figure 1.6 Observations by the *Fermi*–LAT satellite of bubbles of gamma-ray emission symmetrically placed each side of the plane of the Milky Way. Credit: NASA/DOE/Fermi LAT/D. Finkbeiner et al.

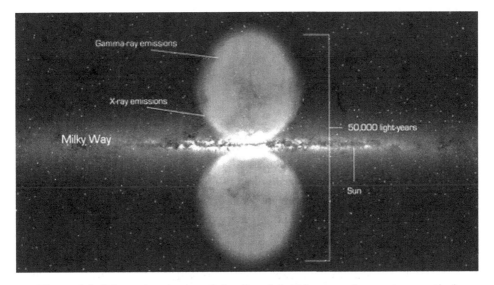

Figure 1.7 Schematic picture of the Fermi bubbles superimposed on optical observations of the plane of the Milky Way. Credit: NASA Goddard.

1.5 The SMBH Environment

We have seen that dynamical friction ensures that most supermassive black holes are in the central bulges of spiral galaxies, or the centres of elliptical galaxies. Both of these environments are roughly spherical systems – often collectively called spheroids – which are probably the outcomes of mergers with other galaxies. In a spiral the mergers were minor – that is, with smaller galaxies, whereas an elliptical probably resulted from a major merger of two galaxies of similar masses. The stellar motions in both types of spheroid are generally characterized by velocity distributions similar to the Maxwellian distribution of the particles of a hot gas. The velocity dispersions σ are spatially constant, as expected theoretically (Lynden-Bell, 1967). The resemblance to a constant-temperature gas means this kind of distribution is called isothermal. To remain in dynamical equilibrium under its mutual self-gravity, the stellar mass density obeys

$$\rho = \frac{\sigma^2}{2\pi G r^2}. \tag{1.13}$$

The stellar mass inside radius R is then

$$M(R) = 4\pi \int_0^R \rho(r)r = \frac{2\sigma^2 R}{G}. \tag{1.14}$$

The SMBH's own gravity controls stellar motions only in the region where $M(R) \lesssim M$, so inside a radius

$$R_{\text{inf}} \simeq \frac{GM}{\sigma^2}, \tag{1.15}$$

called its sphere of influence. In a typical galaxy with black hole mass $M = 10^8 M_8 M_\odot$, $\sigma = 200\sigma_{200}$ km s^{-1}, we find

$$R_{\text{inf}} \simeq \frac{8M_8}{\sigma_{200}^2} \text{ pc.} \tag{1.16}$$

The radius of influence R_{inf} is much smaller than the typical scale $R_b \gtrsim 5$ kpc of a spiral bulge, or the even larger scale of an elliptical galaxy. Other components of the galaxy, such as gas, presumably have velocity dispersions comparable with the stars, and so do not feel the black hole gravity outside the sphere of influence R_{inf} either.

Within R_{inf} stars move in elliptical orbits under the black hole's influence, but they also exchange energy with each other by repeated gravitational encounters, adjusting their orbits accordingly. After one relaxation time, stars of a similar mass should be distributed with number density $\rho \propto R^{-7/4}$, out to a radius $\sim 0.2 R_{\text{inf}}$ (Bahcall & Wolf, 1976). This implies a sharply rising stellar 'cusp' near the SMBH. Similar results hold if the restriction to similar stellar masses is lifted. But so far no clear example of a cusp like this has been found observationally in any galaxy. For example, in our Galaxy the cusp should contain about 10^7 stars and extend out to about 0.4 pc (since $R_{\text{inf}} \simeq 2$–3 pc). This would be easy to resolve from Earth, but the observed distribution is flat or even declining towards the SMBH. It may be that the relaxation time here is in reality comparable to the age of the Galaxy, so that no cusp has yet had time to form, or that the cusp is present, but in a population of stars that is so far unobserved. We shall see in Chapter 6 that there may be other reasons why the cusp is apparently not present.

Other dynamical effects of the central SMBH are directly observed. The SMBH can disrupt stellar binary systems orbiting it. The more massive star of a pair is captured into a closer orbit around the SMBH, while the gravitational energy this liberates goes into propelling the lighter star away from the SMBH at very high speed (Hills, 1988). Examples of these hypervelocity stars have been observed in our Galaxy with speeds approaching 2,000 km s^{-1}, and velocities pointing directly away from the Galactic Centre.

Even more violently, SMBH can completely disrupt stars whose orbits get too close to it – for example, stars falling inwards on near-parabolic orbits. In these tidal disruption events (TDEs) the tides exerted by the SMBH overcome the stars' own self-gravity. This happens if a star's radius R_* is large enough to fill its tidal lobe, of size $R_{\rm tide} \simeq 0.5(M_*/M)^{1/3}a$, where M_* is its mass and a its distance of closest approach to the SMBH. We shall see in the next chapter that there is a minimum value for a for stable orbits not falling directly into the hole, of the form $a_{\rm min} \gtrsim zR_g \propto M$, with $z \sim$ few. The parameter z here depends on the spin of the black hole. If the condition

$$R_* = R_{\rm tide} \simeq 0.5(M_*/M)^{1/3}a_{\rm min} \gtrsim 0.5(M_*/M)^{1/3}zR_g \qquad (1.17)$$

fails, the star is never disrupted: it either orbits the hole ($a > a_{\rm min}$) without being torn apart by tidal forces, or if $a = a_{\rm min}$, simply falls into it 'silently', that is, is swallowed whole by it without disruption. Rearranging (1.17) gives the condition for disruption as

$$M \lesssim 5 \times 10^7 {\rm M}_\odot \rho_*^{-1/2} z^{-3/2}, \qquad (1.18)$$

where $\rho_* \simeq 3M_*/4\pi R_*^3 \sim 1/(M_*/{\rm M}_\odot)^2$ g cm^{-3} for a low-mass main-sequence star. We see that tidal disruption is only possible for normal stars if the SMBH is fairly low, that is, $M \lesssim 10^7 {\rm M}_\odot$. If disruption occurs, we will see (Section 5.8) that about half of the star's gas accretes on to the hole, while the other half is ejected. Observations have revealed a number of likely TDEs, and the associated SMBH masses are low, agreeing with (1.18). This, and the Soltan relation, tell us that TDEs cannot be the main growth channel for the most massive holes observed at low redshift. But it leaves open the possibility that TDEs may be important in growing SMBH with smaller masses, both in the local Universe, and possibly in the early growth of all SMBH at high redshift. Stars that do not quite fall close enough to be fully disrupted by the SMBH tides, or are the captured component of a binary disrupted by the Hills mechanism (Hills, 1988), are probably the cause

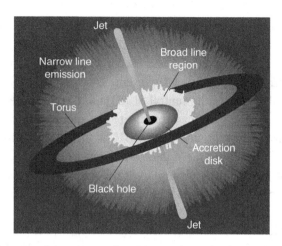

Figure 1.8 Cartoon of the surrounding of an active galactic nucleus (AGN). Note that the central accretion disc and the associated jet may be misaligned from the torus. The entire structure has in general no spatial alignment with the large-scale structure of the host galaxy. Credit: NASA.

of the quasi-periodic eruptions (QPEs) seen in X-ray emission from the nuclei of low-mass galaxies. Here a small star (white dwarfs are favoured by observational selection) is trapped in a highly eccentric orbit about the black hole. At pericentre the star fills its tidal lobe and gives a burst of mass transfer to the black hole, powering the QPE. The emission of gravitational radiation keeps the orbiting star close enough for this to continue (see Section 5.9).

Since our own SMBH is not currently active, it may not be a good guide to what happens when there is significant accretion, as in active galactic nuclei. The brightest of these objects are quasars, where the accreting nucleus outshines the host galaxy so strongly that the host is difficult to observe. Seyfert galaxies also have accreting nuclei, but their lower luminosities allow more observational insight into the galaxy. Seyferts are classified into Type I and Type II through the characteristics of their bright emission lines coming from highly ionized gas. Seyfert I's are luminous in the X–rays and ultraviolet, and show both broad (velocity widths $\Delta v \sim 10^4\,\mathrm{km\ s^{-1}}$) and narrow ($\Delta v \sim 10^2\,\mathrm{km\ s^{-1}}$) emission lines, while Seyfert II's show only narrow lines. Since higher velocities must come from gas orbiting close to the SMBH, an appealing picture (Antonucci & Miller, 1985; Antonucci, 1993) is that these two types of galaxy are essentially the same, but viewed from different inclinations to an obscuring torus of matter, probably containing significant amounts of dust, around the region producing the broad lines (see Figure 1.8).

Figure 1.9 The X-ray jet from the galaxy Cen A. Credit: NASA/SAO/R. Kraft et al. (2002).

Another striking property of accreting SMBH whose origin is still not fully settled is that many of them produce jets, usually observed in the radio. These are high-speed (often with velocities close to c), narrowly collimated outflows (see Figures 1.9 and 1.10), and can extend over scales of Mpc. Using the standard rule of thumb that the terminal velocity of an outflow is very close to the escape velocity at the point where it was launched, their relativistic velocities strongly suggest that they come from the close vicinity of the SMBH. Jets are not unique to AGN – every class of accreting objects is able to produce them, with terminal velocities again close to the escape value from the surface of the accretor. So stellar-mass black holes and neutron stars in accreting binaries also make relativistic jets, while protostars forming out of accreting discs of matter make much slower jets (see Figure 1.11). Despite the very different accretors, the similarities between the jets is obvious, suggesting that a very similar mechanism makes jets from all accretors. As we shall see in Chapter 7, there is a general expectation that jets from accreting objects arise from the combination of infalling matter, magnetic fields and rotation. Radio pulsars, which are rapidly spinning magnetic neutron stars, are not accreting, but also make jets, which are aligned with the pulsar spin axis. It may be that the plasma that we know must be present in their magnetospheres flows in ways that mimic accretion flows in all other jet sources.

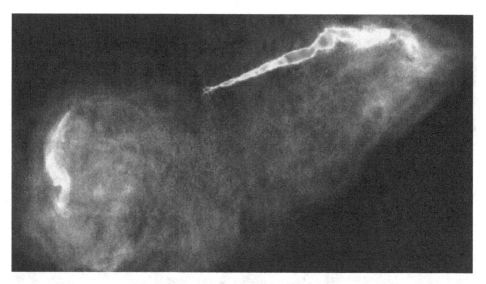

Figure 1.10 The radio jets from the SMBH in the galaxy M87 and their effects on the surrounding gas, observed by the Very Large Array. The left-hand jet is partly obscured. Credit: NASA.

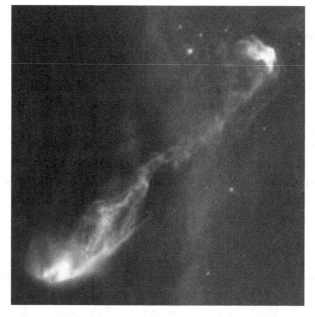

Figure 1.11 The jets from the protostar HH-47. Despite the very different spatial scales (pc/Mpc) and masses of the accreting objects $(1M_\odot / 10^9 M_\odot)$, the similarity with the Cen A jet is striking. Credit: STScI & NASA.

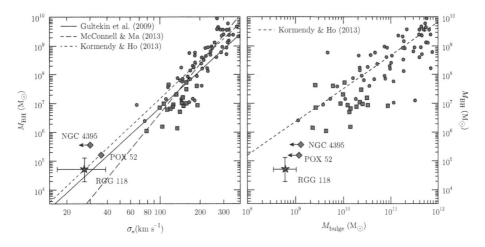

Figure 1.12 The observed relations between black hole mass M and (left) host galaxy velocity dispersion σ, and (right) black hole mass and host bulge stellar mass M_b. Credit: Baldassare et al. (2015).

A marked characteristic of AGN is that although the jet appears to propagate close to the axis of the AGN torus, the orientation of this combined jet–torus structure is completely random when compared with the large-scale structure of the host galaxy (e.g. Nagar & Wilson, 1999, Kinney et al., 2000).

An important feature of AGN spectra is that they show that the gas near the AGN is strongly metal-enriched (Shields, 1976; Baldwin & Netzer, 1978; Hamann & Ferland, 1992; Ferland et al., 1996; Dietrich, Appenzeller & Wagner, 1999, 2003a, 2003b; Arav et al., 2007). This presumably means that this gas has been recycled through several generations of massive stars. We shall see how this may have come about in Chapter 7.

1.6 Supermassive Black Holes and Galaxies

The observations summarized in this chapter give a rich picture of the connections between SMBH and their hosts. In their brief active intervals we know that SMBH must accrete gas, and so are likely to show signatures of disc accretion. These include jets, emission lines from gas moving with a range of speeds, and obscuration by a dusty torus. Even non-active SMBH can show themselves in tidal disruption events, and there are many effects related to the central SMBH seen in our own Galaxy, such as hypervelocity stars, the Fermi bubbles, and a mass of detailed information from direct observation of the Galactic Centre.

But one class of observation above all suggests that SMBH are not simply interesting in their own right, but play a determining role in how all galaxies form and evolve. Since the beginning of this century we have known (Ferrarese & Merritt, 2000; Gebhardt et al., 2000) that the black hole mass M is tightly related to two properties of its host bulge (or spheroid) – its total stellar mass M_b, and its stellar velocity dispersion (Figure 1.12).

We can roughly summarize these relations as

$$M \simeq 10^{-3} M_b \qquad (1.19)$$

and

$$M \simeq 3 \times 10^8 \sigma_{200}^4 M_\odot, \qquad (1.20)$$

where $\sigma_{200} = \sigma/(200 \, \text{km s}^{-1})$.

The connection between black hole and spheroid properties implied by both scaling relations appears very surprising at first sight, since we know that the direct gravitational influence of the SMBH is important only inside R_{sph}. With a size of a few parsecs, this is far smaller than the spheroid scale ~ 1–$10 \, \text{kpc}$, so for most purposes the host spheroid must be completely unaware of the SMBH. We shall try to answer these questions in Chapter 6.

2

Black Holes and General Relativity

2.1 Introduction

We have so far discussed black holes in quasi-Newtonian terms, but it is obvious that we need to use general relativity (GR) for a first-principles description. In GR, gravity appears differently from all other forces (electromagnetism and the nuclear forces), and we give a very brief introduction here. This does not attempt to describe the full technical aspects of the theory, but aims to clarify how GR is used in astrophysical practice. This generally reduces to using results derived from GR in an otherwise Newtonian treatment. We outline in Section 2.5 the limitations and potential pitfalls of this approach.

In GR, gravity is specified by the spacetime metric tensor $g_{\mu\nu}$, which governs how physical effects propagate in space and time. The suffixes μ, ν refer to four spacetime coordinates uniquely specifying any given event. An example of these are the usual time and Cartesian space coordinates $x^{\mu} = (t, x, y, z)$. All coordinate systems specify 10 independent values for the array $g_{\mu\nu}$, as this obeys $g_{\mu\nu} = g_{\nu\mu}$.

In the immediate spacetime vicinity of any spacetime event the metric can always be re-expressed by a change of coordinates to the Minkowski metric of special relativity. This is also known as 'flat space', as the equivalence principle asserts that spacetime curvature, or local gravity, is unimportant on very small scales. This metric specifies the distance ds between two events at (t, x, y, z) and $(t + dt, x + dx, y + dy, z + dz)$ as

$$ds^2 = -c^2 dt^2 + dx^2 + dy^2 + dz^2 \equiv g_{\mu\nu} dx^{\mu} dx^{\nu}, \tag{2.1}$$

where c is the speed of light in vacuo. As always in GR we use the summation convention that repeated indices are interpreted as a sum over all four coordinates.[1]

---

[1] Raised and lowered indices are related by $x^{\mu} = g^{\mu\nu} x_{\nu}$, where $g^{\mu\nu}$ is the matrix inverse of $g_{\mu\nu}$, that is, $g^{\lambda\mu} g_{\mu\nu} = \delta^{\lambda}_{\nu}$ where δ is the matrix with zeroes except on the diagonal, where it has entries $-1,1,1,1$.

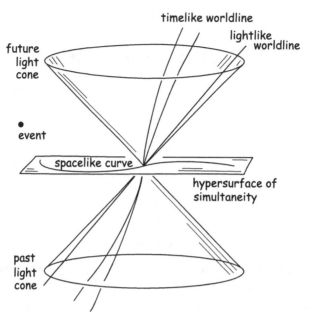

Figure 2.1 Schematic picture of light cone structure. Physical causality is restricted to the interior (timelike curves) or the surface (lightlike or null curves). Only events within the past light cone can influence the event at the origin, and this event can only influence events within the future light cone. Credit: John D. Norton, Einstein for Everyone, https://sites.pitt.edu/~jdnorton/teaching/HPS_0410/index.html.

It is very important to realize that coordinates themselves have in general *no* physical significance, and are simply labels distinguishing different spacetime events. The spacetime interval between two events is specified only by using all four coordinates and computing ds^2, as defined in (2.1). If ds^2 is positive, we call the interval *spacelike*; if it is negative we call this *timelike*; and if $ds^2 = 0$ the interval is *lightlike*, or *null*. These conditions define a four-dimensional structure in spacetime, usually represented in three dimensions as a double cone, called the *light cone*, surrounding the future-pointing and past-pointing timelike paths from any given point (see Figure 2.1). The paths themselves are often called worldlines. All physical interactions that have affected physical variables at a given event must have worldlines lying within or on its past light cone, and conditions at this event can only physically influence events in or on its future light cone. Except in flat space, the shape (but not the topology) of the light cone differs between different events.

Given the metric $g_{\mu\nu}$, the motion of a test particle under gravity is specified by *geodesics*, defined as the spacetime paths that minimize the spacetime interval between two timelike (or null) separated events, that is,

$$S = \int \sqrt{-g_{\mu\nu}\mathrm{d}x^\mu \mathrm{d}x^\nu}, \tag{2.2}$$

where the minus sign appears because the worldlines of all physical particles must be timelike or null, depending on whether the particle has non-zero rest mass.

The metric tensor $g_{\mu\nu}$ is determined by the physical content of the Universe through the Einstein field equations

$$R_{\mu\nu} - \frac{1}{2}Rg_{\mu\nu} = \frac{8\pi G}{c^4}T_{\mu\nu}. \tag{2.3}$$

Here $R_{\mu\nu}$ is the Ricci tensor, derived from the metric $g_{\mu\nu}$ and its derivatives, $R = R^\mu_{\ \mu}$, and $T_{\mu\nu}$ is the energy-momentum tensor, specified by the energy and momenta associated with all the other forces. (All of these quantities also have 10 independent values.)[2]

The constant G is Newton's gravitational constant. Since G and/or c appear in principle in every formula in GR, it is an almost universal practice to adopt 'geometrized units', in which $G = c = 1$, when working in GR. This has huge advantages in writing formulae, but makes it easy to lose sight of an important point, namely that gravity is far weaker than all other forces. In the form (2.3) we can immediately see this, as the constant on the right-hand side is

$$\kappa = \frac{8\pi G}{c^4} = 2.1 \times 10^{-48}\ \mathrm{dyn}^{-1}. \tag{2.4}$$

We can also write this constant as $8\pi c/L$, where $L = c^5/G \simeq 3.7 \times 10^{59}\ \mathrm{erg\ s}^{-1}$ is a luminosity. L appears in all formulae for gravitational radiation power.

The very small size of this constant shows how extremely weak gravity is compared with all other forces. For example, the gravitational effect of an electric charge expressed by (2.4) is utterly negligible compared with its effects on other charges or its reaction to electromagnetic fields. This means in practice that the two sides of (2.3) often lead almost separate lives. Changes in local physical fields usually have almost no effect on the spacetime metric $g_{\mu\nu}$. Conversely, a mathematical identity means that a direct consequence of (2.3) is

$$T^{\mu\nu}_{\ \ ;\nu} = 0, \tag{2.5}$$

where the semi-colon specifies the *covariant derivative* – essentially a kind of four-dimensional vector derivative. Here this has the form of a divergence, and indeed

---

[2] We neglect the cosmological constant term $\Lambda g_{\mu\nu}$ on the left-hand side as this is small in all non-cosmological applications.

the physical content of (2.5) is the conservation laws of energy and momentum for the fields specifying $T_{\mu\nu}$. The metric tensor $g_{\mu\nu}$ appears implicitly in these equations, as it specifies the precise form of the covariant derivatives. Then one often solves the conservation equations $T^{\mu\nu}{}_{;\nu} = 0$ for the matter fields assuming a fixed form of the metric $g_{\mu\nu}$. This is usually described as solving the equations on a fixed background. So in astrophysical applications, detailed knowledge of GR is often not needed, and it is enough to supplement essentially Newtonian treatments with relevant results from GR.

But it is important to realize the limitations of this approach. Many procedures that are routine in Newtonian descriptions are simply meaningless in GR, even when applied to seductively similar-looking formulae. Section 2.5 discusses these and other difficulties in detail, and is particularly recommended to those readers unfamiliar with GR.

2.2   Event Horizons and Causal structure

General relativity is a fundamentally non-linear theory where exact solutions of the field equations describing the spacetime geometry produced by any given matter distribution are limited to a few special situations of high symmetry. And as we shall see, the status of approximations in GR is unclear because the concept of 'small' perturbations is difficult to make covariant. This greatly complicates attempts to describe a real star in GR: if the star does not have complete spherical symmetry (e.g. if it is rotating), even a Newtonian description has to use multipole expansions, and procedures like this are problematic in full GR. So we might imagine that the full GR description of black holes would be even more complex.

Remarkably, this is not so. Black holes are the simplest macroscopic objects in the Universe – all black holes are fully described by a family of exact solutions of the GR field equations (Kerr, 1963). In practice just two physical parameters – mass and angular momentum – characterize the gravitational effect of *all* realistic black holes encountered in the Universe.[3] This remarkable conclusion follows from a long chain of GR results (often called the 'no-hair' theorems, because of the theoretical physicist John Wheeler's dictum that 'a black hole has no hair'). These prove the result rigorously, but their physical content is that gravitational radiation carries away all irregularities in the matter distribution on the light travel time R_g/c,

---

[3] In principle the hole may have a net electromagnetic charge, which is also a parameter, but in reality any charge must be vanishingly small in geometrized units and the spacetime metric is always effectively uncharged Kerr to a very high approximation. But because electromagnetic interactions are far stronger than gravitational (i.e. since $\kappa \ll 1$; cf. (2.4)), even a charge that is utterly negligible in *gravitational* terms in this way may have a significant effect on the motion of charged test particles. See Section 7.3 for a discussion of a specific case of this kind.

so the black hole always quickly settles to a stationary state which must have very high symmetry.

We start by considering the first exact solution of the GR field equations (published almost immediately after they were first announced). This is the Schwarzschild (1916) metric, describing the gravitational field exterior to a non-rotating point mass M at the origin. This is a vacuum, so here $T_{\mu\nu} = 0$. The metric is the unique spherically symmetric and static solution of the vacuum field equations $R_{\mu\nu} = 0$, and is given by

$$ds^2 = -(1 - 2M/r)dt^2 + \frac{dr^2}{(1 - 2M/r)} + r^2(d\theta^2 + \sin^2\theta \, d\phi^2). \qquad (2.6)$$

As usual in the GR literature, this modern form (2.6) uses geometrized units, in which the constants G, c are each set $= 1$. We adopt these units whenever we need to discuss GR directly, as in this section, but we revert to physical units everywhere else. To recover physical units from geometrized ones we remember that $GM/c^2, GM/c^3$ are a length R_g and a time R_g/c, respectively. The quantity $R_s = 2R_g$ is the Schwarzschild radius, which appears twice in the Schwarzschild metric (2.6).[4] Numerically we have

$$R_g = 1.5 \left(\frac{M}{M_\odot} \right) \text{km} \qquad (2.7)$$

so that a stellar-mass black hole has a gravitational radius of a few kilometres, while a supermassive black hole has

$$R_g \simeq 1.5 \times 10^{13} M_8 \text{ cm}, \qquad (2.8)$$

which can be of the order of the Sun–Earth distance (1 AU) or more, depending on the value of M_8.

The form (2.6) is static and has spherical symmetry, as one would expect. The metric coefficients are independent of all the coordinates other than r, except for the $\sin^2\theta$ appearing in the polar-coordinate form of the usual Cartesian element of solid angle. At distances far from the mass point (as $r \to \infty$) the metric becomes the Minkowski metric of special relativity, expressed in spherical polar coordinates, and spacetime is flat, as is again physically reasonable.

But the most striking feature of (2.6) is that the coefficient of dr^2 becomes infinite at $r = R_s$. This 'Schwarzschild singularity' caused years of confusion among GR theorists, and delayed the recognition of the concept of a black hole. We now know

---

[4] In physical units, equation (2.6) is

$$ds^2 = (1 - 2GM/c^2r)c^2dt^2 - dr^2/(1 - 2GM/c^2r) + r^2(d\theta^2 + \sin^2\theta \, d\phi^2).$$

that despite the mathematical singularity in this coefficient, nothing physically singular happens at R_s – as one test, if we work out the Riemann curvature tensor $R_{\mu\nu\lambda\sigma}$ specifying the gravitational field controlling the motion of test particles, we find that

$$R_{\mu\nu\lambda\sigma}R^{\mu\nu\lambda\sigma} \propto \frac{M^2}{r^6}, \tag{2.9}$$

which is perfectly finite at $r = R_s$. As expected, this expression becomes infinite only as $r \to 0$, just as a Newtonian gravitational field would.

We can remove the mathematical singularity at R_s by defining a new coordinate v through the transformation

$$dt = dv - \frac{dr}{1 - 2M/r}, \tag{2.10}$$

that is,

$$v = t + r_* \tag{2.11}$$

where

$$r_* = \int \frac{dr}{1 - 2M/r} = r + 2M \ln|r/2M - 1| + \text{constant}. \tag{2.12}$$

The metric becomes

$$ds^2 = -(1 - 2M/r)dv^2 + 2dvdr + r^2(d\theta^2 + \sin^2\theta d\phi^2), \tag{2.13}$$

which is regular for all $r > 0$. These are called (ingoing) Eddington–Finkelstein coordinates, and their existence shows that the whole region $r > 0$ is a perfectly regular spherically symmetric spacetime.[5] By removing the infinity at $r = 2M$ they also allow us a real insight into the physics of a gravitating point mass, as we can now study the way light rays travel in this spacetime for all $r > 0$. Since no physical effect can propagate faster than light, this approach narrows down the physical causes of any given event. The study of light propagation fixes the *causal structure* of a spacetime.

Light rays propagating under the gravity of the point mass M are given by setting $ds^2 = 0$ in (2.13), and these rays are radial if $d\theta = d\phi = 0$. The two roots of $ds^2 = 0 = d\theta = d\phi$ define two families of light rays. Photons with

$$dv = 0 \tag{2.14}$$

---

[5] As the name suggests, Eddington actually wrote down the non-singular form (2.13) (in 1924). But he and other contemporaries drew no conclusions from its regularity at $r = 2M$. This was just one of several missed opportunities to realize that the 'Schwarzschild singularity' was no such thing, before understanding began to arrive in the mid-1960s. The probable cause was a prevailing view that GR was mathematically intractable and of little importance to other branches of physics, leaving the subject largely neglected at the time.

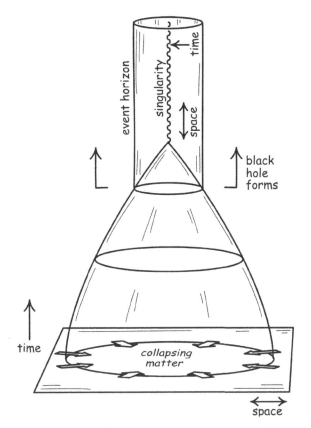

Figure 2.2 The formation of an event horizon as a star collapses under its own gravity. Credit: John D. Norton, Einstein for Everyone, https://sites.pitt.edu/~jdnorton/ teaching/HPS_0410/index.html.

fall inwards: proper time increases as r decreases, since $t + r_* =$ constant along them. The other 'outgoing' family satisfies

$$dv = \frac{2dr}{1 - 2M/r} \tag{2.15}$$

so that $t - r_* =$ constant along them, which is equivalent to requiring

$$v = 2r_* + \text{constant} \tag{2.16}$$

along them. These rays do propagate to larger r provided that they start from positions where $r > 2M$, since from (2.15), $dv > 0$ along them. But rays starting from positions with $r < 2M$ have $dv < 0$, and (from (2.16)) must move to smaller r. So *in the region $r < 2M$ both families of light rays fall inwards.*

This is the defining feature of black holes: a region of spacetime from which light (and so any other physical effect) cannot propagate to infinity. The locus $r = 2M$

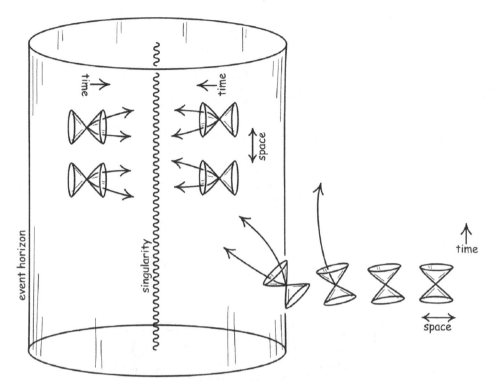

Figure 2.3 Causal structure near a black hole (the upper part of Figure 2.2). The future light cones of all events on the null surface called the *event horizon* (specified by $r = 2M$ in the Schwarzschild metric ((2.6), (2.13))) are tangent to it. No physical future-directed worldline can escape this region. Inside the event horizon, all future light cones point towards the singularity. Credit: John D. Norton, Einstein for Everyone, https://sites.pitt.edu/~jdnorton/teaching/HPS_0410/index.html.

is a null (lightlike) surface, since $dv = 0$ there. This is the *event horizon* defining the trapped region (see Figure 2.2). Another way of describing these black hole properties is that for $r > 2M$ the light cones contain directions pointing to both large and smaller r, but for $r < 2M$ always point to smaller r. Light can propagate along the null surface $r = 2M$ as the local light cones are tangent to it – see Figure 2.3.

2.3 Black Hole Spacetimes

Understanding the causal structure of the Schwarzschild solution opened the way to a general picture of black hole spacetimes. Theorists derived a succession of fundamental results within a few years. All-importantly, although this work was

informed by the insights gained in discussing exact solutions of the field equations of GR with high symmetry, particularly the concept of an event horizon, the
results themselves did not assume *any* symmetries, but simply a rigorous version
of the central idea that gravity is always attractive.[6] Technically this is achieved
by requiring certain positivity properties of projections of the energy–momentum
tensor which is the source of the gravitational field – the right-hand side of the
Einstein field equations. This idea can be used to show two crucial things.

First, the existence of an event horizon – a 'closed trapped surface', such as the
region $r < 2M$ in the Schwarzschild solution – implies that somewhere within
it, matter must collapse to extreme densities, as occurs at the genuine 'curvature'
singularity at $r = 0$ in the Schwarzschild metric (see (2.9)). Here matter densities
formally try to become infinite, and current physics – including GR itself – must
break down. The generality of the methods of proof show that existence of this
singularity has nothing to do with symmetry. A singularity always appears, however
asymmetric the matter distribution is, provided only that an event horizon forms.

The second fundamental result is that if the spacetime exterior to the event horizon settles to some time-independent end state, this must be axisymmetric. Again,
this result is quite unaffected by the lack of any symmetries in the original matter
distribution that collapsed to form the black hole event horizon.

This result strongly suggests that all such black hole end states must be characterized by their masses and angular momenta (spins) alone. The spin itself must
be time-independent (this kind of situation in GR is called 'stationary', rather
than 'static', which excludes steady rotation). All other properties of the collapsing matter forming the hole – its mass multipoles – have disappeared. Physically,
gravitational waves must have radiated away these irregularities during the collapse. An axisymmetric system cannot emit gravitational radiation, and so a black
hole with this symmetry is the end point of the collapse. The free-fall velocity of
collapse approaches c near a black hole, so this radiation must be emitted on a
timescale $R_g/c \sim 500 M_8$ s (a few milliseconds for a stellar-mass black hole). The
same things must happen whenever the hole accretes more matter, which may of
course be another black hole. A short burst of gravitational radiation leaves behind
a new stationary and axisymmetric black hole with a larger mass and a new angular momentum reflecting what was added. Recent LIGO–Virgo observations give
direct tests of this in the case where two stellar-mass black holes merge.

A general point follows from this – *all* black hole interactions with external matter try to re-establish stationary and axisymmetric conditions. So from a
quasi-Newtonian viewpoint, *forces* must act between the hole and the interacting

---

[6] This conceptual breakthrough earned Roger Penrose a half-share of the 2020 Nobel Prize in Physics.

matter trying to bring these symmetries about. We shall discuss examples of these 'symmetry forces' in Section 2.4.

The next step in the chain of results constituting the modern picture of black holes was the proof that there is a unique family of solutions of the vacuum GR field equations with the three properties of stationarity, axisymmetry, and the possession of a regular event horizon. The solutions of this family are completely specified by their masses and spins.[7]

Given the complexity of the field equations, it is a supreme piece of good fortune that these metrics are given by a family of exact analytic solutions. The space-time metric of this Kerr (1963) family is (in what are called Boyer–Lindquist coordinates)

$$ds^2 = -(1 - 2Mr/\Sigma)dt^2 - (4Ma'r \sin^2\theta/\Sigma)dtd\phi + (\Sigma/\Delta)dr^2$$
$$+ \Sigma d\theta^2 + (r^2 + a'^2 + 2Ma'^2 r \sin^2\theta/\Sigma)\sin^2\theta d\phi^2 \qquad (2.17)$$

where

$$\Delta = r^2 - 2Mr + a'^2, \qquad (2.18)$$
$$\Sigma = r^2 + a'^2 \cos^2\theta. \qquad (2.19)$$

The quantity a' fixes the modulus of the black hole's specific angular momentum (and so $a' \geq 0$) and has the same dimensions as M in geometrized units. The magnitude of the black hole's physical angular momentum is

$$J = R_g ca' = \frac{GM}{c}a' = \frac{GM^2}{c}a, \qquad (2.20)$$

where the dimensionless quantity $a = a'/M$ is generally used in the astrophysical literature rather than a'.[8]

For $a' = 0$, the Kerr metric (2.17) reduces to Schwarzschild (cf. (2.6)), as we would expect. The event horizon defining the one-way membrane characteristic of black holes is at the larger root of the equation $\Delta = 0$, that is,

$$r = r_+ = M + (M^2 - a'^2)^{1/2}, \qquad (2.21)$$

---

[7] In principle the hole can also have an electric charge, although in practice this is never large enough to have a significant gravitational effect on the spacetime metric: in geometrized units the charge is always negligible. But, as emphasized after equation (2.4), because of the extreme weakness of gravity compared with all other forces it is possible (and may indeed happen) that a gravitationally negligible charge on the black hole can have purely *electromagnetic* effects.

[8] In physical units, Equation (2.17) is

$$ds^2 = -(1 - 2GMr/c^2\Sigma)c^2 dt^2 - (4GMa'r \sin^2\theta/c\Sigma)dtd\phi + (\Sigma/\Delta)dr^2$$
$$+ \Sigma d\theta^2 + [(r^2 + a'^2) + 2GMa'^2 r \sin^2\theta/c^2\Sigma]\sin^2\theta d\phi^2.$$

where the coefficient of dr^2 tends to infinity. For $a' \geq M$ one can show that there is no event horizon, so this is no longer a black hole. The lack of an event horizon would allow observers to 'see' the physical singularity at $r = 0$, where the gravitational field becomes infinite. This 'naked singularity' is generally regarded as unphysical – evidently conventional GR breaks down there, and must be replaced by some newer theory. A real naked singularity would be able to contaminate the surrounding spacetime with effects unconstrained by current physics. It would then be difficult to understand why conventional physics appears to provide a good description of most of the Universe.

Fortunately, we will see that there are good reasons to believe that no physical process can ever spin a black hole up to make a' equal to M, avoiding this catastrophe. This 'cosmic censorship' principle then allows us to use current physics on macroscopic scales everywhere in the Universe except where we cannot observe, that is, the interior of any event horizon. Accordingly, we adopt the conventional view that the angular momentum parameter obeys $a' < M$ (analogous to a rotational breakup limit).

Over the range $0 < a' < M$ we have $2M > r_+ > M$. Just as in the non-rotating (Schwarzschild) case, we can remove the coordinate singularity at the horizon location by transforming from Boyer–Lindquist coordinates to Kerr coordinates, which are analogues of the Eddington–Finkelstein coordinates used in the Schwarzschild case.

The larger root of the equation $(\Sigma - 2Mr) = 0$, where the coefficient of dt^2 tends to infinity, defines what is called the *ergosphere* of the black hole:

$$r = r_{\text{ergo}} = M + (M^2 - a'^2 \cos^2 \theta)^{1/2}. \tag{2.22}$$

This is outside the event horizon except at the poles $\theta = 0$ where the two radii are equal. All physical (so timelike or null) orbits within r_{ergo} are forced to move in the positive ϕ direction – they are dragged around in the direction of the hole's spin. This is an explicit example of the dragging of inertial frames by rotation in GR, known as the Lense–Thirring effect. Outside r_{ergo} both prograde and retrograde orbits are possible, but if they are inclined with respect to the black hole spin, they must precess in the positive ϕ direction at the Lense–Thirring frequency

$$\omega = -\frac{g_{t\phi}}{g_{\phi\phi}} \simeq \frac{2ac}{R_g}\left(\frac{R_g}{r}\right)^3. \tag{2.23}$$

Here $g_{tt}, 2g_{t\phi}$ are the coefficients of $dt^2, dtd\phi$ in (2.17), expressed in physical units.

2.4 Extracting Black Hole Spin Energy

The dragging effect on orbits inside the ergosphere opens the possibility of extracting energy from the spin of the black hole, formally because there exist physical
orbits whose energies are negative with respect to infinity. The Penrose process
shows how this can work in principle: a body breaking into fragments inside the
ergosphere may put one or more of them on negative-energy orbits, which all eventually plunge through the event horizon into the black hole. Since orbital energy
is a conserved quantity (the Kerr metric is stationary, so there is a timelike Killing vector – see Section 2.5), the sum of the energies of the remaining fragments
must exceed the original energy of the body before it broke up. The extra energy
can only have come from the black hole spin. Since the ergosphere lies outside
the event horizon, this extracted spin energy can in principle be transported to the
exterior spacetime.

There have been a number of suggestions of how extracting black hole spin
energy might work in practice. It appears remarkably difficult to find astrophysically reasonable ways of doing this (see Bardeen, Press & Teukolsky, 1972, who
discuss possible particle and radiation reaction processes and conclude that all are
'unlikely in any astrophysically plausible context'[9]). One idea that has been popular is that magnetic fields might manage this (Blandford & Znajek, 1977). We will
see in Section 7.3 that this too is questionable.

There is one known case where spin energy is clearly extracted, although its
practical importance for astrophysical applications is unclear. We saw in Section
2.3 that the result that all stationary black holes must be axisymmetric means that
from a Newtonian viewpoint there must be forces acting between a spinning black
hole and its physical surroundings. An example of these 'symmetry forces' occurs
when we consider a spinning black hole immersed in an external magnetic field. If
the magnetic field is not parallel to the hole spin, axisymmetry is violated and the
hole–field system must evolve in time until it is restored. There are evidently two
limits, depending on whether the field or the hole have the greater inertia.

In the first case the field must be fixed by currents in sources far from the
hole. Near the hole the field must be distorted from uniformity to obey the correct boundary condition (no outgoing electromagnetic radiation) at the black hole
event horizon. A long GR calculation (King & Lasota, 1977) shows that the hole
acts as if it were a Newtonian body subject to a torque

$$\mathbf{T} = \frac{2G^2}{3c^5} M(\mathbf{J} \times \mathbf{B}) \times \mathbf{B} \qquad (2.24)$$

---

[9] The main difficulty appears to be in arranging nearby matter to have near-neighbouring trajectories with
velocities differing by a significant fraction of c.

where $\mathbf{J}, \mathbf{B}$ are the hole angular momentum and the (uniform) magnetic field strength far from the hole. This is an explicit example of a symmetry force.

The form of (2.24) means that there is no precession as the hole aligns. The angular momentum component $\mathbf{J}_\parallel$ parallel to $\mathbf{J}$ remains fixed, and the total angular momentum $J = |\mathbf{J}|$ decreases on the timescale

$$t_h = \frac{3c^5}{2G^2 M B^2},\tag{2.25}$$

extracting rotational energy from the hole.

The surprisingly simple form (2.24) has a straightforward interpretation. With the substitutions $\mu = R_g^3 \mathbf{B}$, $\omega = \mathbf{J}/aMR_g^2$ (so $\omega = c/R_g$) where $R_g = GM/c^2$ is the gravitational radius, we get

$$\mathbf{T} = \frac{2\omega^2}{3c^3}(\omega \times \mu) \times \mu,\tag{2.26}$$

which is the formula for the effective reaction torque produced by radiation from a magnetic dipole of moment μ rotating with angular velocity ω. The effect of immersing the black hole in a fixed misaligned field is evidently equivalent to inducing an effective dipole moment within its ergosphere. This is forced to rotate in the direction of the hole spin, producing the dipole emission and the torque. This process carries energy and angular momentum from the hole spin to infinity. Some of this is in the form of gravitational radiation (as the mass and spin of the hole are changed) and some as electromagnetic radiation (the emission along the spin axis is circularly polarized, which carries off angular momentum).

In practice the timescale (2.25) is far too long in a realistic interstellar magnetic field for this magnetic alignment effect to be important.[10] But the opposite limit, in which the magnetic field has much lower inertia than the hole, can produce significant dipole emission. However, here the energy is extracted from accretion, as the black hole spin remains unchanged – see Section 7.3.

Since energy is equivalent to mass, extracting spin energy must mean that the black hole mass M *decreases*. There is a simple and elegant way of quantifying this. We can define the area of the black hole event horizon as the area of any timelike two-dimensional slice through it. Because the horizon is a null (lightlike) surface, one can show that the resulting area is independent of whatever particular slicing we take. So in the Schwarzschild case (equation (2.6)) we take $dt = 0, r = 2M$. Evidently the two-dimensional sphere this defines has the simple metric

$$ds^2 = 4M^2(d\theta^2 + \sin^2\theta\, d\phi^2),\tag{2.27}$$

---

[10] It could conceivably be important in the collapse of a rapidly rotating but misaligned core of a massive star in a gamma-ray burst, if the core has an equipartition-strength magnetic field.

which clearly has total area

$$A = 16\pi M^2 = 4\pi R_g^2, \tag{2.28}$$

which in physical units is

$$A = 16\pi \left(\frac{GM}{c^2}\right)^2. \tag{2.29}$$

A similar calculation for the Kerr solution (2.17) generalizes (2.28) to

$$A = 8\pi M^2 [1 + (1 - a^2)^{1/2}] \tag{2.30}$$

where we have again used the dimensionless spin parameter $a = a'/M$. The importance of the event horizon area A comes from a global result by Hawking (1971), which states that

> in any non-quantum physical interaction involving black holes, the total area of the event horizon(s) involved can never decrease.

Note that this result – usually called the area theorem – does not assume any symmetries for the black holes concerned: they are (by definition) not stationary as they are interacting, and similarly not spherically or even axisymmetric. The result only uses the fact that gravity is always attractive.[11]

The form (2.30) is interesting in combination with Hawking's result. If we spin up the black hole (e.g. by letting it accrete matter from prograde orbits), a increases, and this would violate Hawking's result unless there was a simultaneous increase of the black hole mass M. This of course is what we expect: spin has energy, which has mass. Conversely, spin-down can *decrease* M, but not so much that A decreases. Spinning the hole all the way from $a = 1$ to $a = 0$ can at most decrease the mass from M to a value $M/\sqrt{2}$. As a result the maximum rotational energy extraction from a spinning black hole of mass M is

$$Mc^2(1 - 1/\sqrt{2}) \simeq 0.29Mc^2. \tag{2.31}$$

---

[11] The 'never decreasing' property of A is very reminiscent of entropy, as expressed in the second law of thermodynamics, and indeed this turns out to be no accident: work by Hawking and by Bekenstein (1973) shows that the entropy of a black hole is proportional to A, with a large constant in front which one can think of as allowing for the many ways in which a black hole of given mass and spin could have formed from ordinary matter. The counterpart of temperature turns out to be the surface gravity, which is the acceleration, as exerted at infinity, required to keep an object at the event horizon. This is always inversely proportional to the black hole mass M, with an additional spin dependence for Kerr holes, and characterizes the quantum-mechanical Hawking radiation from black holes. This radiation *is* able to make the horizon area decrease, and in principle can eventually cause the hole to evaporate. Although fascinating for illuminating the connections between gravity, thermodynamics, and quantum mechanics, these properties do not so far appear to have any astrophysical importance: a $1M_\odot$ black hole has a temperature of 6×10^{-8} K, and a black hole mass $\lesssim 10^{15}$ g $\sim 10^{-18} M_\odot$ is needed for a surface temperature exceeding even the low-redshift microwave background value 2.7 K.

Another way of formulating this is to make a^2 the subject of the formula (2.30) and use (2.20) to express it in terms of the hole's angular momentum J. This gives

$$M^2 = M_{\text{irr}}^2 + \frac{4\pi J^2}{Ac^2},$$
(2.32)

where

$$M_{\text{irr}}^2 = \frac{Ac^4}{16\pi G^2}.$$
(2.33)

The form (2.32) shows explicitly that the black hole mass depends on two components, one specified by the black hole angular momentum, which can be made to decrease, or even vanish entirely, and one which cannot in the absence of quantum-mechanical interactions decrease at all. M_{irr} is called the *irreducible mass* of the hole.

LIGO–Virgo observations of gravitational radiation emitted in mergers of black holes give strong and explicit support to it – the black hole formed in the merger always has a larger area than the sum of the areas of the two merging holes.

2.5 Limitations of Quasi-Newtonian Treatments

We have discussed properties of black holes in this chapter by taking results in GR and often setting them in a quasi-Newtonian context. This is a widely adopted way of incorporating them into astrophysical treatments. But it is clear that there must be limitations to this approach. In this section we discuss several instances where quasi-Newtonian treatments may fail to capture the physics correctly, and some where they can be positively misleading.

We can take three important points like this from the discussion of the last two sections. First, the black hole property is *global*, that is, it is defined by the asymptotic behaviour of light rays at large distances from the hole, and not by local physics near the event horizon. In particular, there is no reason for local physics to be any different or more extreme near the event horizon – it does not 'know' whether any particular light ray will reach future null infinity. Certainly, gas does not necessarily heat up and radiate more strongly in its vicinity unless the local equations of motion indicate that dissipation increases there.

Second, we should not be surprised by coordinate singularities, as seen in the form (2.6) at $r = 2M$, or the fact that a particular choice of coordinates does not describe the whole spacetime (e.g. the null surface $r = 2M$). For example, we can take the flat-space Minkowski metric

$$ds^2 = -dt^2 + dx^2 + dy^2 + dz^2$$
(2.34)

and (perversely) define a new coordinate p by setting $z = t/(1-p)$, transforming the metric to

$$ds^2 = -dt^2 + dx^2 + dy^2 + \frac{dt^2}{(1-p)^2} + \frac{2t\,dt\,dp}{(1-p)^3} + \frac{t^2\,dp^2}{(1-p)^4}, \tag{2.35}$$

which is singular as $p \to 1$. Of course, nothing special happens near $p = 1$: spacetime is still flat, with zero gravitational field everywhere. So a coordinate singularity is not sufficient to indicate an interesting feature of a spacetime. A coordinate singularity is not necessary at such locations either, as the completely regular form (2.13) of the Schwarzschild metric shows. So it is simply a minor mathematical coincidence that something interesting – the event horizon – does happen at the mathematical singularity of Schwarzschild's original choice of coordinates in (2.6).

Third, coordinates in general have no physical meaning in GR. They are simply labels distinguishing spacetime points from each other, and even this fails at mathematical singularities. They can change character – the coordinate r in both the Schwarzschild and Kerr metrics has the standard appearance of a radial coordinate in spherical polars at large r but becomes null (lightlike) at the event horizon. Coordinates do not in general measure invariant properties such as proper times or distances, except possibly in cases where the spacetime has well-defined symmetries and the coordinates are carefully chosen. Symmetries are defined in an invariant coordinate-free way by the existence of what are called Killing vectors.[12] Killing vectors specify globally conserved quantities – energy if the spacetime has a timelike Killing vector, angular momentum components if it has rotational symmetries, or a linear momentum component if the Killing vector specifies a translational symmetry. Because Minkowski space has all these symmetries, these conservation laws hold in special relativity. But symmetries like this are very unusual in GR, in the sense that writing down an arbitrary metric does not usually define any symmetries at all, and so no conserved quantities. A direct consequence of this is that the concept of perturbations of spacetimes with few or no symmetries is not in general meaningful. Unless the coordinates are defined by inherent symmetries (i.e. Killing vectors), 'small' coordinate increments can be made arbitrarily large (or vice versa) by changing coordinates (e.g. compare the form (2.35)). Worse, coordinates can become singular in various limits in GR.

Figure (2.4) shows a dramatic example of this. For extreme spin rates $a \to 1$, three physically completely distinct circular equatorial orbits, the marginally bound and marginally stable (ISCO) particle orbits, as well as the photon orbit, all have the same radial coordinate $r \to M$, and so appear to lie on the event horizon. But

---

[12] A Killing vector ξ_ν satisfies the condition $\xi_{\nu;\mu} + \xi_{\mu;\nu} = 0$. In a spacetime with no symmetries the covariant derivatives here produce a condition that no vector can satisfy.

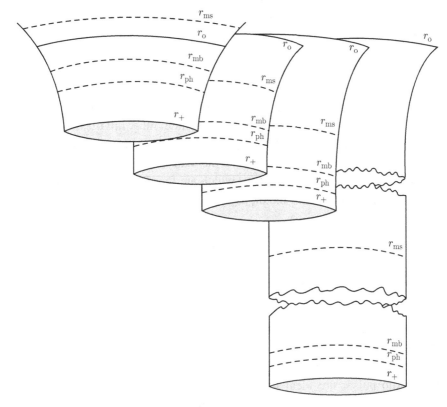

Figure 2.4 Embedding diagrams of the 'plane' $\theta = \pi/2, t = $ constant for rotating black holes with $a \to 1$. The Boyer–Lindquist radial coordinate r determines only the circumference of the 'tube'. When $a \to M$, the marginally stable (ISCO) (r_{ms}) and marginally bound (r_{mb}) particle orbits, as well as the photon orbit (r_{ph}), all have the same circumference and coordinate radius, although, as the embedding diagram shows clearly, they are in fact completely distinct. Credit: Bardeen, Press and Teukolsky (1972).

we know that the horizon is a null hypersurface, and no timelike curves (particle orbits) can lie in it. As the paper by Bardeen, Press and Teukolsky (1972) shows, all three orbits are actually outside the horizon, and all at distinct proper distances from each other – in fact the proper distance between the marginally bound and stable particle orbits becomes infinite (in a spacelike direction) as $a \to 1$.

The coordinate problem obviously affects numerical treatments, as these directly appeal to the concept of expansions in small quantities. Interpreting the results of numerical calculations in GR is inherently far more difficult than in Newtonian or special-relativistic treatments. This is true even if we assume that the metric is fixed, for example as one of the Kerr family, if we are considering gas of negligible mass orbiting or accreting on to a black hole. Numerical calculations of the motion of gas close to the horizon require very careful consideration to extract correct physics.

Another manifestation of the coordinate problem is that it is not in general obvious what the physical meaning (if any) of parameters appearing in a general metric are. For example, to define the mass or angular momentum of an isolated body, the metric coefficients and tensors defined using them must fall off in well-defined ways at large distance – the spacetime represented by the metric must be 'asymptotically flat'. One useful simplification is that the 'causal structure' – the ensemble of light ray paths – are identical in metrics which are 'conformal', that is, differ only by a (possibly variable) overall factor, as

$$ds^2 = g_{\mu\nu}dx^\mu dx^\nu$$

and

$$ds^2 = \Omega^2 g_{\mu\nu}dx^\mu dx^\nu,$$

since light rays are defined by the condition $g_{\mu\nu}dx^\mu dx^\nu = 0$. So, for example, an event horizon is defined by identical coordinate conditions in both metrics.

Major efforts have gone into handling these difficulties in two astronomical contexts. First, close stellar-mass binary systems where one component is a rapidly spinning radio pulsar are ideal for testing the weak-field approximation to GR, where we can be reasonably confident that all the gravitational perturbations really are small because we know that special relativity works to enormous accuracy in much of the Universe. The agreement with observations is quite remarkable (see Figures 2.5 and 2.6).

Second, the search for gravitational waves has stimulated huge efforts to relate the strong-field regime of the merger of two black holes (or a black hole and a neutron star) to the weak-field limit in which the waves are detected. This approach again appears to be justified, given the impressive agreement with the first LIGO detections. But it is clear that these successes largely result because of the close connection to the weak-field limit and its symmetries and conservation laws, and we cannot be so confident in other contexts.

As a final example of the problems that can affect quasi-Newtonian treatments, we consider what happens near the event horizon, and in particular its relation to boundary conditions. This can have significant effects on numerical treatments. A simple and illuminating problem is to describe the accretion of a particle with charge q. To keep the discussion as simple as possible we assume that the black hole is non-rotating (Schwarzschild) and has no electric charge at the start of the calculation. We can anticipate that accreting this particle will cause the black hole to become a charged (Reissner–Nordström) black hole. But in naive Newtonian

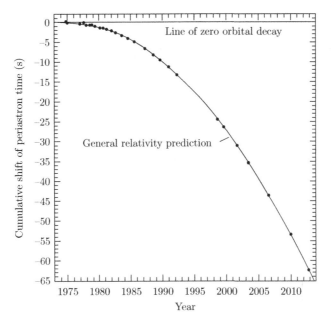

Figure 2.5 Shift in the time of periastron passage (Einstein precession) of the Hulse–Taylor binary pulsar PSR B1913+16 caused by gravitational wave damping. The parabola is the GR prediction. The error bars on the observed points are too small to show. Credit: Weisberg and Huang (2016).

terms this appears surprising, as one might imagine the particle has to fall into the hole to give radial fieldlines, but then cannot have any influence on the spacetime outside the hole. In calculating what happens is we imagine the particle placed at various distances from the black hole, assuming it is held stationary. This is a surprisingly accurate description of infall, since as viewed by a distant observer, the time as a function of radial distance r of a freely falling particle is

$$\tau(r) = -\frac{4}{3}x^3 - 4x + 2\ln\left(\frac{x+1}{x-1}\right) \tag{2.36}$$

where $x = (r/2M)^{1/2}$. We see that $\tau \to \infty$ as $r \to 2M$, so that the particle appears to hover ever more slowly above the horizon, and never reaches it. The fieldlines of the charge are given by solving the time-independent Maxwell equations in the Schwarzschild background. We see from Figures 2.7 and 2.8 that far from the charge, the fieldlines are radial about it. Between the charge and the horizon the fieldlines curve closer together, but away from this region they spread out. We can interpret this as the charge inducing an apparent distribution of charge on the horizon – of opposite sign to the original charge q – where the fieldlines curve closer, and of the same sign as q as they curve away. The total induced charge

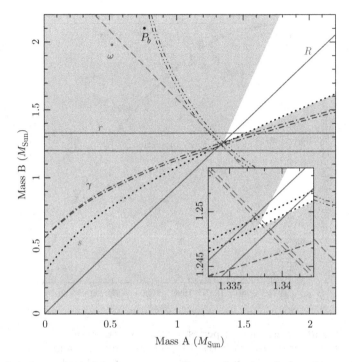

Figure 2.6 General relativity mass–mass diagram based on timing observations of the double pulsar J0737–3039, where both neutron stars are pulsars. The curves represent various observational constraints which must be satisfied (doubled to show the error ranges where these can be resolved). The inset shows the region around the fitted masses at 10× scale. The entire shaded region is *excluded* because it would require the orbital inclination i to satisfy $\sin i > 1$. Only masses in the small double track across the point of the white area in the inset are permitted by GR, fixing the masses of the two neutron stars to accuracies approaching $10^{-3} M_\odot$. Credit: Kramer et al. (2006).

adds up to zero, of course. As the charge q gets very close to the horizon (where the distant observer sees it 'hover'), the induced opposing charges crowd closely together, while the distributed like charges spread almost evenly over the horizon. In the limit this tends to a dipole of zero strength where the original charge is very close to the horizon, with the original charge q spread uniformly over the horizon. A distant observer sees this as a Reissner–Nordström black hole. Accordingly, the correct boundary condition is that the hole simply gains the charge q, just as it gains the mass of the particle, and would gain its angular momentum if it fell in from the ISCO.

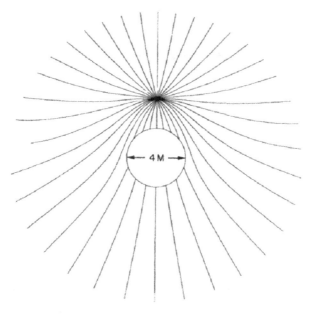

Figure 2.7 Fieldlines from a point charge at $R = 4M$ in the Schwarzschild metric. Credit: Hanni and Ruffini (1973).

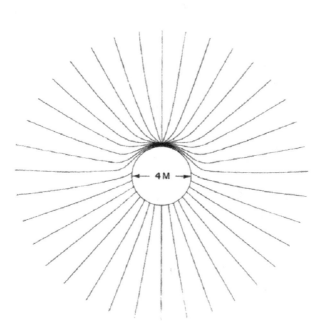

Figure 2.8 Fieldlines from a point charge at $R = 2.2M$ in the Schwarzschild metric. Credit: Hanni and Ruffini (1973).

3

Astrophysical Gases

3.1 Introduction

Essentially, all accreting matter near black holes is gaseous, like most of the baryonic matter in the Universe. In this chapter we give a brief introduction to the gas dynamics we use in the rest of the book. (For a fuller description see Chapter 3 in Frank, King & Raine, 2002, Chapter 3, or Pringle & King, 2007.)

A gas is a collection of particles which interact through short-range forces. In most astrophysical cases these are Coulomb collisions between electrons and ions, but for a fluid description to apply all we need to know is that the mean free path λ between collisions is much shorter than all other scales of interest. Then the net effect of the collisions is very often to randomize the motions of the gas particles around a mean velocity $\mathbf{v}$. In the reference frame moving with $\mathbf{v}$, the particles have Maxwell–Boltzmann velocity distributions described by a temperature T. On lengthscales $L \gg \lambda$ we can treat the fluid as a continuous medium with a velocity $\mathbf{v}$, temperature T and mass density ρ, which are all functions of position and time. We can study how these quantities change over space and time by using the conservation laws for mass, momentum and energy, and this is gas dynamics. Sometimes the gas-dynamical equations do not apply, for example if there are large changes in $\mathbf{v}$, ρ and T on lengthscales $\sim \lambda$. In other cases the detailed behaviour of the gas is strongly influenced by the particle interactions themselves, and by the magnetic fields the particle motions may generate, and we have to consider more detailed treatments, such as magnetohydrodynamics or plasma physics. We do not attempt a comprehensive treatment of these subjects here, and in some cases simply write down the equations rather than giving derivations. The interested reader can find these in many books (see, e.g. Choudhuri 2010; Shu, 2011).

3.2 Gas Dynamics

We start with the simplest treatment of gas dynamics, where we neglect direct effects of the finite mean path λ, such as diffusion and viscosity. We will consider these in Section 3.3 onwards.

As in any fluid, mass conservation is expressed by

$$\frac{\partial \rho}{\partial t} + \nabla.(\rho \mathbf{v}) = 0, \tag{3.1}$$

which is often called the continuity equation. Momentum conservation is given by the Euler equation

$$\rho \left(\frac{\partial \mathbf{v}}{\partial t} + \mathbf{v}.\nabla \mathbf{v} \right) = -\nabla P + \mathbf{f}. \tag{3.2}$$

Here $\mathbf{f}$ is the force density (force per unit mass) of all the forces (e.g. gravity, radiation, and magnetic forces) acting on the gas, with the exception of those caused by variations in the gas pressure P. This is usually related to density and temperature through the perfect gas law

$$P = \frac{\rho k T}{\mu m_{\mathrm{H}}}, \tag{3.3}$$

where k is Boltzmann's constant, $m_H \simeq m_p$ is the mass of the hydrogen atom (with m_p the proton mass) and μ is the mean mass per gas particle in units of m_{H} (so $\mu = 1$ for neutral hydrogen, $= 1/2$ for ionized hydrogen, and somewhere between for all the gases we will consider). In some cases radiation pressure is important, so that for a gas close to thermal equilibrium we have

$$P = \frac{\rho k T}{\mu m_{\mathrm{H}}} + \frac{4\sigma}{3c} T^4, \tag{3.4}$$

where σ is the Stefan–Boltzmann constant.

The third conservation law is energy:

$$\frac{\partial}{\partial t} \left(\frac{1}{2}\rho v^2 + \rho \epsilon \right) + \nabla. \left[\left(\frac{1}{2}\rho v^2 + \rho \epsilon + P \right) \mathbf{v} \right] = \mathbf{f}.\mathbf{v} - \nabla.\mathbf{F}_{\mathrm{rad}} - \nabla.\mathbf{q}. \tag{3.5}$$

Here ϵ is the thermal energy per unit mass, and can change as the pressure does work on the fluid by compressing it, or the gas expands. The term $\mathbf{F}_{\mathrm{rad}}$ is the energy flux from radiation emitted or absorbed by the gas, and $\mathbf{q}$ is the heat flux arising from temperature gradients within it.

This is the most complicated of the conservation laws, and in principle we would need to add further conservation laws, for example for radiation energy (the radiative transfer equation), but we can very often avoid this by adopting simple forms for $\mathbf{F}_{rad}$ and $\mathbf{q}$. To a good approximation, most astrophysical gases are monatomic, so that

$$\epsilon = \frac{3kT}{2\mu m_H}. \tag{3.6}$$

If the gas is optically thin to radiation we have

$$- \nabla . \mathbf{F}_{rad} = 4\pi \int j_\nu d\nu, \tag{3.7}$$

where j_ν erg cm^{-3} s^{-1} is the emissivity of the gas at frequency ν. In the opposite limit of a very optically thick gas, we have instead

$$\mathbf{F}_{rad} = - \left(\frac{16\sigma}{3\kappa_R \rho} \right) T^3 \nabla T, \tag{3.8}$$

where κ_R is the Rosseland mean opacity defined by

$$\frac{1}{\kappa_R} = \frac{\int \frac{1}{\kappa_\nu} \frac{\partial B_\nu}{\partial T} d\nu}{\int \frac{\partial B_\nu}{\partial T} d\nu}, \tag{3.9}$$

where κ_ν is the monochromatic opacity, and B_ν is the Planck function

$$B_\nu[T(R)] = \frac{2h\nu^3}{c^2(e^{h\nu/kT(R)} - 1)} \text{ erg s}^{-1} \text{ cm}^{-2} \text{ sr}^{-1}. \tag{3.10}$$

We note that κ_ν, κ_R both have dimensions 1/(length): in general, $\kappa_\nu^{-1}, \kappa_R^{-1}$ are of order the effective photon mean free path in the gas.

In hydrostatic equilibrium under gas pressure we have $\mathbf{v} = 0$. Given a hydrostatic solution $\rho = \rho_0(r)$, $P = P_0(r)$ we can consider small perturbations $\rho = \rho_0 + \rho'$, $P = P_0 + P'$, and $\mathbf{v} = \mathbf{v}'$ where all the primed quantities are assumed small. In general we have a relation

$$P = K\rho^\gamma \tag{3.11}$$

between the pressure and density perturbations. Here K is a constant, and $\gamma = 5/3$ or 1 in the two cases of adiabatic flow (i.e. with no heat exchange with the surroundings, so that the right-hand side of (3.5) is zero), or isothermal flow, where the terms on the right-hand side of (3.5) are assumed to keep the temperature T constant everywhere. In practice both cases occur – adiabatic flow holds if the flow is so rapid that there is no time for interaction with the surroundings, and isothermal flow when cooling is very fast.

Then linearizing (3.1) and (3.2), we get

$$\frac{\partial \rho'}{\partial t} + \rho_0 \nabla . \mathbf{v}' = 0$$
$$\frac{\partial \mathbf{v}'}{\partial t} + \frac{1}{\rho_0} \nabla P' = 0. \tag{3.12}$$

From (3.11), $P = P(\rho)$, so $\nabla P' = (dP/d\rho)_0 \nabla \rho'$, where the subscript indicates that the derivative is taken for the zero-order solution $P = P_0, \rho = \rho_0$. Then the second of (3.12) becomes

$$\frac{\partial \mathbf{v}'}{\partial t} + \frac{1}{\rho_0} \left(\frac{dP}{d\rho} \right)_0 \nabla \rho' = 0. \tag{3.13}$$

Eliminating $\mathbf{v}'$ between this equation and the first of (3.12) by operating with ∇. and $\partial/\partial t$, respectively, and subtracting, we get finally

$$\frac{\partial^2 \rho'}{\partial t^2} = c_s^2 \nabla^2 \rho', \tag{3.14}$$

where

$$c_s^2 = \frac{dP}{d\rho}. \tag{3.15}$$

We see that (3.14) is the wave equation, with wave speed c_s. Clearly all the other small quantities $P', \mathbf{v}'$ also obey wave equations with speed c_s. So small perturbations (e.g. of pressure) propagate through the gas as waves, with sound speed

$$c_s = \left(\frac{5P}{3\rho} \right)^{1/2} \left(\frac{5kT}{3\mu m_H} \right)^{1/2} \propto \rho^{1/3} \tag{3.16}$$

and

$$c_s = \left(\frac{P}{\rho} \right)^{1/2} \left(\frac{kT}{\mu m_H} \right)^{1/2} \tag{3.17}$$

in the adiabatic and isothermal cases. Numerically, we have

$$c_s \simeq 10(T/10^4 \text{ K})^{1/2} \text{ km s}^{-1} \tag{3.18}$$

in both cases. The sound speed c_s is a fundamental quantity defined at all points in a gas, and it is always of order the mean thermal speed of the ions (see (3.6)). In terms of physical causality, it plays a similar role in a gas to the speed of light in other contexts, as it limits the speed of pressure changes. So if the flow is supersonic, that is, $v > c_s$, pressure gradients have little effect as they cannot respond in time to influence the flow. Conversely in subsonic flow, that is, $v < c_s$, pressure can adjust rapidly and the flow passes through a sequence of states in near-hydrostatic balance. These properties follow from estimating the terms in the Euler equation (3.2). We have

$$\frac{|\rho \mathbf{v}.\nabla}{\mathbf{v}|} |\nabla P| \sim \frac{v^2/L}{P/\rho L} \sim \frac{v^2}{c_s^2} = \mathcal{M}^2, \tag{3.19}$$

where

$$\mathcal{M} = \frac{v}{c_s} \tag{3.20}$$

is the Mach number. For supersonic flow $v/c_s > 1$, so we can neglect pressure gradients to a first approximation. We will see in Section 3.3 that the density dependence $c_s \propto \rho^{1/3}$ of the adiabatic sound speed is important.

In *steady* flows ($\partial/\partial t = 0$) with a force density given by $\mathbf{f} = -\nabla\phi$ (e.g. under gravity), operating on the Euler equation (3.2) with $\mathbf{v}$. shows that $(\mathbf{v}.\nabla)\mathcal{B} = 0$, where

$$\mathcal{B} = \int \frac{\mathrm{d}P}{\rho} + \frac{v^2}{2} + \phi. \tag{3.21}$$

This is Bernoulli's theorem, and $\mathcal{B}$ is the Bernoulli integral, which is constant along streamlines of the flow. We can evaluate $\mathcal{B}$ explicitly if the polytropic relation (3.11) holds:

$$\mathcal{B} = \frac{\gamma}{\gamma - 1}\frac{P}{\rho} + \frac{v^2}{2} + \phi = \frac{c_s^2}{\gamma - 1} + \frac{v^2}{2} + \phi \tag{3.22}$$

if $\gamma \neq 1$, and

$$\mathcal{B} = K \ln P + \frac{v^2}{2} + \phi \tag{3.23}$$

for the isothermal case $\gamma = 1$.

To get a feel for the subject, it is instructive to consider a case frequently (if not almost exclusively) treated in the standard fluid dynamics literature, although it is almost never applicable for an astrophysical fluid. An *incompressible* fluid has $\rho =$ constant throughout. Water, and air at room temperature, are incompressible to good approximations. From (3.15) this is formally equivalent to assuming infinite sound speed c_s: physically, the approximation is that the pressure can always adjust at effectively infinite speed to keep ρ constant everywhere in the fluid. Since the volume of a fluid element always remains constant, pressure can do no work on the fluid, and so disappears from the energy equation. The continuity equation (3.1) becomes simply $\nabla.\mathbf{v} = 0$, and the pressure only remains in the term ∇P in the Euler equation. Mathematically it is now tempting to eliminate P entirely by taking the curl of this equation. Clearly, this must introduce the *vorticity*

$$\omega = \nabla \times \mathbf{v}, \tag{3.24}$$

which measures the tendency of the fluid to rotate on small scales. Using the identity

$$\mathbf{v} \times \omega = \frac{1}{2}\nabla v^2 - \mathbf{v}.\nabla\mathbf{v}, \tag{3.25}$$

we can eliminate the term $\mathbf{v}.\nabla\mathbf{v}$ from the Euler equation. If also $\mathbf{f} = -\nabla\phi$, as is true for motion under gravity, for example, the Euler equation becomes

$$\frac{\partial\mathbf{v}}{\partial t} + \nabla\left(\frac{v^2}{2} + \phi + \mathbf{v}.\nabla\mathbf{v}\right) + \frac{\nabla P}{\rho} - \mathbf{v}\times\omega = 0, \tag{3.26}$$

and now taking the curl gives the physical content of the Euler equation as

$$\frac{\partial\omega}{\partial t} = \nabla\times(\mathbf{v}\times\omega). \tag{3.27}$$

The pressure has disappeared from the fluid dynamics equations completely. This mathematical convenience accounts for the prominent role of vorticity in much of the fluid dynamics literature. One can solve an incompressible fluids problem without any knowledge of the pressure – if necessary we can 'read it off' after the problem is solved. For example, in a steady state with force density $\mathbf{f} = -\nabla\phi$, the Bernoulli integral for incompressible flow is

$$\mathcal{B} = \frac{P}{\rho} + \frac{v^2}{2} + \phi, \tag{3.28}$$

giving $P(\mathbf{r})$ in terms of the solution for $v(\mathbf{r})$ and the initial conditions.

Since astrophysical fluids are very rarely even approximately incompressible, vorticity is hardly ever mentioned in discussing them. But in Section 3.5 we will see that there is a remarkable analogy with the logical and formal mathematical structure of the usual magnetohydrodynamic (MHD) approximation used to treat (compressible) astrophysical fluids with significant magnetic fields. In MHD *three* quantities disappear from the main equations – the electric field, and both the electric charge and current density. But in both the incompressible fluid and MHD cases it is important to remember that the mathematical convenience introduced by these powerful approximations comes with the risk that they cannot hold in all cases, and may be very misleading when they do not.

3.3 Shocks

In astrophysics we very often deal with cases where a supersonic gas flow encounters an obstacle. We have seen in (3.19) that pressure forces are negligible in these flows, so there is evidently no gentle way of causing a supersonic flow to decelerate. To see what happens we consider a simple paradigm.

In the rest frame of the supersonic gas, an obstacle of sufficient inertia is like a piston pushing into it at supersonic speed. To simplify things we consider a tube lying along the x-axis filled with gas initially at rest (see Figure 3.1). At time $t = 0$ a supersonic piston begins to push into it. We can assume that the pressure changes

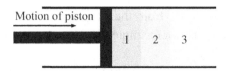

Figure 3.1 Compression of a gas by a piston in a long cylinder. Credit: Frank, King, and Raine (2002).

induced by the piston are so rapid that they behave adiabatically. Then a pressure wave travels ahead of the piston at the adiabatic sound speed c_s, compressing the gas slightly. Since $c_s \propto \rho^{1/3}$ (equation (3.16)), this means that the compressed gas moves even faster into the uncompressed gas ahead of it, causing the density profile $d\rho/dx$ ahead of the piston to steepen continuously (see Figure 3.2).

This cannot continue indefinitely, as once $\rho/(d\rho/dx)$ becomes of order the mean free path λ the fluid approximation begins to break down. The internal thermal motions within the gas begin to carry significant amounts of gas mass and bulk momentum, dissipating the associated energy in collisions. The gas ahead of this narrow region (beyond the sound travel time) has not yet received the 'news' of the violent pressure change heading towards it, and remains in its original undisturbed state. The limit of this process is a sharp transition moving into the gas to larger x, changing its state across a length $\Delta x \sim \lambda$ from cool and stationary to hot, compressed and rapidly moving.

The sharp changes in the gas state on lengthscales $\sim\lambda$ imply that processes such as viscosity become important. We explicitly neglected these in writing the conservation equations describing the gas motions. We can nonetheless continue to use these simpler equations if we idealize this region as a *shock* – a mathematical discontinuity between fluid properties on each side – and connect these by conservation laws in a steady state. We apply these in the frame of the shock discontinuity, so that cool gas (quantities labelled '1' in Figure 3.3) enters the shock and leaves as hot compressed gas (quantities labelled '2'). Evidently, the continuity equation implies that the mass flux is constant across the shock, that is,

$$\frac{d}{dx}(\rho v) = 0, \qquad (3.29)$$

so that the mass flux

$$J = \rho v \qquad (3.30)$$

is conserved. The Euler equation is

$$\rho v \frac{dv}{dx} + \frac{dP}{dx} = f_x \qquad (3.31)$$

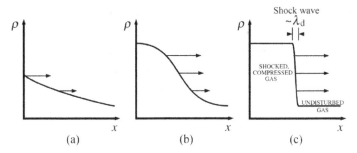

Figure 3.2 The formation of a shock wave in a gas. (a) The denser gas tends to overtake the dilute gas ahead of it as its adiabatic sound speed is greater; (b) as a result, the density gradient and the tendency to overtake both increase, leading (c) to a shock wave of thickness of order the mean free path of the gas particles. Credit: Frank, King, and Raine (2002).

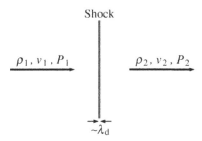

Figure 3.3 The calculation of the shock jump (or Rankine–Hugoniot) conditions. Credit: Frank, King, and Raine (2002).

or, using (3.29),

$$\frac{d}{dx}(P + \rho v^2) = f_x,$$ (3.32)

where f_x is the x-component of the force density. This must be continuous across the shock, so

$$(P_1 + \rho_1 v_1^2) - (P_2 + \rho_2 v_2^2) = \lim_{dx \to 0} \int_{-dx}^{dx} f_c dx = 0,$$ (3.33)

showing that the momentum flux

$$I = P + \rho v^2$$ (3.34)

is another conserved quantity. Finally, the energy equation gives

$$\frac{d}{dx}\left[v \left(\frac{1}{2}\rho v^2 + \rho \epsilon + P \right) \right] = f_x v$$ (3.35)

if we assume radiative losses, thermal conduction, and so on are negligible in the shock region. This amounts to assuming adiabatic conditions, and these conditions, together with the assumption that we can neglect the work $(f_x v)$ done against external forces in the shock region, characterize *adiabatic shocks*. This is not the only possible case, as we shall see below. Then using $\rho v = $ constant and $\epsilon = 3P/2\rho$, we find that the quantity

$$E = \frac{1}{2}v^2 + \frac{5P}{2\rho} \tag{3.36}$$

is also continuous across an adiabatic shock.

Enforcing continuity across the shock of the quantities defined by (3.30), (3.34), and (3.36) gives the three Rankine–Hugoniot relations determining the jumps in ρ, v, and P for an adiabatic shock. We have

$$\frac{I}{Jv} = \frac{P}{\rho v^2} + 1, \tag{3.37}$$

and defining the Mach number $\mathcal{M}$ by

$$\mathcal{M}^2 = \frac{v^2}{c_s^2}, \tag{3.38}$$

we get

$$\frac{I}{Jv} = \frac{3}{5\mathcal{M}^2} + 1. \tag{3.39}$$

Using (3.30) and (3.34) in (3.36), we get

$$E = \frac{v^2}{2} + \frac{5}{2}\left(\frac{Iv}{J} - v^2\right), \tag{3.40}$$

or

$$v^2 - \frac{5I}{4J}v + \frac{E}{2} = 0. \tag{3.41}$$

This is a quadratic equation for v, derived from the Rankine–Hugoniot relations, so its two roots must be v_1, v_2. Then $v_1 + v_2$ is given by the sum of the roots of the quadratic, that is,

$$v_1 + v_2 = \frac{5I}{4J}, \tag{3.42}$$

so that from (3.39),

$$1 + \frac{v_2}{v_1} = \frac{5}{4}\left[\frac{3}{5\mathcal{M}_1^2} + 1\right], \tag{3.43}$$

where $\mathcal{M}_1$ is the upstream Mach number. In many astrophysical situations the upstream (pre-shock) velocity is highly supersonic, so that $\mathcal{M}_1 \gg 1$. We call this a *strong shock*. For strong shocks, (3.43) gives

$$\frac{v_1}{v_2} = \frac{1}{4}, \qquad (3.44)$$

and velocity drops by a factor of four across the shock. From (3.30) we have

$$\frac{\rho_1}{\rho_2} = 4, \qquad (3.45)$$

that is, a strong shock compresses a monatomic gas by a factor of four. From (3.39) we get

$$I = \rho_1 v_1^2 \left(1 + \frac{3}{5}\mathcal{M}_1^2\right) \simeq \rho_1 v_1^2, \qquad (3.46)$$

so ahead of a strong shock the thermal pressure P_1 is negligible compared with the ram pressure $\rho_1 v_1^2$. The right-hand side of (3.34) then gives

$$P_2 = \rho_1 v_1^2 - \rho_2 v_2^2 = \rho_1 v_1 (v_1 - v_2), \qquad (3.47)$$

and using (3.44) we get

$$P_2 = \frac{3}{4}\rho_1 v_1^2. \qquad (3.48)$$

The gas pressure after a strong shock is 3/4 of the pre-shock ram pressure. The meaning here is clear – the effect of the dissipation in the shock is to randomize the ordered kinetic energy of the supersonic pre-shock gas and convert it into heat. The post-shock gas is subsonic, partly because it has been slowed, but also, importantly, because the gas temperature has increased markedly. Its adiabatic sound speed is

$$c_s = \left(\frac{5P_2}{3\rho_2}\right)^{1/2} = \left(\frac{5\rho_1 v_1^2}{16\rho_1}\right)^{1/2} = \frac{\sqrt{5}}{4} > v_2 = \frac{v_1}{4}. \qquad (3.49)$$

The randomizing effect of the shock implies that the post-shock gas (quantities labelled '2') has higher entropy ($\propto \ln[P/\rho^{5/3}]$) than the pre-shock gas, as we have implicitly assumed – this is easily verified.

The post-shock gas temperature follows from the perfect gas law (3.3) as

$$T_2 = \frac{\mu m_H P_2}{k\rho_2} = \frac{3}{16}\frac{\mu m_H}{k}v_1^2, \qquad (3.50)$$

again emphasizing that the shock has turned much of the pre-shock kinetic energy into heat.

We have derived these relations for the case of a plane, strong, adiabatic shock. They can easily be generalized beyond this, for example for finite pre-shock Mach number. In the one-dimensional argument given in this section, the steepening of density gradients into shocks was inevitable. In other cases both dissipation and particularly geometrical expansion may prevent this.[1]

In an adiabatic shock the energy losses are neglected. In the opposite limit we can imagine cases where radiative losses are so large that the post-shock gas quickly cools to a similar temperature as it had before the shock. The energy equation (3.36) is replaced simply by

$$\frac{P_1}{\rho_1} = \frac{P_2}{\rho_2} = c_s^2, \tag{3.51}$$

where c_s is the sound speed on both sides of the shock. Then, using (3.34) we have

$$\rho_1(c_s^2 + v_1^2) = \rho_2(c_s^2 + v_2^2), \tag{3.52}$$

and from (3.29) in the form $\rho_1 = \rho_2 v_2/v_1$, this becomes

$$(v_2 - v_1)c_s^2 = v_1 v_2(v_2 - v_1). \tag{3.53}$$

Since $v_1 \neq v_2$, this means that

$$v_1 v_2 = c_s^2. \tag{3.54}$$

We see immediately that

$$v_2 = \frac{c_s}{\mathcal{M}_1}, \tag{3.55}$$

$$\frac{v_1}{v_2} = \mathcal{M}_1 \frac{v_1}{c_s} = \mathcal{M}_1^2 \tag{3.56}$$

and

$$\rho_2 = \rho_1 \frac{v_1}{v_2} = \rho_1 \mathcal{M}_1^2. \tag{3.57}$$

A strong isothermal shock ($\mathcal{M}_1 \gg 1$) produces extreme compression.

We derived the jump conditions for adiabatic and isothermal shocks *in the frame of the shock*. Often shocks move relative to the gas, and we then have to apply a velocity transformation, cancelling it before considering the jump conditions. For example, if the piston considered at the beginning of this section is pushed into the gas-filled cylinder at speed u and the shock moves into the gas at speed v_s, then in the shock frame, gas enters the shock from larger x with velocity $-v_s$, and leaves it with a $u - v_s$. Applying the adiabatic condition (3.44), we have

$$\frac{-v_s}{u - v_s} = 4, \tag{3.58}$$

so that $v_s = 4u/3$, that is, the shock runs ahead of the piston with a larger velocity.

---

[1] This is obvious (and important!) in everyday life – fortunately, a shout does not turn into a shock wave.

A particularly important and characteristic shock pattern arises when two gas flows with different speeds collide. In general the gas particle mean free paths are so short that the two gases do not interpenetrate at the boundary between them, and this is treated as a 'contact discontinuity' – a surface at rest in both fluids, with momentum transfer but no mass or energy transfer across it. So thermal pressure and velocity are continuous there. On each side of a contact discontinuity, shocks locally parallel to it run into the two gas flows, slowing the 'intruding' gas and accelerating the 'target' gas velocities. If both shocks are isothermal, the entire pattern of the two shocks and the contact discontinuity between them move effectively as a narrow interface between the two gases. We will see examples of this later in the book (e.g. Figure 6.7).

3.4 Plasmas

So far we have neglected electromagnetic fields in considering gas motions. But most astrophysical gases are ionized and highly conducting, and we will find that magnetic fields can be important in accretion discs and jets close to black holes. A full description – as given by plasma physics – couples the Maxwell equations governing the electromagnetic fields with separate equations of continuity, motion and energy for ions and electrons, each with velocity fields $v_{i,e}(\mathbf{r}, t)$ and temperatures $T_{i,e}(\mathbf{r}, t)$. This is often extremely complex in practice. Fortunately, in many cases we can make powerful simplifying assumptions.

First, experience suggests that the Universe is charge-neutral overall: the number densities of ions and electrons rarely remain strongly different for long, and integrate to give zero net charge over macroscopic volumes. Physically this happens because electrostatic (Coulomb) forces are long range. Any charge imbalance simultaneously attracts cancelling opposite charges and repels the local excess of like charges. So a 1 per cent positive charge excess in a sphere of radius r and number density N attracts electrons near the edge of the sphere with acceleration

$$\dot{v} = \frac{e|\mathbf{E}|}{m_e} \simeq \frac{4\pi r^3}{3m_e} \frac{N}{100} \frac{e^2}{[4\pi\epsilon_0]r^2}, \tag{3.59}$$

where $-e, m_e$ are the electron charge and mass. With typical parameters $r = 3$ cm, $N = 10^{17}$ cm^{-3} we find $\dot{v} \sim 10^{17}$ cm s^{-2}, so electrons would move to neutralize the net charge on a timescale $t \sim (r/\dot{v})^{1/2} \sim 3 \times 10^{-9}$ s. The high speed, $v \sim r/t \sim 10^9$ cm s^{-1}, implies that the electrons overshoot the site of the charge imbalance and cause the local charge density to oscillate. For a hydrogen plasma with very slightly different ion and electron charge densities $N_i = N, N_e = N + N_0(\mathbf{r}, t)$, where $N_0 \ll N$, and $N_0 = 0$ outside a small region, Maxwell's equations give

$$\nabla \cdot \mathbf{E} = -\frac{4\pi}{[4\pi\epsilon_0]} N_1 e. \tag{3.60}$$

This electric field causes the charges to move. We can neglect the motion of the ions (protons) because of their far greater inertia. The electrons must follow the Euler equation (3.2), where we can neglect the $(\mathbf{v}_e \cdot \nabla)\mathbf{v}_e$ since we are considering a small perturbation from rest. Since the mass density of the electron fluid is $N_e m_e$ and the electric force density is $-N_e e\mathbf{E}$, this equation becomes

$$m_e \frac{\partial \mathbf{v}_e}{\partial t} = -e\mathbf{E}. \tag{3.61}$$

The continuity equation for conservation of electron fluid mass is

$$\frac{\partial N_e}{\partial t} + \nabla \cdot (N_e \mathbf{v}_e) = 0, \tag{3.62}$$

which to first order in small quantities is

$$\frac{\partial N_1}{\partial t} + N_0 \nabla \cdot \mathbf{v}_e = 0. \tag{3.63}$$

We eliminate $\mathbf{v}_e$ from equations (3.61) and (3.63) by taking the divergence of the first and subtracting the time derivative of the second:

$$\frac{1}{N_0} \frac{\partial^2 N_1}{\partial t^2} - \frac{e}{m_e} \nabla \cdot \mathbf{E} = 0, \tag{3.64}$$

and from (3.60) we get

$$\frac{\partial^2 N_1}{\partial t^2} + \left[\frac{4\pi}{[4\pi\epsilon_0]} \frac{N_0 e^2}{m_e} \right] N_1 = 0, \tag{3.65}$$

which shows that the charge imbalance oscillates at the *plasma frequency*

$$\omega_p = \left[\frac{4\pi}{[4\pi\epsilon_0]} \frac{N_0 e^2}{m_e} \right]^{1/2}. \tag{3.66}$$

With N_0 in units of cm^{-3}, we have

$$\omega_p = 5.7 \times 10^4 N_0^{1/2} \text{ rad s}^{-1}, \tag{3.67}$$

or

$$\nu_p = \frac{\omega_p}{2\pi} = 9.0 \times 10^3 N_0^{1/2} \text{ Hz}. \tag{3.68}$$

Plasmas are opaque to radiation of frequencies $\nu < \nu_p$ as the plasma oscillations are faster than the changes in the applied electromagnetic fields, so the plasma electrons move rapidly to 'short out' the radiation, which is reflected. The Earth's ionosphere ($N_0 \simeq 10^6$ cm^{-3}) reflects radio waves of frequencies lower than ~ 10 MHz.

The other implication of the plasma frequency is that from (3.63) with $\partial/\partial t \sim \omega_p$, $\nabla \sim 1/l$ we find a typical lengthscale

$$l \sim v_e/\omega_p, \tag{3.69}$$

which limits the size of any region of charge imbalance. Plasma electrons are subject to thermal fluctuations even without applied radiation, so there is a fundamental shielding scale, the *Debye length* λ_D, given by setting $v_e \sim (kT_e/m_e)^{1/2}$ in this relation:

$$\lambda_D = \left[\frac{[4\pi \epsilon_0]kT_e}{4\pi N_0 e^2} \right]^{1/2}. \tag{3.70}$$

With N_0 in units of cm^{-3} and T_e in Kelvin this gives

$$\lambda_D \simeq 7(T_e/N_0)^{1/2} \text{ cm}. \tag{3.71}$$

This is far smaller than macroscopic scales of interest in most astrophysical contexts.[2]

3.5 Magnetohydrodynamics

The expression (3.71) means that we can very often regard plasmas as charge neutral at every point, since λ_D is small compared with the lengthscales of interest. The *magnetohydrodynamic approximation* makes the additional assumption of a connection

$$\mathbf{j}' = \sigma_c \mathbf{E}' \tag{3.72}$$

between the current density and the electric field in the co-moving fluid frame – a kind of Ohm's law. In the simplest version, *ideal* MHD, the fluid is assumed to have a high enough density of charge carriers that it has effectively perfect conductivity. This is often quite a good approximation in astrophysics, although deviations from it are important in some cases, as we will see. In ideal MHD with conductivity $\sigma_c \to \infty$, finite current density requires (from (3.72))

$$\mathbf{E}' = 0. \tag{3.73}$$

[2] But exceptions can occur if the plasma has very low density, as we shall see in Section 3.7.

The corresponding fields $\mathbf{E}, \mathbf{B}$ in the observer's frame follow from the standard special-relativistic Lorentz transformations of electromagnetic fields between reference frames with relative velocity $\mathbf{v}$, that is,

$$
\begin{aligned}
E'_\parallel &= E_\parallel \\
\mathbf{E}'_\perp &= \gamma(\mathbf{E}_\perp + \mathbf{v} \times [c]\mathbf{B}/c) = 0 \\
B'_\parallel &= B_\parallel \\
\mathbf{B}'_\perp &= \gamma(\mathbf{B}_\perp - \mathbf{v} \times [c]\mathbf{E}/c),
\end{aligned}
\tag{3.74}
$$

where $\gamma = (1 - v^2/c_2)^{-1/2}$ and the suffixes $\parallel, \perp$ denote the components parallel and orthogonal to $\mathbf{v}$. With (3.73) the first two relations of (3.74) show that the observer measures an electric field

$$
\mathbf{E} = \mathbf{E}_\perp = -[c]\mathbf{v} \times \frac{1}{c}\mathbf{B}.
\tag{3.75}
$$

This immediately implies

$$
\mathbf{E} \cdot \mathbf{B} = 0,
\tag{3.76}
$$

so that there is no electric field component locally parallel to the magnetic field. This property is called *degeneracy*, and is valid in all reference frames if it holds in any one, as $\mathbf{E} \cdot \mathbf{B}$ is a Lorentz invariant. This means that at every point there is a reference frame in which one of $\mathbf{E}, \mathbf{B}$ is zero (cf. (3.74)). Since the other Lorentz invariant $[c^2]B^2 - E^2$ is positive, it is $\mathbf{E}$ that vanishes. As the frame in which it does so varies at each point in any plasma where $\mathbf{v}$ varies with position, Equation (3.74) shows that we cannot formally set $\mathbf{E} = 0$ everywhere in a global reference frame. But we will see that $\mathbf{E}$ is completely absent from the equations of ideal MHD.

In most cases we can neglect the permeability and dielectric properties of astrophysical fluids, so Maxwell's equations are

$$
\begin{aligned}
\frac{\partial \mathbf{B}}{\partial t} &= -\left[\frac{1}{c}\right]c\nabla \times \mathbf{E} \\
\nabla \cdot \mathbf{B} &= 0 \\
\nabla \times \mathbf{B} &= \left[\frac{\mu_0}{4\pi}\right]4\pi\mathbf{j} + \frac{1}{[c]}\frac{1}{c}\frac{\partial \mathbf{E}}{\partial t} \\
\nabla \cdot \mathbf{E} &= \left[\frac{1}{4\pi\epsilon_0}\right]4\pi\rho_e.
\end{aligned}
\tag{3.77}
$$

Taking the divergence of the third equation and the time derivative of the fourth, and eliminating the mixed derivatives of $\mathbf{E}$, we find

$$
\nabla \cdot \mathbf{j} + \frac{\partial \rho_e}{\partial t} = 0,
\tag{3.78}
$$

which expresses charge conservation. Equation (3.75) and the first of (3.77) give the *ideal MHD induction equation*:

$$\frac{\partial \mathbf{B}}{\partial t} = \nabla \times (\mathbf{v} \times \mathbf{B}). \tag{3.79}$$

The divergence of this equation implies

$$\frac{\partial}{\partial t}(\nabla \cdot \mathbf{B}) = 0, \tag{3.80}$$

and since any MHD flow must start with magnetic fields obeying $\nabla \cdot \mathbf{B} = 0$ everywhere, this shows that the induction equation preserves this condition.

So far we have placed no restriction on the fluid velocity $\mathbf{v}$, so the induction equation is valid for relativistic flows in ideal MHD. The motions in an accretion disc are non-relativistic, that is, $v \ll c$, except possibly very close to a central black hole. Then in applying ideal MHD to a disc, we retain terms of order v/c, but neglect $O(v^2/c^2)$ and higher. From (3.74) and (3.75) we find

$$\mathbf{B}' = \mathbf{B}[1 + O(v^2/c^2)], \tag{3.81}$$

so the magnetic field $\mathbf{B}$ is the same in all rest frame. Equation (3.75) and the third equation of (3.77) give

$$\left[\frac{1}{4\pi \epsilon_0}\right] 4\pi \mathbf{j} - \frac{[c]}{c}\frac{\partial}{\partial t}\mathbf{v} \times \mathbf{B} = [c]c\nabla \times \mathbf{B}. \tag{3.82}$$

The second term on the left-hand side comes from the displacement current in Maxwell's equations, and is negligible compared with the right-hand side for $v \ll c$. To see this, let V, L, and $\tau = L/V$ be the typical velocity, length and timescales in a given situation. Then the second term on the left-hand side is of order (V^2/c^2) compared with the right-hand side for non-relativistic disc motions. This is reasonable, since the displacement current is what makes Maxwell's equations compatible with wave propagation at speed c, and it is small in non-relativistic applications. Dropping this term we have

$$\left[\frac{1}{4\pi \epsilon_0}\right] 4\pi \mathbf{j} = [c]c\nabla \times \mathbf{B}, \tag{3.83}$$

or

$$\mathbf{j} = \frac{c}{4\pi}\left[\frac{4\pi}{\mu_0 c}\right]\nabla \times \mathbf{B} \tag{3.84}$$

so that

$$\nabla \cdot \mathbf{j} = 0, \tag{3.85}$$

which from (3.78) gives

$$\frac{\partial \rho_e}{\partial t} = 0. \qquad (3.86)$$

Any MHD flow must start with $\rho_e = 0$, so charge neutrality is guaranteed.[3] Equation (3.85) means that ideal MHD currents do not have sources or sinks, and we do not need to work out how these currents close. We shall discuss this in more depth later in this section.

To see how magnetic fields affect the motion of the fluid in ideal MHD we work out the Lorentz force density (force per unit volume) $f_L \propto \mathbf{j} \times \mathbf{B}$, which gives

$$\mathbf{f}_L = \frac{1}{4\pi} \left[\frac{4\pi}{\mu_0} \right] (\nabla \times \mathbf{B}) \times \mathbf{B}. \qquad (3.87)$$

Although this may look unfamiliar, it is important to remember that this is the Lorentz force density only for the particular case of an electrically neutral and infinitely conducting fluid. We add this force law to the fluid equations of motion (Euler) and mass conservation (continuity) and combine this with the induction equation governing the evolution of the magnetic field in ideal non-relativistic MHD:

$$\rho \frac{\partial \mathbf{v}}{\partial t} + \rho(\mathbf{v} \cdot \nabla)\mathbf{v} = -\nabla P + \frac{1}{4\pi} \left[\frac{4\pi}{\mu_0} \right] (\nabla \times \mathbf{B}) \times \mathbf{B},$$

$$\frac{\partial \rho}{\partial t} + \nabla.(\rho \mathbf{v}) = 0, \qquad (3.88)$$

$$\frac{\partial \mathbf{B}}{\partial t} = \nabla \times (\mathbf{v} \times \mathbf{B}).$$

Then with the usual thermodynamic equations and the equation of state, we have a full description of the system.

Remarkably, this only requires us to introduce the single extra vector field $\mathbf{B}$, while the full set of Maxwell equations (3.77) are reduced to the induction equation (3.79) for $\mathbf{B}$. These drastic simplifications come about because the assumptions of ideal MHD are very restrictive, and mean that we must be very careful when adopting the ideal MHD approximation, which can sometimes be seriously misleading (for example, we shall see in Section 3.7 that it does not hold at low plasma densities). The two inhomogenous Maxwell equations involving ρ_e and $\mathbf{j}$ are not needed: charge neutrality is guaranteed (by assumption), and Ampère's law is reduced to the status of telling us what current densities correspond to the magnetic field we have calculated in solving for the MHD flow (cf. (3.84)). The complete absence of the electric field $\mathbf{E}$ from the equations is not surprising given the discussion following (3.76)).

---

[3] This is true within the plasma, but we shall find that surface charge densities can appear at boundaries, for example with a surrounding near-vacuum – see the discussion following Equation (3.115).

This simplicity is a direct consequence of the ideal MHD approximation – in making it, we tacitly assume that the currents can always adjust instantaneously to produce any magnetic field specified by the ideal MHD equations, since perfect conductivity means that charges can move at effectively infinite velocities, simultaneously restoring complete charge neutrality and 'shorting out' the electric field **E** everywhere.

There is a close analogy with the effect of assuming incompressibility in a non-magnetic fluid (see Section 3.2). There we can eliminate the fluid pressure from the equation of motion by taking its curl, and solve the Equation (3.27) for the vorticity $\omega = \nabla \times \mathbf{v}$, which has precisely the same form as the induction equation (3.79). Then if we need the value of the pressure at any point we can simply read it off at the end of the calculation by using the Bernoulli integral. The tacit assumption here is that the pressure can always take the required form because it adjusts at the sound speed, which is effectively infinite. This means that thinking in terms of pressure is not useful if the fluid is assumed incompressible.

In exactly the same way, thinking in terms of currents is not helpful in ideal MHD, and can be directly counterproductive in terms of understanding. For example, currents do *not* in general move with the fluid. Formally this is because there is no continuity equation relating them to the fluid velocity. From a more physical microscopic viewpoint, the plasma ions carry the positive charges and almost all the mass and inertia of the plasma, and so move with mean velocity very close to **v**. The plasma electrons carry the negative charges and move far more rapidly, by a factor of $\sim (m_i/m_e)^{1/2} \sim 43$ – it is this mobility that may allow us to assume complete charge neutrality in the ideal MHD equations, and, as we see, it is not directly related to the fluid velocity.

The ideal MHD approximation has a wide range of validity in astrophysics. But it has limitations, and as cautioned after (3.88), we should be careful in adopting it. The conductivity is in reality never infinite, so in non-ideal MHD new effects appear (see Section 3.6). Moreover, positive and negative charges can move independently to some degree in certain situations, particularly where the density is low and/or the plasma is permeated by strong magnetic fields.[4] Eddington (1926) and Bally & Harrison (1978) noted that this effect means that, for example, all stars in hydrostatic equilibrium carry a slight net positive charge (which is negligible for almost all purposes) distributed over their atmospheres.

Ideal MHD has a further characteristic property which greatly simplifies discussion of its effects. Expanding the cross products in the induction equation (3.79) and using $\nabla \cdot \mathbf{B} = 0$, we get

---

[4] For example, charge separation occurs in electrolytes (Debye & Hückel, 1923) and in ionized astrophysical plasmas (Salpeter, 1954), and it very significantly affects the rates of nuclear burning in stars (e.g. Clayton, 1968).

$$\frac{\partial \mathbf{B}}{\partial t} = -\mathbf{B}\nabla \cdot \mathbf{v} - (\mathbf{v} \cdot \nabla)\mathbf{B} + (\mathbf{B} \cdot \nabla)\mathbf{v}. \tag{3.89}$$

We define the magnetic flux

$$\Phi(\mathrm{L}) = \int_S \mathbf{B}.\mathrm{d}\mathbf{S} \tag{3.90}$$

of a loop L consisting of fluid particles. Here S is any surface spanning L, with $\mathrm{d}\mathbf{S} = \mathbf{n}\mathrm{d}S$ and $\mathbf{n}$ the unit normal to it. Because $\nabla.\mathbf{B} = 0$, the divergence theorem means that Φ is the same for any choice of S. Then one can show that the flux through any loop moving with the flow is conserved, that is,

$$\frac{\partial \Phi}{\partial t} + (\mathbf{v}.\nabla)\Phi = 0. \tag{3.91}$$

This is Alfvén's theorem, and shows that in ideal MHD, magnetic fieldlines move with the fluid (they are often described as 'frozen-in'). As a direct result, the magnetic field strength may be strongly amplified as the fluid moves in a converging or shearing flow. (We do not give proofs here, but these appear in Frank, King & Raine, 2002 (Section 3.7) and in any MHD text.)

The crucial point is that charged particles are closely tied to magnetic fieldlines. This is obvious for infinite conductivity, but also holds for a different reason if the particles gyrate many times around fieldlines before they collide with another particle, that is, the cyclotron frequency

$$\omega_B = [c]\frac{qB}{mc} \tag{3.92}$$

is much larger than the Coulomb collision frequency (see Frank, King & Raine, 2002, Section 3.3). This happens in hot and/or rarified plasmas, which are often called *collisionless*. The equations of MHD hold here also, even though one might think that a two-fluid plasma treatment was necessary. The flows and fields in the solar wind agree closely with expectations from MHD, for example, even though the Coulomb collision length is of order an astronomical unit.

As with any powerful approximation, some essential features of the full theory are suppressed in ideal MHD. We have seen already that the displacement current is neglected, so a first-principles treatment of the production and interaction of matter and electromagnetic radiation is impossible. For many astrophysical treatments this is not a serious problem, as we can use emissivities and absorption coefficients as in radiative transfer.

We can get a better insight into the dynamical effects of the Lorentz force in ideal MHD by transforming (3.87) using standard vector algebra. This gives

$$\mathbf{f}_L = -\frac{1}{8\pi}\left[\frac{4\pi}{\mu_0}\right]\nabla B^2 + \frac{1}{4\pi}\left[\frac{4\pi}{\mu_0}\right](\mathbf{B}\cdot\nabla)\mathbf{B}. \tag{3.93}$$

For many applications it is useful to write the Lorentz force density (3.93) as the divergence of the magnetic stress tensor: we rewrite its *i*th (Cartesian) component using

$$[(\nabla\times\mathbf{B})\times\mathbf{B}]_i = \epsilon_{ijk}\epsilon_{jlm}\frac{\partial B_m}{\partial x_l}B_k$$

$$= (\delta_{kl}\delta_{im} - \delta_{km}\delta_{il})\frac{\partial B_m}{\partial x_l}B_k \tag{3.94}$$

$$= \frac{\partial}{\partial x_k}(B_iB_k - \frac{1}{2}B^2\delta_{ik}),$$

where ϵ_{ijk} is the completely antisymmetric symbol, δ_{ij} is the Kronecker delta, and we sum over repeated indices and use $\nabla.\mathbf{B} = 0$ in the final form. The stress tensor is

$$M_{ij} = \left[\frac{4\pi}{\mu_0}\right]\left(\frac{1}{8\pi}B^2\delta_{ij} - \frac{1}{4\pi}B_iB_j\right), \tag{3.95}$$

and (3.94) means that

$$\mathbf{f}_L = -\nabla.\mathbf{M}. \tag{3.96}$$

Then the force on fluid in a volume V bounded by a surface S is

$$\int_V \mathbf{f}_L\,dV = \int_S \mathbf{n}.\mathbf{M}\,dS, \tag{3.97}$$

where $\mathbf{n}$ is the outward normal on S. The force this volume of fluid exerts on its surroundings is the negative of this expression, and so it exerts a *surface force* per unit area

$$\mathbf{f}_S = \mathbf{n}.\mathbf{M} = \left[\frac{4\pi}{\mu_0}\right]\left(\frac{1}{8\pi}B^2\mathbf{n} - \frac{1}{4\pi}B_n\mathbf{B}\right) \tag{3.98}$$

on its surroundings, with $B_n = \mathbf{B}.\mathbf{n}$ the outward normal component of $\mathbf{B}$.

To see what this means in physical terms we consider a rectangular box of ideal MHD fluid with a uniform magnetic field $\mathbf{B} = (0, 0, B)$ in the z-direction, assuming that the field vanishes outside this box. The box of magnetic field exerts a force

$$\mathbf{F} = -\int \mathbf{f}_M \cdot d\mathbf{S} \tag{3.99}$$

on its surroundings, where d**S** is the element of surface area of the box. On faces perpendicular to the field, that is, with normals in the x and y directions, we find forces

$$F_x = \left[\frac{4\pi}{\mu_0}\right]\left(\frac{B^2}{8\pi} - \frac{B_x B_z}{4\pi}\right) = \left[\frac{4\pi}{\mu_0}\right]\frac{B^2}{8\pi} \tag{3.100}$$

per unit area, with a similar expression for F_y, while on the face in the positive z-direction we find

$$F_z = \left[\frac{4\pi}{\mu_0}\right]\left(\frac{B^2}{8\pi} - \frac{B_z B_z}{4\pi}\right) = -\left[\frac{4\pi}{\mu_0}\right]\frac{B^2}{8\pi} \tag{3.101}$$

per unit area. The x, y faces orthogonal to the field experience a magnetic *pressure*, while the faces along the fieldlines feel a *tension*, each proportional to the energy density of the magnetic field. In this sense an ideal MHD fluid is rather more like an elastic solid than a fluid. In particular, waves (called Alfvén waves) can propagate along the field with speed

$$v_A = \left(\left[\frac{4\pi}{\mu_0}\right]\frac{B^2}{4\pi\rho}\right)^{1/2}. \tag{3.102}$$

In an ideal MHD fluid the Alfvén speed v_A plays a role similar to the sound speed in a gas, and, in particular, changes in the magnetic field propagate at this speed.

Electromagnetic energy flow is described by the Poynting vector

$$\mathbf{S} = \left[\frac{4\pi}{\mu_0 c}\right]\frac{c}{4\pi}\mathbf{E} \times \mathbf{B}. \tag{3.103}$$

Using (3.75), we have

$$\mathbf{S} = \left[\frac{4\pi}{\mu_0}\right]\frac{1}{4\pi}\mathbf{B} \times (\mathbf{v} \times \mathbf{B}), \tag{3.104}$$

and expanding the vector triple product, we get

$$\mathbf{S} = \left[\frac{4\pi}{\mu_0}\right]\mathbf{v}_\perp\frac{B^2}{4\pi}, \tag{3.105}$$

where $\mathbf{v}_\perp$ denotes the component of $\mathbf{v}$ in the plane orthogonal to $\mathbf{B}$.

This expression suggests an interpretation of **S** as moving magnetic energy around with the velocity $\mathbf{v}_\perp$. But the coefficient of $\mathbf{v}_\perp$ is *twice* the energy density. The extra factor of two enters in the same way that the energy transport in a moving fluid does not simply involve the movement of local heat energy, but must also account for pressure (PdV) work. In a magnetic field, energy density and pressure are each separately equal to $[4\pi/\mu_0](B^2/8\pi)$.

We cannot simply add the magnetic effects to the fluid ones (e.g. in a combined Bernoulli integral), as they do not move together in general. Also, interpreting the

Poynting flux can be more complex than one might think. In particular, the polarity of the fieldlines matters: patches of field with the same polarity (the same sign of $\mathbf{B} \cdot \mathbf{n}$, where $\mathbf{n}$ is the normal to a surface in the plasma) repel each other. The system has to supply energy to push them closer, so the Poynting flux is directed *into* the surface bounding the volume they occupy. But for opposite polarities the two field patches attract, with fieldlines linking the field patches, so the Poynting flux is directed *out of* the volume as they approach each other.

Equations (3.88) allow us to classify ideal MHD flows in terms of typical fluid quantities. We use the definition $c_s^2 = P/\rho$ of the isothermal sound speed to write the first equation (motion) as

$$\mathcal{M}^2 \left(\frac{\partial \tilde{\mathbf{v}}}{\partial \tilde{t}} + \tilde{\mathbf{v}} \cdot \tilde{\nabla} \mathbf{v} \right) \sim -\tilde{\nabla} \ln \rho + \frac{2}{\beta} (\tilde{\nabla} \times \tilde{\mathbf{B}}) \times \tilde{\mathbf{B}}. \qquad (3.106)$$

Here $\mathcal{M}$ is the Mach number of the flow velocity v, and

$$\beta = \left[\frac{\mu_0}{4\pi} \right] \frac{8\pi P}{B^2} \qquad (3.107)$$

is the ratio of gas pressure to magnetic pressure, while quantities with a tilde are of order unity. Clearly, for $\beta \gg 1$ we recover the usual Euler equation of gas dynamics.[5]

In flows with $\beta \lesssim 1$, magnetic effects are important. If this kind of flow is supersonic, then we must have $\mathcal{M} \sim (\beta)^{-1/2}$, or

$$v \sim v_A, \qquad (3.108)$$

so the gas moves with velocities comparable to the Alfvén velocity. Gas clouds forming stars are often in this state.

If a small-β flow is instead subsonic, so that the inertia terms on the left-hand side are also small, then we must have

$$(\tilde{\nabla} \times \tilde{\mathbf{B}}) \times \tilde{\mathbf{B}} = O(\beta) \sim 0. \qquad (3.109)$$

There are two ways of arranging this: if

$$\nabla \times \mathbf{B} = 0, \qquad (3.110)$$

then

$$\mathbf{B} = -\nabla \phi, \qquad (3.111)$$

which is called a *potential* field. The atmospheres of magnetic stars can often be modelled like this.

---

[5] One might reasonably object that β would have been better defined as the reciprocal of the right-hand side of (3.107), so that the two similar-sounding quantities β and B^2 would be proportional, but attempts to rectify this usage have failed and the definition here is irreversibly standard.

If instead $\nabla \times \mathbf{B} \neq 0$, we must have[6]

$$(\nabla \times \mathbf{B}) \times \mathbf{B} = 0, \tag{3.112}$$

which is called a *force-free* field. (The name describes the fact that gas forces are negligible compared with magnetic ones. Unfortunately, the designation 'force-free' is also used in a rather different context – see Section 3.7.) From (3.84) we can write (3.112) as

$$\mathbf{j} \times \mathbf{B} = 0, \tag{3.113}$$

meaning that all currents run along magnetic fieldlines. Small displacements do no work against the magnetic field, so this state is extremal for the magnetic energy. It is straightforward to show that the force-free condition characterizes the minimum magnetic energy state if the ends of the fieldlines are kept fixed by the external medium. The field exerts a non-zero stress on this medium, which must be able to counter this stress to maintain the field. This requirement holds for all magnetic fields, because their energy density $\propto B^2$ is always positive. They always tend to expand away from each other and simultaneously exert a tension along the field-lines. In this book the anchoring medium is often an accretion disc (see Chapter 4), and it is important to check that this can support a postulated magnetic field.

Force-free conditions highlight a subtlety in discussing charge density in ideal MHD. We did not use the fourth Maxwell equation

$$\nabla \cdot \mathbf{E} = \left[\frac{1}{4\pi \epsilon_0} \right] 4\pi \rho_e \tag{3.114}$$

in deriving the equations (3.88) of ideal non-relativistic MHD. The expression (3.75), $\mathbf{E} = -[c]\mathbf{v} \times (1/c)\mathbf{B}$, used there is relativistically correct in ideal MHD.[7] Then, using (3.114), we get

$$\left[\frac{1}{4\pi \epsilon_0} \right] 4\pi \rho_e = -[c]\frac{1}{c}\nabla \cdot \mathbf{E} = [c]\frac{1}{c}\{\mathbf{B} \cdot (\nabla \times \mathbf{v}) - \mathbf{v} \cdot (\nabla \times \mathbf{B})\}. \tag{3.115}$$

The right-hand side of this equation is non-zero in general, so it seems that we have charge densities appearing in a theory (ideal MHD) that starts by assuming charge neutrality. The resolution of this apparent paradox is that ρ_e is not the charge, but the charge *density*. Charges are Lorentz invariant, but a charge density involves division by a volume, which is not. The second term in (3.115) is proportional to the current $\mathbf{j} \propto \nabla \times \mathbf{B}$. If there is a net current $\mathbf{j}$, positive and negative charges stream in opposite directions with respect to the gas velocity $\mathbf{v}$ in the local rest frame of the plasma. The Lorentz contractions of these two streams differ, so charge densities

---

[6] This is an example of 'terms so large they are zero'.

[7] But *not* when the diffusion terms of non-ideal MHD appear (see Section 3.6), since these imply initial signal speeds that formally exceed c.

appear in the MHD equations because we cannot usefully write them in the local rest frame, as this differs at each point. Similarly, the rotational nature of the first term means that although we can choose a frame where **E** vanishes at a given point, **E** becomes non-zero as we move away from it, and a charge density appears in that frame.

This argument shows that inside the conducting MHD fluid, charge densities result from frame transformations, and vanish in the locally co-moving and co-rotating frame at each point. But at a fluid boundary, for example with a vacuum, we cannot transform the charge densities away and they imply surface charges.

A simple analogy illustrates this. Consider what happens if we move a conducting sphere into a uniform magnetic field **B** with velocity **v**. In practice even a highly conducting sphere will eventually become permeated by the field (in the language of Section 3.6, it must have finite *diffusivity*), so after a time the field is uniform inside the sphere also. In its co-moving reference frame the sphere now feels an external electric field $\mathbf{E} = [c]\mathbf{v} \times \mathbf{B}/c$, but this vanishes inside it because of its high conductivity. The result is that the sphere acquires a surface charge density

$$\sigma_s = [4\pi\epsilon_0]\frac{3vB\cos\theta}{4\pi Rc}, \tag{3.116}$$

where θ is the polar angle with respect to the magnetic field. This creates an internal uniform electric field which exactly cancels the external field **E** inside the sphere, and in the observer's frame outside the sphere a dipole field which is negligible at large distance. The cosine dependence of σ_s shows that the total charge on the sphere is zero, and the polarization induced by the magnetic field has separated positive and negative charges in opposite directions on its surface.

From this we see that for the MHD approximation to apply, the plasma must be able to separate charges enough to give a minimum local charge density (3.115) which can move to supply the current density **j** without exceeding the speed of light. From (3.115) one can show that this requires a number density

$$N > N_0 = [4\pi\epsilon_0 c]\frac{\Omega B}{4\pi ec} \simeq 0.55\Omega B \,\text{cm}^{-3} \gtrsim 2\times10^9\frac{vB}{cl}\,\text{cm}^{-3} \tag{3.117}$$

where

$$\Omega \gtrsim \frac{v}{l} \tag{3.118}$$

is the typical value of $|\nabla \times \mathbf{v}|$, and v, l, B are the characteristic velocity, lengthscale, and magnetic field at each point.[8]

---

[8] The quantity N_0 appears as the Goldreich–Julian charge density in the theory of pulsar magnetospheres.

3.6 Non-ideal MHD

The discussion of magnetic fields in Section 3.5 considered only *ideal* MHD, where the conductivity is assumed to be infinite. In reality this can never be exactly true, so we should ask what effects appear for finite conductivity. We assume a linear 'Ohm's law' relating current density and electric field in the fluid frame, that is,

$$\mathbf{j}' = \sigma_c \mathbf{E}', \tag{3.119}$$

where the conductivity σ_c is now finite. Then we have

$$\frac{\mathbf{j}}{\sigma_c} = \mathbf{E}' = \mathbf{E} + [c]\frac{\mathbf{v}}{c} \times \mathbf{B}, \tag{3.120}$$

giving the new form of the induction equation (3.79) as

$$\frac{\partial \mathbf{B}}{\partial t} = \nabla \times (\mathbf{v} \times \mathbf{B} - \eta \nabla \times \mathbf{B}), \tag{3.121}$$

where

$$\eta = \left[\frac{4\pi}{c^2 \mu_0}\right] \frac{c^2}{4\pi \sigma_c}, \tag{3.122}$$

where η has the dimensions of (length)^2/time. This equation is obviously parabolic, that is, $\mathbf{B}$ obeys a diffusion equation. If η is constant, (3.121) gives

$$\frac{\partial \mathbf{B}}{\partial t} = \nabla \times (\mathbf{v} \times \mathbf{B}) + \eta \nabla^2 \mathbf{B}, \tag{3.123}$$

showing explicitly that the magnetic field diffuses on a timescale

$$t_{\text{mag}} \sim \frac{R^2}{\eta} \tag{3.124}$$

analogous to the viscous timescale (4.33) we shall meet in Chapter 4. In a non-ideal MHD plasma, magnetic fieldlines may slip with respect to the gas flow at a rate depending on the ratio of the 'momentum diffusivity' (kinematic viscosity) ν to the magnetic diffusivity η

$$\mathcal{P}_m = \frac{\nu}{\eta}, \tag{3.125}$$

which is called the (magnetic) Prandtl number.[9]

This slippage means that the freezing-in approximation is no longer valid: the flux of magnetic fieldlines across a surface moving with the fluid is not constant.

---

[9] The original Prandtl number is the ratio of the momentum diffusivity (i.e. kinematic viscosity) to the thermal diffusivity (sound speed times mean free path). The adjective 'magnetic' is often omitted in the astrophysics literature, where there is little ambiguity in doing so. Note that in some papers (e.g. Lubow, Papaloizou, & Pringle, 1994) $\mathcal{P}_m$ is defined as the inverse of the right-hand side of (3.125), that is, η/ν. In practice this inverse definition has little effect on the discussion, because the interesting case is $\mathcal{P}_m = O(1)$. Another variant is that the term 'resistivity' is sometimes used for the diffusivity η (e.g. Balbus & Henri, 2008).

We can nevertheless often think of individual lines as having an identity while moving through the fluid if some region of the gas does have high conductivity. The fieldlines are effectively frozen-in there, allowing us to label them.

The fieldline slippage described by (3.123) can have important consequences for accretion discs, as we will see in Chapter 4.

3.7 Force-Free Electrodynamics

In and near accretion flows the mass density is often extremely low, and we can safely ignore particle inertia and interparticle collision terms.

Under these conditions one can write the equation of motion of the charge carriers as

$$\rho_e \mathbf{E} + \frac{[c]}{c} \mathbf{j} \times \mathbf{B} = 0 \qquad (3.126)$$

rather than (3.113) or ultimately (3.88).

This has some resemblance to force-free ideal MHD, again with 'terms so large they are zero', but is clearly very distinct from it, since in ideal MHD we have $\mathbf{j} \times \mathbf{B} = 0, \mathbf{E} = 0$. The assumption (3.126) characterizes *force-free electrodynamics*, or FFE. An immediate consequence is that

$$\rho_e \mathbf{E} \cdot \mathbf{B} = 0, \qquad (3.127)$$

so that there can be non-zero charge density at points where $\mathbf{E} \cdot \mathbf{B} = 0$. The physical implication is that the charge carriers spiral very rapidly along magnetic fieldlines to arrange themselves so that they feel no net Lorentz force, by 'shorting out' the local component of $\mathbf{E}$ parallel to $\mathbf{B}$. Positive and negative charge carriers evidently move in opposing directions along magnetic fieldines.

The quantity $\mathbf{E} \cdot \mathbf{B}$ is Lorentz (and GR) invariant, and so is the combination $F = [c^2]B^2 - E^2$. Depending on the sign of F (fixed by boundary conditions, and almost always positive), (3.127) ensures that at each point there is a reference frame in which the local $\mathbf{E}$ (usually) or $\mathbf{B}$ (rarely) can be set to zero. As for charge densities in MHD, there is in general no global frame in which this holds throughout the gas.

It is important to realize that the equilibrium state where (3.127) holds does not in general correspond to zero net charge at each point. This is in stark contrast with ideal MHD, where charge neutrality at every point is the defining assumption. It can happen that in the presence of electromagnetic fields the charge carriers in a plasma move independently, rather than as a fluid – this is called *charge separation*. This behaviour is common in several physical and astrophysical situations, such as electrolytes (Debye & Hückel, 1923) and in ionized astrophysical plasmas (Salpeter, 1954). Charge separation significantly affects the rates of nuclear fusion in stars (e.g. Clayton, 1968). It is also important in some accretion flows, as we shall see in Section 7.3.

4

Disc Accretion on to Black Holes

4.1 Orbits around Black Holes

We argued in Section 1.1 that supermassive black holes must power active galactic nuclei, the brightest persistent objects in the Universe, by releasing energy through gas accretion. We have also seen that SMBH lie at the centres of their host galaxies, which therefore presumably supply this gas. Most galaxies contain large amounts of orbiting gas, but we know that the SMBH has a negligible gravitational effect on its host stars and gas outside a very small region of size $\sim R_{\text{inf}} = 2GM/\sigma^2$, where M is the SMBH mass and σ is the velocity dispersion in the central region of the galaxy. The total mass of this region is $\sim M$, but is mostly in the form of stars, so the infall of gas towards the hole cannot simply result from its gravity alone. Just how galaxies arrange that significant masses of gas get close enough to the black hole to be accreted by it is probably the main current uncertainty in understanding the connection between SMBH and their hosts. We will return to this question in Chapter 5. In this chapter we consider how gas behaves once it is within the SMBH sphere of influence R_{inf}.

Although gravity is by far the weakest of the four fundamental forces, it is generally the most important on macroscopic scales because it is long-range (unlike the strong and weak nuclear forces) and always attractive (unlike electromagnetism). As a result most astrophysics deals either with gravity and inertia alone (e.g. celestial mechanics, galaxy dynamics) or with the contest between gravity, inertia, and forces that are electromagnetic in origin, such as gas or radiation pressure, or viscosity (e.g. stellar evolution, accretion, feedback). The strong and weak nuclear forces enter only through their effects in powering stars through nuclear burning and in driving supernova explosions.

In Section 1.1 we noted that at sufficient distance a black hole's gravitational field is the same as that of any other gravitating object of the same mass. So our basic knowledge of Newtonian orbits around a point mass applies to all orbits around black holes that lie within R_{inf} but do not get close to the hole, that is, within distances of order a few times R_g, where orbital speeds begin to approach c and relativistic effects appear. In general, any matter attracted towards the hole must have a non-zero net angular momentum with respect to it – it is extremely unlikely that matter should be 'aimed' directly at the hole. Although at large distance the angular momentum may be negligible, so that inflow is essentially radial, it becomes all-important for the dynamics close to the hole. A particle in a Newtonian orbit about a point mass M has the energy equation

$$\frac{v^2}{2} - \frac{GM}{r} = E,$$ (4.1)

where v, E are the velocity and specific energy of the particle. In polar coordinates in the orbital plane we have $v^2 = \dot{r}^2 + r^2 \dot{\theta}^2$, where the angular velocity $\dot{\theta}$ is related to the specific angular momentum J as

$$r^2 \dot{\theta} = J.$$ (4.2)

Then (4.1) becomes

$$\frac{1}{2} \dot{r}^2 + V(r) = E,$$ (4.3)

where we can regard

$$V(r) = \frac{J^2}{2r^2} - \frac{GM}{r}$$ (4.4)

as an effective potential. This equation shows explicitly how the importance of the particle's angular momentum grows strongly as r decreases, and reveals the following main features of Newtonian orbits:

(a) For $J = 0$, orbiting particles escape to infinity ($r \to \infty$) if $E > 0$, or instead fall into the gravitating mass at $r = 0$ if $E < 0$.
(b) For $J \neq 0$, orbiting particles again escape to infinity ($r \to \infty$) if $E > 0$, but remain bound (r varies between two finite values) if $E < 0$.
(c) Particles have circular orbits (fixed r) if $\dot{r} = 0$, so that $V(r) = E < 0 =$ constant and $dV/dr = 0$. These orbits have the lowest energy E for fixed specific angular momentum J. They are stable only if $d^2V/dr^2 > 0$; that is, circular orbits occur at minima of the effective potential $V(r)$.

These relations illustrate a characteristic feature of all gravitationally bound systems. Circular orbits have $v = (GM/r)^{1/2}$, and (4.1) becomes

$$E = -\frac{GM}{2r} = -\frac{v^2}{2}.$$ (4.5)

Particles orbiting with lower (i.e. more negative) total energy E move *faster* (and orbit at smaller r). This is the virial theorem, which governs the relation between energy and speed in all gravitationally bound systems. For a star made of hot gas a similar result shows that as it loses heat energy by radiating it away, it shrinks sufficiently that the gas actually heats up. As this usually increases the energy loss, it is the reason that stars evolve at all, through a sequence of long-lasting equilibria where various forms of nuclear burning supply their luminosities, instead of simply cooling down.

The properties (1)–(3) accurately describe distant ($r >> R_g$) orbits about black holes. To find what happens for closer orbits we have to integrate the full GR geodesic equations of motion. The Kerr geometry has only two obvious symmetries, specified by Killing vectors corresponding to stationarity and axisymmetry. These require that an orbiting particle conserves its energy, and its angular momentum component parallel to the black hole spin. With the rest mass of the orbiting particle this gives three constants of the motion, which would not fix the orbits uniquely. But another piece of good luck means that the orbits (geodesics) in the Kerr solution are integrable despite this. There is a fourth constant,[1] related to the particle's total angular momentum (unusually, specified by what is called a Killing tensor, rather than a Killing vector), which allows one to characterize all orbits uniquely. Unlike in the Newtonian regime, the orbits do differ depending on their orientation with respect to the black hole spin. Fortunately, in many cases orbits that get close to the hole are in practice in or close to the spin (equatorial) plane, for reasons we will discuss in detail (see Section 5.3), so we confine the discussion to these aligned orbits at this point.

At large r the full GR orbits closely resemble the Newtonian ones, as we can deduce from the closeness of the dashed and dash-dotted curves in Figure 4.1. In particular, circular orbits in the equatorial plane $\theta = \pi/2$ appear at minima of an effective potential like $V(r)$. But something must change for orbits with $r \sim R_g$, as (4.25) then formally predicts $v \sim c$. To prevent this, the minima of the GR effective potential become progressively shallower as the orbital radius r decreases (see Figure 4.1). At a certain radius, specified as a multiple of R_g which depends on the hole's angular momentum parameter a, and on whether the orbit is prograde or retrograde with respect to the hole spin, the minima disappear altogether. Then any circular orbit is unstable to perturbations pushing it radially inwards, and falls into the event horizon. The critically stable orbit is called the ISCO (innermost stable circular orbit). As the name suggests, equatorial circular orbits inside the ISCO are unstable, and particles following them plunge into the black hole, growing its mass.

---

[1] Usually called the Carter constant, after its discoverer Brandon Carter (Carter, 1968).

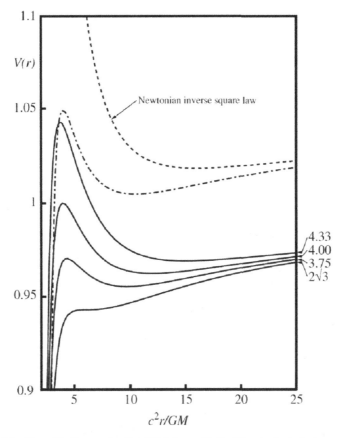

Figure 4.1 The effective potential $V(r)$ from (4.4). The dash-dotted curve is the pseudo-Newtonian potential $\Phi = -GM/(r-2R_g)$ of Paczyński and Wiita (1982). Credit: APIA3.

We will see in Chapter 5 that gas orbiting the hole is likely to spiral inwards through a sequence of circular obits of decreasing radius. All gas reaching the ISCO falls into the hole, so the mass inflow rate $\dot{M}$ to the ISCO, together with the ISCO properties, then determine all the main features of the accretion process. The specific angular momentum J and specific binding energy E of particles orbiting at the ISCO specify how the hole spin evolves, and the total energy which must have been released by the accretion disc as electromagnetic radiation, that is, the accretion yield. Convenient expressions for these quantities are

$$J_{\text{ISCO}} = \frac{2R_g c}{3\sqrt{3}}[1 + 2(3z - 2)^{1/2}], \qquad (4.6)$$

$$E_{\text{ISCO}} = \left[1 - \frac{2}{3z}\right]^{1/2} c^2, \qquad (4.7)$$

where z is defined by

$$z = \frac{R_{\text{ISCO}}}{R_g} \tag{4.8}$$

and is related to a via

$$a = \frac{1}{3}z^{1/2}[4 - (3z - 2)^{1/2}]. \tag{4.9}$$

From (4.7) we find the dimensionless accretion yield

$$\eta = 1 - \left[1 - \frac{2}{3z}\right]^{1/2}. \tag{4.10}$$

It is often useful to compare the specific angular momentum J_{ISCO} with the Kepler value $J_K(R_{\text{ISCO}}) = (GMR_{\text{ISCO}})^{1/2} = z^{1/2}R_g c$. Using (4.6), this gives

$$\frac{J_{\text{ISCO}}}{J_K(R_{\text{ISCO}})} = f(z) = \frac{2(1 + 2(3z - 2)^{1/2})}{3\sqrt{3}z^{1/2}}. \tag{4.11}$$

The function $f(z)$ is virtually constant at $f(z) \simeq 1.4$ for $9 \geq z \gtrsim 3$, decreasing slightly to 1.36 and 1.16 for $z = 2, 1$. If a black hole with initial mass and spin parameters (M_1, z_1) grows its mass by accreting matter from the ISCO, Bardeen (1970) shows that these evolve as

$$\frac{z}{z_1} = \left(\frac{M_1}{M}\right)^2. \tag{4.12}$$

The connection between the ISCO radius zR_g and a, M is

$$z = \{3 + Z_2 \mp [(3 - Z_1)(3 + Z_1 + 2Z_2)]^{1/2}\}, \tag{4.13}$$

where

$$Z_1 = 1 + (1 - a^2)^{1/3}[(1 + a)^{1/3} + (1 - a)^{1/3}] \tag{4.14}$$

and

$$Z_2 = (3a^2 + Z_1^2)^{1/2}, \tag{4.15}$$

and the upper/lower signs refer to prograde/retrograde orbits, respectively, that is, rotating in the same or opposed sense as the black hole spin.

The ratio $z = R_{\text{ISCO}}/R_g$ decreases from 9 for a retrograde orbit in maximal Kerr (formally $a = -1$) through 6 for $a = 0$ (i.e. Schwarzschild), to 1 for prograde orbits in maximal Kerr ($a = 1$) (see Table 4.1). Similar to Newtonian orbits, the accretion efficiencies of the relatively wide ISCO orbits ($9 > z > 6$) are low. Prograde accretion, where $6 > z > 1$, gives higher accretion yields. The accretion efficiency can be as high as $\eta \sim 0.4$, although this requires prograde accretion on to a hole with the highest spin $a \simeq 1$. It is conventional to use $\eta = 0.1$ (corresponding

Table 4.1 *ISCOs and accretion yields*

	a	z	η
Maximal retrograde	−1	9	0.038
Schwarzschild	0	6	0.057
Maximal prograde	1	1	0.422

The dimensionless ISCO radius $z = R_{ISCO}/R_g$ and accretion efficiency η (Equation (4.10)) for various black hole spin parameters a.

to $a \sim 0.7$) in rough estimates, as this is less than a factor of two away from the correct value for all but the most extreme spin rates $a > 0.9$.

Prograde mass accretion from the ISCO grows the hole mass and spin, while retrograde accretion grows M but reduces the spin. Table 4.1 shows that spinning a maximally rotating Kerr hole ($a = -1$) down to Schwarzschild ($a = 0$) with retrograde accretion requires a black hole mass increase by a factor $\sqrt{3}$ from the initial value. At this point continued accretion in the same sense is now prograde with respect to the hole spin. If this continues it will spin the hole up to $a = 1$ after a further factor $\sqrt{3}$ mass increase.

4.2 Thin Accretion Discs

So far we have only considered particle orbits near black holes, rather than gas flows around them. The important difference is that since a gas flow has finite radial extent, its azimuthal velocity is likely to vary with radius across it, implying internal shear. Any radial communication between gas particles, for example any process we would call a 'viscosity', must tend to resist the shear and try to reduce it ultimately to zero.[2] As the gas at smaller r is likely to have higher velocity than that further out, this implies a 'viscous' torque transporting angular momentum outwards, so ultimately causing matter to spiral inwards to smaller r. This process simultaneously produces dissipation as shear energy is converted to heat, which the gas will try to radiate away.

This argument strongly suggests that a gas accretion flow with angular momentum (realistically, every gas accretion flow) tends to radiate, ultimately taking the energy for this from gravity as gas moves inwards. Angular momentum must be transported outwards in this process, but the relative rate is generally far slower

---

[2] Technically, 'viscosity' here means any internal force proportional to the fluid velocity gradient, as assumed by both Newton and Stokes. See Pringle and King (2007) for a discussion.

than the energy loss. Accordingly, gas tends to spiral slowly in towards the black hole through a sequence of orbits which have the lowest energy possible for their angular momenta. As we saw in point (3) in Section 4.1 these orbits are circles in the plane orthogonal to the angular momenta. The resulting flattened, differentially rotating gas configuration is called an *accretion disc*.

This reasoning clearly applies to realistic gas flows around any gravitating central mass, not simply supermassive black holes. Most of the progress in studying accretion discs has until recently come from studying those around white dwarfs, neutron stars and stellar-mass black holes, rather than SMBH, for reasons discussed in the Preface. There is a rich astrophysical literature about such discs going back at least to the early 1970s, most of which applies directly to SMBH discs. The treatment here closely parallels that in APIA3, and we refer the reader to that source for details.

Most of the disc flow is in the Newtonian regime, but GR appears in two ways. First, the inner edge of the disc is usually at the ISCO appropriate to the black hole spin, where the disc gas must fall directly into the hole. The boundary conditions at the ISCO connect the Newtonian and GR regimes, and describe how the disc flow affects the mass and spin of the hole. Second, gas orbiting in discs around spinning black holes experiences forced precessions through the Lense–Thirring (LT) effect (see (2.23)) unless the orbit lies in the equatorial plane. In deriving the disc equations here we assume completely equatorial discs. Since the LT effect falls off as the cube of the radial distance, this is effectively no restriction for most of the disc. Even for 'misaligned' discs not in this plane at large radius, we will see (Section 5.3) that the net effect of LT frame-dragging is often to warp the local disc plane into the black hole equatorial plane.

In many discs the gas flows are confined so closely to the plane defined by their angular momenta that we can regard them as two-dimensional for most purposes. We will see that this corresponds to the well-defined physical condition that the gas is able to radiate away efficiently the gravitational binding energy it releases. This suggests that thermal gas pressure is unlikely to be important, and we will verify this later (see Section 4.6). Then at each cylindrical radius R the gas feels only gravity, so in the Newtonian regime simply follows Keplerian rotation. Its angular velocity is

$$\Omega(R) = \Omega_K(R) = \left(\frac{GM}{R^3}\right)^{1/2}, \tag{4.16}$$

so that the circular velocity is

$$v_\phi = R\Omega_K(R) = \left(\frac{GM}{R}\right)^{1/2}. \tag{4.17}$$

Rather than the mass density ρ it is convenient to use the surface density

$$\Sigma(R,t) = \int_{-\infty}^{\infty} \rho \, dz \sim \rho H, \qquad (4.18)$$

where z is the vertical cylindrical coordinate and H is the disc scaleheight, that is, the characteristic lengthscale over which $\rho(z)$ decreases by a factor ~ 2 as we move away from the disc mid-plane $z = 0$.

As we suggested at the start of this section, interactions between shearing gas layers – 'viscous torques' – are fundamental in accretion discs, because they cause gas to transport angular momentum outwards and so allow it to spiral inwards and liberate gravitational energy. These torques result from any random motions about the mean streaming motion v_ϕ of the gas. Clearly, thermal motions are like this, and they are the usual source of viscosity in laboratory fluids. But we will find this conventional 'molecular' viscosity is totally inadequate for driving observed disc accretion, which must instead involve relatively large-scale turbulent motions. Whatever their origins, the random motions have similar mass fluxes in both directions, and so (like diffusion in laboratory fluids) have a negligible effect on the surface density Σ. But the difference in v_ϕ between adjacent gas streams means that the motions can in principle transfer angular momentum across a shear. The discussion in APIA3 shows that the resulting viscous torque exerted by an outer gas ring on an inner one is

$$G(R) = 2\pi R \nu \Sigma R^2 \Omega', \qquad (4.19)$$

where $\Omega' = d\Omega/dR$, and we write

$$\nu \sim \lambda \tilde{v}, \qquad (4.20)$$

where $\lambda, \tilde{v}$ are the typical lengthscale and velocity of the random radial motions. This form agrees with expectations: if Ω decreases outwards, as it does for Kepler rotation ($\Omega \propto R^{-3/2}$), the torque is negative. This means that the outer ring is trying to slow the inner ring, and angular momentum is being transported outwards. For rigid rotation, $\Omega' = 0$, and the torque vanishes.

Of course, each ring of gas has both an inner and an outer face. So it is subject to competing torques, losing angular momentum across its outer face $R + dR$ and gaining it across the inner face at R. The net viscous torque on it (trying to speed it up) is

$$G(R + dR) - G(R) = \frac{\partial G}{\partial R} dR. \qquad (4.21)$$

This torque is in the sense of the angular velocity $\Omega(R)$, so it does work at the rate

$$\Omega \frac{\partial G}{\partial R} dR = \left[\frac{\partial}{\partial R}(G\Omega) - G\Omega' \right] dR. \qquad (4.22)$$

But here the term

$$\frac{\partial}{\partial R}(G\Omega)dR \tag{4.23}$$

is simply the rate of viscous transport ('convection') of rotational energy across the disc. None of this shows up as dissipated energy and so as radiation. The rate of that is given by the other term, $-G\Omega'dR$, which gives the dissipated energy per unit area $2\pi R dR$ as

$$D(R) = \frac{G\Omega'}{4\pi R} = \frac{1}{2}\nu\Sigma(R\Omega')^2. \tag{4.24}$$

This form is reassuringly positive definite, and we see that dissipation can vanish only for the unlikely case of rigid rotation, $\Omega' = 0$. We note that for Keplerian rotation

$$\Omega(R) = \Omega_K = (GM/R^3)^{1/2}, \tag{4.25}$$

which often holds, we have

$$D(R) = \frac{9}{8}\nu\Sigma\frac{GM\dot{M}}{R^3}. \tag{4.26}$$

Now we can write down conservation equations for the accretion disc. We assume that the gas moves predominantly in the ϕ-direction, but drifts slowly inwards with radial velocity $v_R < 0$. Conservation of mass requires

$$\frac{\partial}{\partial t}(2\pi R\Delta R\Sigma) = v_R(R,t)2\pi R\Sigma(R,t) - v_R(R+\Delta R,t)$$

$$\times 2\pi(R+\Delta R)\Sigma(R+\Delta R) \simeq 2\pi\Delta R\frac{\partial}{\partial R}(R\Sigma v_R).$$

In the limit $\Delta R \to 0$ we get the mass conservation equation

$$R\frac{\partial\Sigma}{\partial t} + \frac{\partial}{\partial R}(R\Sigma v_R) = 0. \tag{4.27}$$

Similarly, there is a conservation equation for the angular momentum component parallel to the disc axis, except that we must include the effects of the viscous torques $G(R)$. Then

$$\frac{\partial}{\partial t}(2\pi R\Delta R\Sigma R^2\Omega) = v_R(R,t)2\pi R\Sigma(R,t)R^2\Omega$$

$$- v_R(R+\Delta R,t)2\pi(R+\Delta R)\Sigma(R+\Delta R)$$

$$\times (R+\Delta R)^2\Omega(R+\Delta R) + \frac{\partial G}{\partial R}\Delta R$$

$$\simeq 2\pi\Delta R\frac{\partial}{\partial R}(R\Sigma v_R R^2\Omega) + \frac{\partial G}{\partial R}\Delta R.$$

In the limit $\Delta R \to 0$ we get

$$R\frac{\partial}{\partial t}(\Sigma R^2 \Omega) + \frac{\partial}{\partial R}(R\Sigma v_R R^2 \Omega) = \frac{1}{2\pi}\frac{\partial G}{\partial R}. \tag{4.28}$$

The similarity between Equations (4.27) and (4.28) let us simplify the second equation to

$$R\Sigma v_R (R^2 \Omega)' = \frac{1}{2\pi}\frac{\partial G}{\partial R}. \tag{4.29}$$

(We have assumed $\partial\Omega/\partial t = 0$, which must hold for orbits in a fixed gravitational potential, as here.) We eliminate v_R from this equation using (4.19), so

$$R\frac{\partial\Sigma}{\partial t} = -\frac{\partial}{\partial R}\left[\frac{1}{2\pi(R^2\Omega)'}\frac{\partial G}{\partial R}\right]. \tag{4.30}$$

If we now assume Keplerian rotation (4.25) in (4.19) we get finally

$$\frac{\partial\Sigma}{\partial t} = \frac{3}{R}\frac{\partial}{\partial R}\left[R^{1/2}\frac{\partial}{\partial R}(\nu\Sigma R^{1/2})\right], \tag{4.31}$$

while the combination of (4.27), (4.28), and (4.19) gives the radial drift velocity as

$$v_R = -\frac{3}{\Sigma R^{1/2}}\frac{\partial}{\partial R}[\nu\Sigma R^{1/2}]. \tag{4.32}$$

Equation (4.31) governs time-dependent disc accretion. It is non-linear, because ν is in general a function of Σ, R and t. Unless some external agency unconnected with the disc is influencing it, the time dependence of ν must come entirely from the fact that the surface density is a function of $\Sigma(R, t)$, so that $\nu = \nu(R, \Sigma)$. The single time derivative and two space derivatives show that (4.31) is a diffusion equation. Writing the derivatives as ratios we see that the diffusion timescale is

$$t_{\text{visc}} \sim \frac{R^2}{\nu}, \tag{4.33}$$

and similarly, (4.32) implies

$$v_R \sim -\frac{\nu}{R}. \tag{4.34}$$

Equation (4.31) is non-linear, and so far we have no clear picture of the process defining the viscosity ν. Then it is not obvious that the diffusion it describes acts to spread disc gas out and reduce spatial density gradients – as familiar in other cases such as heat transport – or instead to cause it to clump together in some kind of instability (antidiffusion). We can check this by setting $\Sigma = \Sigma_0(R) + \Delta\Sigma$, where Σ_0 is a steady-state solution of (4.31) and $\Delta\Sigma$ is an axisymmetric perturbation at each R. Setting $\mu = \nu\Sigma$, and $\nu = \nu(R, \Sigma)$, we have a corresponding perturbation $\Delta\mu = (\partial\mu/\partial\Sigma)\Delta\Sigma$. Using (4.31) we find

$$\frac{\partial}{\partial t}(\Delta\mu) = \frac{\partial\mu}{\partial\Sigma}\frac{3}{R}\frac{\partial}{\partial R}\left[R^{1/2}\frac{\partial}{\partial R}(R^{1/2}\Delta\mu)\right]. \tag{4.35}$$

As we would expect, $\Delta\mu$ obeys a similar diffusion equation, but now with a diffusion coefficient taking the sign of $\partial\mu/\partial\Sigma$, which can be either positive or negative, depending on the structure of the unperturbed steady disc. With a positive sign, spatial maxima in μ (i.e. rings where $\partial^2\mu/\partial R^2 < 0$) imply $\partial(\Delta\mu)/\partial t < 0$. So matter spreads out diffusively away from peaks in surface density, smoothing these out and trying to reach a steady state. But if the unperturbed disc has a region with

$$\frac{\partial\mu}{\partial\Sigma} < 0, \tag{4.36}$$

matter is fed *towards* surface density maxima, and the disc formally tends to break up into rings. Section 4.3 gives a physical interpretation of this. We will see in Section 4.7 that discs can sometimes have regions where $\partial\mu/\partial\Sigma < 0$, but in practice clumping does not happen, as the disc finds a very different steady state on a timescale shorter than viscous.

Even without a detailed form for ν we can get a good idea of the character of solutions of (4.31) by considering the simple case $\nu = $ constant. Then $\partial\mu/\partial\Sigma = \nu > 0$, so diffusion spreads gradients out in the usual way. We can solve (4.31) by separation of variables. The simplest possible case is a ring of gas placed in a Kepler orbit at some initial radius R_0:

$$\Sigma(R, t = 0) = \frac{m}{2\pi R_0}\delta(R - R_0), \tag{4.37}$$

where $\delta(R - R_0)$ is a Dirac delta function. This gives (Lynden-Bell & Pringle, 1974)

$$\Sigma(R, t) = \frac{m}{\pi R_0^2}\tau^{-1}x^{-1/4}\exp\left[-\frac{(1 + x^2)}{\tau}\right]I_{1/4}(2x/\tau), \tag{4.38}$$

where $I_{1/4}$ is a modified Bessel function and $x = R/R_0, \tau = 12\nu t R_0^{-2}$ are dimensionless radius and time variables.

This initially spreads into a disc on a characteristic viscous timescale

$$t_0 = \frac{R_0^2}{12\nu} \tag{4.39}$$

(see Figure 4.2).

As the disc evolves the spatial gradients of Σ become shallower, and local spreading is on longer timescales $\sim l^2/\nu$, with $l > R_0$. Although some gas initially spreads to larger R, it eventually begins to move to smaller R once $R \gg R_0 t/2t_0$.

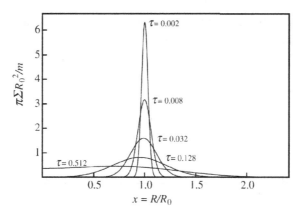

Figure 4.2 A ring of matter of mass m placed in a Kepler orbit at $R = R_0$ spreads out under the action of viscous torques. The surface density Σ, given by (4.38), is shown as a function of $x = R/R_0$ and the dimensionless time variable $\tau = 12\nu t R_0^{-2}$, where ν is the constant kinematic viscosity. Credit: APIA3.

At times $t \gg t_0$ almost all of the disc mass m_0 has accreted to the origin $R = 0$, while all the original angular momentum is carried off to large R by an ever-smaller fraction of the original mass. All of this behaviour is characteristic of all disc evolution. Under stable viscous evolution alone (i.e. with $\partial\mu/\partial\Sigma > 0$ everywhere) mass diffuses inwards and angular momentum outwards.

4.3 Steady Thin Discs

Given the tendency of viscosity to smooth out density gradients, a disc supplied with gas at a constant rate naturally settles to a steady condition. In SMBH discs we will see that a steady mass supply is unlikely. Despite this, the viscous smoothing means that each region of the disc attempts to reach a *locally* steady structure given by the local mass rate $\dot M(R, t)$ at any given time.[3] From (4.18) with $\partial/\partial t = 0$ a locally steady disc has

$$R\Sigma v_R = \text{constant.} \tag{4.40}$$

Since this is the integral of the mass conservation equation, it gives the steady mass accretion rate

$$\dot M = 2\pi R\Sigma(-v_R), \tag{4.41}$$

[3] Many conceptual errors in treating SMBH (and other) accretion discs originate from the tacit (and often inadvertent) assumption of a *globally* steady state.

which is constant with radius and time through the disc. The angular momentum conservation equation (4.28), with $\partial/\partial t = 0$, gives

$$2\pi R\Sigma v_R R^2 \Omega = G(R) + C. \tag{4.42}$$

Here C is a constant, which specifies how much angular momentum the disc feeds to the hole. Using (4.41) and (4.19), we get

$$-\dot{M}R^2\Omega = v\Sigma(2\pi R^3 \Omega') + C. \tag{4.43}$$

As before, we can assume that the innermost gas orbits lie in the spin plane of the hole. Then the hole cannot exert any torque on them, and we can assume that all the disc angular momentum reaching the ISCO simply feeds the hole, that is, $C = \dot{M}J_{\text{ISCO}}$. (A possible exception to this would occur if the infalling matter was connected to disc gas slightly further out by magnetic fields, but we discount this for the moment.) Then setting $\Omega = \Omega_K(R) = (GM/R^3)^{1/2}$, we find

$$v\Sigma = \frac{\dot{M}}{3\pi}\left[1 - \frac{J_{\text{ISCO}}}{J(R)}\right] \simeq \frac{\dot{M}}{3\pi}\left[1 - f(z)\left(\frac{R_{\text{ISCO}}}{R}\right)^{1/2}\right], \tag{4.44}$$

using (4.11), where we found that $f(z) \simeq$ constant ~ 1.4 for all z. Here $J(R)$ is the true specific angular momentum at R, which tends to the Kepler value $(GMR)^{1/2}$ away from the ISCO, and must smoothly tend to J_{ISCO} as $R \rightarrow R_{\text{ISCO}}$. The second form of (4.44) shows that the term in square brackets is close to unity for $R > R_{\text{ISCO}}$, and reaches zero at $R = R_{\text{ISCO}}$.

Equations (4.44) and (4.24) show that the dissipation per unit disc face area is

$$D(R) = \frac{3GM\dot{M}}{8\pi R^3}\left[1 - \frac{J_{\text{ISCO}}}{J(R)}\right], \tag{4.45}$$

since the disc has two faces. This is independent of the viscosity, because in a steady state the binding energy lost by the disc gas must appear as radiation. The total luminosity of the disc is

$$L = 2\int_{R_{\text{ISCO}}}^{R_{\text{out}}} D(R)2\pi R dR, \tag{4.46}$$

where R_{out} is the outer radius of the disc (the factor 2 in front of the integral comes because the disc radiates over two faces). For $R_{\text{out}} >> R_{\text{ISCO}}$ this gives

$$L_{\text{disc}} \simeq \frac{GM\dot{M}}{2R_{\text{ISCO}}}. \tag{4.47}$$

This is just one-half of the total binding energy, since by the virial theorem, the gas falling into the hole takes with it kinetic energy equal to one-half of the potential energy at infinity. This points to another subtlety – the total emission from the two faces of a disc ring between R and $R + dR$ is

$$2 \times 2\pi R dR \times D(R) \simeq \frac{3GM\dot{M}}{2R^2}\left[1 - \frac{J_{\rm ISCO}}{J(R)}\right] dR. \tag{4.48}$$

But the total binding energy release between $R, R+dR$ is just $GM\dot{M}/2R^2$. The extra energy

$$\frac{GM\dot{M}}{2R^2}\left[1 - \frac{3}{2}\frac{J_{\rm ISCO}}{J(R)}\right] dR \tag{4.49}$$

is 'convected' into the ring from smaller R. So for $R > R_{\rm in} \simeq 9R_{\rm ISCO}/4$ (i.e. most of the disc area) the total radiation exceeds that given by the local release of binding energy – viscous transport of energy released deeper in the potential well is twice as big. The viscous transport simply redistributes the energy release, and cannot change the total, so the excess emission from the outer disc $R > R_{\rm in}$ is compensated by emission of less than the binding energy release for $R < R_{\rm in}$, as the expression (4.49) is negative there.

We can use (4.45) to get an idea of how a disc radiates. Assuming it is optically thick (we will check this assumption for consistency later), we know its emission will be close to a blackbody at the effective temperature $T(R)$ given by equating the blackbody flux to D, that is,

$$\sigma T^4(R) = D(R), \tag{4.50}$$

where σ is the Stefan–Boltzmann constant. From (4.45) we get

$$T(R) = \left\{\frac{3GM\dot{M}}{8\pi R^3 \sigma}\left[1 - \frac{J_{\rm ISCO}}{J(R)}\right]\right\}^{1/4}. \tag{4.51}$$

For $R \gg R_{\rm ISCO}$ we have $T \simeq T_*(R/R_{\rm ISCO})^{-3/4}$, where

$$T_* = \left(\frac{3GM\dot{M}}{8\pi R^3 \sigma}\right)^{1/4} = 2.3 \times 10^5 \dot{M}_{26}^{1/4} M_8^{1/4} R_{14}^{-3/4} \text{ K}. \tag{4.52}$$

Here we have parametrized $\dot{M} = 10^{26}\dot{M}_{26}$, g s^{-1}, $M = 10^8 M_8 M_\odot$, R $= 10^{14} R_{14}$ cm, as appropriate for radius $R = 10R_g$ in a disc around a $10^8 M_\odot$ SMBH accreting near its Eddington rate. The value $T_* \sim 10^5$ K suggests that a large fraction of the accretion energy must be radiated in the ultraviolet and soft X-rays, in the form of what is often called the 'big blue bump'. If we roughly approximate the local spectrum emitted by the optically thick part of the disc as blackbody, the continuum spectrum at each radius R is the Planck function

$$I_\nu = B_\nu[T(R)] = \frac{2h\nu^3}{c^2(e^{h\nu/kT(R)} - 1)} \text{ erg s}^{-1}\text{ cm}^{-2}\text{ sr}^{-1}, \tag{4.53}$$

where k, h are Boltzmann's and Planck's constants. This approximation neglects the effect of the local disc atmosphere, that is, the part of the disc above the photosphere where optical depth is $\lesssim 1$, in redistributing the spectrum over frequency. This is particularly likely at frequencies where the opacity changes rapidly. An observer at distance D whose line of sight is at an angle i to the local disc normal sees a flux

$$F_\nu = \frac{2\pi \cos i}{D^2} \int_{R_{\text{ISCO}}}^{R_{\text{out}}} I_\nu R dR \qquad (4.54)$$

from the disc, where R_{out} is its outer radius. With the blackbody assumption (4.53) this becomes

$$\frac{4\pi h\nu^3 \cos i}{c^2 D^2} \int_{R_{\text{ISCO}}}^{R_{\text{out}}} \frac{R dR}{e^{h\nu/kT(R)} - 1}. \qquad (4.55)$$

Not surprisingly this is a stretched-out blackbody (see Figure 4.3). For frequencies $\nu << kT(R_{\text{out}})/h$ the entire disc radiates the Rayleigh–Jeans limit $2kT(R)\nu^2/c^2 \propto \nu^2$, so the full disc has this form, $I_\nu \propto \nu^2$. For $\nu >> kT_*/h$ the whole disc emits the Wien limit $2hc^{-2}\nu^3 e^{-h\nu/kT(R)}$, and the integral (4.55) is dominated by the hottest (inner) parts of the disc where $T \sim T_*$, so $I_\nu \propto \nu^3 e^{-h\nu/kT_*}$, giving an exponential cutoff. For intermediate frequencies $kT_{R_{\text{out}}}/h << \nu << kT_*/h$ we let $x = h\nu/kT(R) \simeq (h\nu/kT_*)(R/R_*)^{3/4}$, and (4.55) becomes roughly

$$F_\nu \propto \nu^{1/3} \int_0^\infty \frac{x^{5/3}}{e^x - 1} dx \propto \nu^{1/3}, \qquad (4.56)$$

as the upper limit of the integral is $h\nu/kT(R_{\text{out}}) >> 1$ and the lower limit is $h\nu/kT_* << 1$. We note that the $F_\nu \propto \nu^{1/3}$ region is only appreciable provided that $T(R_{\text{out}}) << kT_*$, which requires $R_{\text{out}} >> R_{\text{ISCO}}$.

Importantly, the surface dissipation rate $D(R)$ is independent of the disc viscosity for blackbody discs. Stellar-mass accreting binary systems (particularly cataclysmic variables, where a white dwarf accretes from a low-mass main sequence companion star) allow abundant tests, as eclipses by the companion occult different parts of the disc with different blackbody temperatures. The good agreement of observation and theory here was a major step in verifying that disc theory was a realistic model for accretion processes.

In common with other accreting systems (see APIA3) we expect that AGN discs will emit some radiation at frequencies $\nu \sim kT_{\text{dyn}}/h$, where

$$T_{\text{dyn}} \sim \frac{GMm_p}{2kR_{\text{ISCO}}} \lesssim \frac{m_p c^2}{2k} \sim 10^{12}\,\text{K} \qquad (4.57)$$

is the dynamical temperature corresponding to the binding energy of a proton at the ISCO, in addition to the blackbody-like emission described above. The temperatures T_* and T_{dyn} play for accretion discs the roles of the effective and coronal

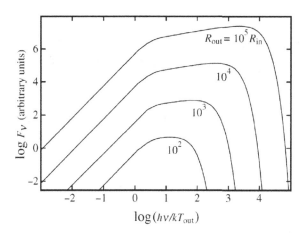

Figure 4.3 The continuum spectrum of a steady optically thick accretion disc radiating locally as a blackbody, for discs with different ratios R_{out}/R_{in}. The frequency is normalized to kT_{out}/h where $T_{out} = T(R_{out})$. Credit: APIA3.

temperatures for stars (T_{dyn} is also the central temperature in stars, which allows nuclear fusion to occur).

4.4 Disc Spectra and Emission Lines

Just as for stars, the radiation from an accretion disc seen at infinity is modified from the pure blackbody form of Figure 4.3 by the effects of the disc atmosphere – the upper layers where the continuum optical depth is $\lesssim 1$. One can attempt to construct more detailed theoretical spectra by considering the effects of these layers. This is more complicated than the analogous procedure for stars. There the energy being radiated is generated by nuclear reactions in the star's centre, far below the atmosphere, which is a purely passive absorber and re-emitter of this continuum. By contrast, in discs we do not yet have a good understanding of where the energy release from viscous dissipation occurs, and there is no obvious reason to assume that the atmosphere here is simply a passive layer reacting to the flux generated far below, as in a star. There is clear observational evidence to the contrary, in that most AGN show strong resonance lines of highly ionized iron *in emission* in X-rays (see, e.g. Figure 1.2).[4] Emission lines must originate in layers of the disc that are *hotter* than those producing the continuum. This suggests either that, as we envisaged in the discussion after (4.49), viscous dissipation in the outer parts of the disc atmosphere is significant, or that significant parts of the disc surface are irradiated

[4] The most abundant element for which resonance lines lie in the X-ray region is iron, so these emission lines come from blends of hydrogen-like and helium-like iron (i.e. ionized to the point where the ions have only either one or two bound electrons).

by emission generated at smaller radii near the SMBH, or both. Because we do not yet know the detailed form of local viscous dissipation, no first-principles description of either case is possible: we do not know the detailed form of the vertical disc structure.

As a result, modelling AGN iron lines requires ad hoc assumptions, for example about the surface emissivity and the disc's geometric profile, and it is not yet definitive. Moreover, rigorous separation of the X-ray emission lines from the underlying steeply sloping continuum is not possible, making their true velocity widths uncertain. Although the lines are broad, as we would expect if they are produced in rapidly rotating and strongly sheared parts of the disc near the ISCO, it is difficult to translate this rigorously into a limit on the ISCO size and so an estimate of the SMBH spin parameter a.

X-ray irradiation of parts of the disc surface by emission at smaller radii is known to occur in low-mass X-ray binary systems (see APIA3, Section 5.10). These systems are close analogues of AGN in many ways, particularly if the accretor is a stellar-mass black hole rather than a neutron star. Irradiation of most of the disc surface by harder radiation from close to the ISCO flattens the effective surface temperature dependence from $T \propto R^{-3/4}$ (4.51) to $T \propto R^{-1/2}$. Since in general $R_{\mathrm{out}} >> R_{\mathrm{ISCO}}$, this gives a significant regime where the disc spectrum is

$$F_\nu \propto \nu^{-1} \tag{4.58}$$

(see APIA3, Equation (5.97)), which is perhaps closer to what is observed (see Figure 1.2) for AGN.

For any form of F_ν, the total disc flux emitted between two frequencies (ν_1, ν_2) is

$$F(1,2) \propto \int_{\nu_1}^{\nu_2} F_\nu \mathrm{d}\nu = \int_{\ln \nu_1}^{\ln \nu_2} \nu F_\nu \mathrm{d}\ln\nu. \tag{4.59}$$

If F_ν does not vary strongly with ν (e.g. a power-law dependence $F_\nu \propto \nu^\gamma$, with $|\gamma| = O(1)$) in a frequency range with $\ln(\nu_2/\nu_1) = $ few, we can crudely approximate this as

$$F(1,2) \sim \nu_2 F_{\nu_2}. \tag{4.60}$$

This is often used to give simple estimates of the disc luminosity from observations, although this may be misleading if there are parts of the disc spectrum that vary strongly with frequency.

A related pitfall is thinking of regions of the disc as being in one-to-one correspondence with certain components of the total disc spectrum. It is true that the hardest radiation (e.g. X-rays) is almost entirely radiated from the hot innermost regions of a disc, but these regions make non-negligible contributions at lower photon energies too – at a frequency ν on the Rayleigh–Jeans tail of the disc spectrum, the contribution from the disc region within radius R is proportional to $\sim R^2 T(R) \sim R^{5/4}, R^{3/2}$ for non-irradiated and irradiated discs, respectively, so the region with one-half of the outer disc radius contributes 42 per cent or 35 per cent of the flux in the two cases.[5]

4.5 Non-thermal Emission

Most emission from discs is thermal in origin, but this is not necessarily so, particularly at every frequency. For a source at a known distance, so that one can convert from flux to luminosity, a simple test is to compute the *brightness temperature* T_{br} of the emission at a given frequency ν. This is defined as the temperature the emitter would need to have to emit the specific luminosity L_ν at ν as a Rayleigh–Jeans emitter, that is,

$$T_{\mathrm{br}} = \frac{c^2 L_\nu}{2k\nu^2}. \tag{4.61}$$

This is evidently a lower limit to the temperature of any thermal emitter producing L_ν, as a hot body cannot emit more than the Planck function B_ν at any frequency, and the Rayleigh–Jeans form is an upper limit to B_ν at all frequencies. Clearly T_{br} itself depends on frequency.

Thermal emission from accretion processes cannot easily produce photon energies kT higher than the rest-mass energy $m_p c^2 \sim 10^{-3}$ erg of a proton. This implies a limit on the physical temperature $T \lesssim m_p c^2/k \sim 10^{13}$ K, and even then the electrons producing the emission must be moving relativistically, with Lorentz factors $\gamma = (1 - v^2/c^2)^{-1/2} \sim m_p/m_e \sim 10^3$. Since $T > T_{\mathrm{br}}$, brightness temperatures

$$T_{\mathrm{br}} \gtrsim 10^{12}\,\mathrm{K} \tag{4.62}$$

are generally a sign that some *non-thermal* process is producing the emission at that frequency. Examples of this are synchrotron emission from electrons gyrating at relativistic speeds around a magnetic field, or inverse Compton emission from scattering of photons by relativistic electrons (see APIA3, Section 9.2 and Appendix). By definition, non-thermal processes appear in situations far from thermal equilibrium, which may be short-lived. Detected radio emission in astronomy is almost always non-thermal, as these processes are inherently far more luminous than thermal emission, compensating for the low energies of individual radio photons.

---

[5] In stellar-mass binary systems with accretion discs, this property means that eclipses of the disc by the companion star are much wider and shallower at long wavelengths than short. See APIA3, Figure 5.8.

4.6 When Is a Disc Thin?

In the last two sections we assumed that the disc was thin, and in Keplerian rotation. To check these assumptions we note that for consistency there must be very little motion in the axial or 'vertical' direction. Then the disc should be hydrostatic in this direction, and the pressure force should have negligible effect on the radial motion, where gravity is balanced by rotational inertia.

The Euler equation governing the fluid velocity $\mathbf{v}$ is

$$\rho \frac{\partial \mathbf{v}}{\partial t} + \rho (\mathbf{v} \cdot \nabla) \mathbf{v} = -\nabla P + \mathbf{f}. \tag{4.63}$$

Here P is the total pressure and $\mathbf{f}$ the force density. The only external force is gravity, so

$$\mathbf{f} = \rho \nabla \left(\frac{GM}{r} \right), \tag{4.64}$$

with $r = (R^2 + z^2)^{1/2}$ the spherical radial coordinate.

Setting $v_z = 0$, the z-component of this equation is

$$\frac{1}{\rho} \frac{\partial P}{\partial z} = \frac{\partial}{\partial z} \left[\frac{GM}{(R^2 + z^2)^{1/2}} \right]. \tag{4.65}$$

A thin disc must have $z \ll R$, so this becomes

$$\frac{1}{\rho} \frac{\partial P}{\partial z} = -\frac{GMz}{R^3}. \tag{4.66}$$

From the definition of the vertical scaleheight we expect $\partial P / \partial z \sim -P/H$. Using $P = \rho c_s^2$, where c_s is the local sound speed in the disc, we find

$$\frac{H}{R} \simeq \frac{c_s}{v_K}, \tag{4.67}$$

where $v_K = (GM/R)^{1/2}$ is the local Kepler velocity. A thin disc requires $H \ll R$, so we need $c_s \ll v_K$, that is, the disc must be cool enough that the local Kepler velocity is highly supersonic.

The R-component of (4.63) is

$$v_R \frac{\partial v_R}{\partial R} - \frac{v_\phi^2}{R} + \frac{1}{\rho} \frac{\partial P}{\partial R} + \frac{GM}{R^2} = 0. \tag{4.68}$$

The pressure term here is $\sim c_s^2/R$, which by (4.67) is negligible compared with the gravity term GM/R^2, and for a disc flow we must have $v_R \ll v_\phi$, so the v_R term is negligible: $v_R(\partial v_R/\partial R)/(v_\phi^2/R) \sim (v_R/v_\phi)^2 \ll 1$. So only the v_ϕ term is left to balance gravity, and we have finally

$$v_\phi = \left(\frac{GM}{R}\right)^{1/2} = v_K(R). \tag{4.69}$$

We conclude that

> *an accretion disc is simultaneously thin and Keplerian if and only if it cools sufficiently that $c_s \ll v_K$.*

Another way of saying this is that these thin disc conditions hold if the *Mach number* $\mathcal{M} = v_K/c_s$ of the Keplerian flow obeys $\mathcal{M} \gg 1$. If this condition fails, all the thin disc approximations break down simultaneously – the disc is neither thin nor Keplerian.

We have so far said little about viscosity, which is the agency that makes disc accretion possible. We noted that dimensionally the kinematic viscosity ν is the product of a lengthscale λ and a velocity scale $\tilde{v}$ (see (4.20)). We can think of these as characterizing the properties of the random motions around the mean streaming velocity $\mathbf{v}$ of the disc gas. If the random motions are isotropic we must have $\lambda < H$. It is also likely that these motions have speed $\tilde{v}$ less than c_s – otherwise they would shock in colliding with other matter and so become subsonic. These arguments motivate the α-prescription

$$\nu = \alpha c_s H \tag{4.70}$$

(Shakura & Sunyaev, 1973). Here α is dimensionless, and we expect $0 < \alpha \lesssim 1$. Of course, there is no reason to expect α to be constant, either spatially or in time. But its restricted range suggests that it may be reasonable to neglect its variation in calculating disc properties. This approach has been remarkably successful in giving a plausible picture of disc accretion that one can readily compare with observations. It is, of course, not a substitute for a first-principles treatment of viscosity, and we shall discuss this further in Section 4.12.

The α-prescription gives a simple expression for the viscous timescale (4.33), which becomes

$$t_{\text{visc}} \sim \frac{1}{\alpha}\left(\frac{R}{H}\right)^2\left(\frac{R^3}{GM}\right)^{1/2} \tag{4.71}$$

or

$$t_{\text{visc}} \sim \frac{1}{\alpha}\left(\frac{R}{H}\right)^2 t_{\text{dyn}}, \tag{4.72}$$

where we have identified

$$t_{\mathrm{dyn}} \sim \left(\frac{R^3}{GM} \right)^{1/2} = \frac{R}{v_K} \tag{4.73}$$

as the dynamical timescale of the disc. As in any hydrostatic structure (e.g. a star), t_{dyn} also specifies the timescale for pressure changes to affect the vertical structure of the disc, since the thin disc condition (4.67) implies that

$$t_z = \frac{H}{c_s} \sim t_{\mathrm{dyn}}. \tag{4.74}$$

Just like a star, an accretion disc also has a thermal timescale t_{th} governing changes in the disc structure caused by heat gain or loss. The heat content per unit volume of the disc is $\sim \rho c_s^2$, so the content per unit disc face area is simply $\sim H\rho c_s^2 \sim \Sigma c_s^2$. Then the disc thermal timescale is

$$t_{\mathrm{th}} \sim \frac{\Sigma c_s^2}{D(R)} \sim \frac{R^3 c_s^2}{GM\nu} \sim \frac{c_s^2}{v_K^2} \frac{R^2}{\nu} \sim \left(\frac{H}{R} \right)^2 t_{\mathrm{visc}} \sim \frac{1}{\alpha} t_{\mathrm{dyn}}, \tag{4.75}$$

where we have used (4.26) for $D(R)$, and (4.72) for t_{visc}. So we get a hierarchy of timescales

$$t_{\mathrm{dyn}} \sim \alpha t_{\mathrm{th}} \sim \alpha (H/R)^2 t_{\mathrm{visc}}, \tag{4.76}$$

so that

$$t_{\mathrm{dyn}} \lesssim t_{\mathrm{th}} << t_{\mathrm{visc}}. \tag{4.77}$$

This is the same ordering as in a star, where nuclear burning takes the role played by viscosity, although the disc dynamical and thermal timescales in a disc are much closer to each other than in a star. The ordering (4.77) implies that if a disc evolves on a viscous timescale, it passes through a sequence of steady states in thermal and dynamical equilibrium.

4.7 Disc Stability

We can now look more closely at the instability criterion

$$\frac{\partial \mu}{\partial \Sigma} = \frac{\partial (\nu \Sigma)}{\partial \Sigma} < 0 \tag{4.78}$$

found from using (4.35).[6]

   In steady thin discs we have

$$\dot{M} \simeq 3\pi \nu \Sigma = \text{constant} \tag{4.79}$$

---

[6] This condition is sometimes called the Lightman–Eardley instability condition, as it was first found in a paper showing that discs with radiation pressure dominant were unstable (Lightman & Eardley, 1974).

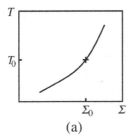 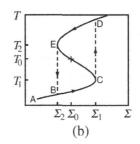

(a)   (b)

Figure 4.4 Effective temperature–surface density diagrams with (a) a unique stable steady solution (T_0, Σ_0), and (b) a case where limit-cycle behaviour occurs as the steady solution (T_0, Σ_0) lies on a region of the curves with $\partial T/\partial \Sigma < 0$. Credit: APIA3.

in most of the disc, which implies $\dot{M} = 3\pi\mu$ with $\mu = \nu\Sigma$. The viscous instability criterion $\partial\mu/\partial\Sigma < 0$ now has a simple interpretation, as it implies

$$\frac{\partial \dot{M}}{\partial R} = 3\pi \frac{\partial \mu}{\partial \Sigma} \frac{\partial \Sigma}{\partial R} \propto -\frac{\partial \Sigma}{\partial R}, \tag{4.80}$$

so gas increasingly flows into a surface density maximum at R_0 from larger R, but is decreasingly able to flow away from the maximum to $R < R_0$.

Further, we can work out how the disc reacts to the instability in many cases. Often the equilibrium relation between $\mu = \nu\Sigma \propto \dot{M} \propto T^4$ and Σ at a given disc radius has a characteristic S-shape, as in Figure 4.4.

This occurs, for example, at ionization fronts, where hydrogen changes from being predominantly neutral to highly ionized. As we see, there is an unstable part of the $\dot{M}(\Sigma)$ curve sandwiched between two stable parts. The equilibrium curve is the locus of thermal balance between viscous heating and radiative cooling. To the right of this curve, heating is stronger than cooling, and to the left, cooling is stronger than heating. In principle three local equilibrium states are possible, where viscous dissipation is balanced by radiative cooling in the S-shaped part of the curve. Let us imagine that the disc region under discussion is currently on the lower stable branch A–C of the curve, and supplied with mass is such a way that it has to increase Σ above the value that would give a steady solution on this lower branch.

Eventually it reaches point C, where to remain in thermal balance it would have to turn back to smaller Σ on the unstable part of the equilibrium curve. But at point A any perturbation taking this part of the disc above the curve grows, because heating is stronger than cooling there. As we have seen, the thermal timescale at each disc radius is shorter than that for viscous evolution, so the disc evolves on this shorter timescale to the point D, on the upper stable branch of the equilibrium curve, where it is (at least locally) hotter and more luminous. A similar but reversed evolution happens on this upper branch if the disc needs to evolve to lower Σ. First it evolves viscously (D $\rightarrow$ E), then thermally to lower temperature (E $\rightarrow$ B).

This discussion shows that if the disc is fed mass at a steady rate which would allow thermal equilibrium only on the unstable part of the curve (EC), it cannot settle to a steady state, and is instead forced to follow this cycle of slow viscous growth of $\dot{M}$ on the lower (cool, low accretion rate) and viscous decrease of $\dot{M}$ on the upper (hot, high accretion rate) branches of the S-curve, with rapid thermal–timescale jumps between them. The prevalence of S-curves in disc theory reflects the fact that there is frequently more than one possible equilibrium accretion rate for a given surface density Σ, depending on the thermal balance. For discs in stellar-mass accreting binaries, this is the basis of the disc outburst model. This works well in describing outbursts in dwarf novae, where the accretor is a white dwarf.

In systems where the accretor is a neutron star or stellar-mass black hole, this model also explains the analogous 'soft X-ray transient' outbursts when the effects of disc irradiation by the accretor are added in (see APIA3, Section 5.10). Here the extra source of heating from the central X-ray source prevents much of the disc (or all of it, in binary systems with short orbital periods and so with small discs) from returning to the cool branch of the S-curve until a large part of the disc mass has been accreted in the high state. These systems have longer outbursts, and therefore much longer low states in which mass accumulates slowly in the disc. AGN discs may behave like this (Burderi, King & Szuszkiewicz, 1998), but the expected timescales are long enough to make it difficult to distinguish this from other forms of variability.

4.8 The Local Structure of Thin Discs

The form (4.71) of the viscous timescale allows estimates of α by comparing with observations of stellar-mass binary systems where mass is transferred via an accretion disc to a compact object (white dwarf, neutron star or black hole). Many of these systems show repeated outbursts, which are now well established as reflecting instabilities in the disc structure, as we have seen in the previous section. The outburst decays are slow (days in dwarf novae, and weeks or months in soft X-ray transients) and easily measured, and reveal the viscous timescale of the whole accretion disc. The orbital period of the binary follows from radial velocity measurements, which also strongly constrain the mass M of the accretor. Kepler's laws then give the binary separation, and tightly constrain the disc size R_d, and so its dynamical time $(R_d^3/GM)^{1/2}$ (which is slightly shorter than the binary period). As the disc luminosity decays on a viscous timescale it passes through a sequence of steady states, so the surface temperature of the disc (found from spectra) constrains the scaleheight (see (4.91)) and thus H/R_d. Then using (4.71) gives an estimate of α. Observations of several hundred accreting binaries of various types consistently give values in the fairly narrow range

$$\alpha \simeq 0.1-0.4 \qquad (4.81)$$

for discs where hydrogen is fully ionized (see King, Pringle & Livio, 2007 for details). Direct estimates from SMBH accretion are more difficult, but consistent with this estimate. As our derivations show, there is no obvious reason to expect a difference.

Given this narrow range, we can use the α-prescription to calculate the properties of thin discs as functions of the accretion rate $\dot{M}$. As we have remarked, we can expect these relations to hold even as $\dot{M}$ changes. In a thin disc, the pressure and temperature gradients are essentially vertical, so at each radius R the disc is like a one-dimensional star with its size given by the disc thickness in the z-direction. The radial dependence enters through the value of the Kepler velocity v_K, and the local viscous dissipation rate $D(R)$, which plays the role taken by central nuclear burning in stars. As in a star, the vertical temperature gradient depends on whether energy transport to the photosphere is predominantly radiative or convective, depending on whether the local specific entropy $S \propto \ln(P/\rho^\gamma)$ increases vertically or not. Here $\gamma \simeq 5/3$ is the adiabatic index. If $dS/dz > 0$, gas parcels attempting to move upwards are not buoyant and so fall back, whereas they can move upwards, carrying their thermal energy, if this gradient is negative. As in stellar structure, we proceed by assuming a positive entropy gradient and so radiative energy transport, and check this assumption for self-consistency afterwards. Then the vertical energy flux through a surface $z = $ constant is

$$F(z) = -\frac{16\sigma T^3}{3\kappa_R \rho}\frac{\partial T}{\partial z}, \qquad (4.82)$$

where $\kappa_R(\rho(z), T(z))$ is the Rosseland mean opacity, for which we usually take the standard Kramers form

$$\kappa_{\text{Kramers}} = 5 \times 10^{24}\rho T_c^{-7/2} \text{ cm}^2 \text{ g}^{-1}. \qquad (4.83)$$

Equation (4.82) implicitly assumes that the disc is optically thick in the z-direction, that is,

$$\tau = \rho\kappa_R H = \Sigma\kappa_R > 1. \qquad (4.84)$$

The energy balance equation is

$$\frac{\partial F}{\partial z} = Q^+, \qquad (4.85)$$

where Q^+ is the local rate of energy production by viscous dissipation. Integrating, we get

$$F(H) = \int_0^H Q^+(z)\mathrm{d}z = D(R), \qquad (4.86)$$

since the net outward energy flux must vanish by symmetry on the mid-plane $z = 0$. From (4.82) this implies

$$F(z) \sim \frac{4\sigma}{3\tau}T^4(z). \qquad (4.87)$$

Assuming that the central temperature $T_c = T(0)$ satisfies $T_c^4 >> T(H)^4$ (where $T(H)$ is close to the surface temperature $T(R)$ at radius R (see (4.51)), this equation becomes

$$\frac{4\sigma}{3\tau}T_c^4 = D(R), \qquad (4.88)$$

which is the vertical energy transport equation for the local disc structure. Finally we take

$$P = \frac{\rho kT_c}{\mu m_p} + \frac{4\sigma}{3c}T_c^4, \qquad (4.89)$$

where the second term is the radiation pressure. The full set of equations is then

$$\rho = \Sigma/H$$
$$H = c_s R^{3/2}/(GM)^{1/2}$$
$$c_s^2 = P/\rho$$
$$P = \frac{\rho kT_c}{\mu m_p} + \frac{4\sigma}{3c}T_c^4$$
$$\frac{4\sigma}{3\tau}T_c^4 = \frac{3GM\dot{M}}{8\pi R^3}\left[1 - \left(\frac{R_{\mathrm{ISCO}}}{R}\right)^{1/2}\right] \qquad (4.90)$$
$$\tau = \kappa\Sigma$$
$$\nu\Sigma = \frac{\dot{M}}{3\pi}\left[1 - \left(\frac{R_{\mathrm{ISCO}}}{R}\right)^{1/2}\right]$$
$$\nu = \alpha c_s H.$$

These equations can be solved algebraically if we assume that Kramers opacity is larger than electron scattering, given by $\kappa_{\mathrm{e.s.}} \simeq \sigma_T/m_p \simeq 0.4\,\mathrm{cm}^2\,\mathrm{g}^{-1}$, with

$\sigma_T = 6.65 \times 10^{-25}\,\mathrm{cm}^2$ the Thomson cross-section, and also that the radiation pressure term $\propto T_c^4$ is small compared with gas pressure. This gives (see APIA3)[7]

$$\Sigma = 5.2 \times 10^6 \alpha^{-4/5} \dot{M}_{26}^{7/10} M_8^{1/4} R_{14}^{-3/4} f^{14/5}\,\mathrm{g\ cm^{-2}}$$

$$H = 1.7 \times 10^{11} \alpha^{-1/10} \dot{M}_{26}^{3/10} M_8^{-3/8} R_{14}^{9/8} f^{3/5}\,\mathrm{cm}$$

$$\rho = 3.1 \times 10^{-5} \alpha^{-7/10} \dot{M}_{26}^{11/20} M_8^{5/8} R_{14}^{-15/8} f^{11/5}\,\mathrm{g\ cm^{-3}}$$

$$T_c = 1.4 \times 10^6 \alpha^{-1/5} \dot{M}_{26}^{3/10} M_8^{1/4} R_{14}^{-3/4} f^{6/5}\,\mathrm{K} \qquad (4.91)$$

$$\tau = 1.9 \times 10^4 \alpha^{-4/5} \dot{M}_{26}^{1/5} f^{4/5}$$

$$\nu = 1.8 \times 10^{18} \alpha^{4/5} \dot{M}_{26}^{3/10} M_8^{-1/4} R_{14}^{3/4} f^{6/5}\,\mathrm{cm^2\ s^{-1}}$$

$$\nu_R = 2.7 \times 10^4 \alpha^{4/5} \dot{M}_{26}^{3/10} M_8^{-1/4} R_{14}^{-1/4} f^{-14/5}\,\mathrm{cm\ s^{-1}},$$

where $f \simeq [1 - (R_{\mathrm{ISCO}}/R)^{1/2}]^{1/4}$. The disc photospheric (surface) temperature is (from (4.51))

$$T(R) = 2.3 \times 10^5 \dot{M}_{26}^{1/4} M_8^{1/4} R_{14}^{-3/4} f. \qquad (4.92)$$

This gas-pressure-dominant form of the disc solution holds where the ratios $\kappa_{\mathrm{Kramers}}/\kappa_{\mathrm{e.s.}}$ and $P_{\mathrm{gas}}/P_{\mathrm{rad}}$ are both >1. The ratios behave similarly, as $\rho/T^{3.5}$ and ρ/T^3, respectively. Both are self-consistently >1 for $R_{14} \gtrsim 10$, with $P_{\mathrm{gas}}/P_{\mathrm{rad}} > 1$ for

$$R \gtrsim R_{\mathrm{rad}} = 3.7 \times 10^{14} \alpha^{4/15} \dot{M}_{26}^{14/15} M_8^{1/3} f^{56/15}\,\mathrm{cm}. \qquad (4.93)$$

We will find in Section 4.9 that SMBH discs can extend to sizes 100 times this limit, so most (if not all) of the disc mass is in the regime described by (4.91). We see that none of the disc properties here depend strongly on α, suggesting that this semi-empirical approach is self-consistent. A striking property is that the aspect ratio H/R of this main part of the disc is almost constant at

$$\frac{H}{R} \simeq 10^{-3}. \qquad (4.94)$$

This arises because the disc equations imply $H/R = c_s/v_K \propto T_c^{1/2}/v_K \propto T^{1/2}\tau^{1/2}/v_K \propto (\tau/R)^{1/8} \sim$ constant.

Just outside the inner edge R_{rad} of this main body of the disc (if it exists) there is a narrow zone in which gas pressure is still dominant, but electron-scattering opacity is larger than Kramers. This produces a slight flattening of the disc radiation spectrum, but the disc structure is otherwise similar to (4.91).

---

[7] Minor differences appear in versions of these solutions appearing in the literature depending on the precise choices of exponents in the opacity coefficients (compare, e.g. the versions in Collin-Souffrin & Dumont, 1990). These differences have no important effects on any of the physical results we discuss in this book. The same holds for the relations resulting for disc regimes with electron scattering larger than Kramers opacity, or radiation pressure larger than gas pressure.

For discs extending within R_{rad} there is a central region with radiation pressure greater than gas pressure. Evidently this requires $R_{rad} > R_g$. Using $\dot{M}_{26} \simeq (\dot{M}/\dot{M}_{Edd})M_8$ and $R_g = 1.5 \times 10^{13}$ cm in (4.93) shows that this holds only if

$$\frac{\dot{M}}{\dot{M}_{Edd}} > 3.2 \times 10^{-2} \left(\frac{R_{ISCO}}{R_g}\right)^{14/15} \alpha^{-2/5} M_8^{-2/7} f^{-4}. \tag{4.95}$$

Since $R_{ISCO} \geq R_g$ and $f^{-4} \to \infty$ as $R \to R_{ISCO}$, we get a central region with dominant radiation pressure if and only if $\dot{M} > \dot{M}_{Edd}$. We discuss this region further in Section 4.10.

4.9 Disc Self-Gravity

We have so far implicitly assumed that self-gravity is unimportant in the disc compared with the SMBH gravity. We can now check this for self-consistency. At the disc surface the z-component of the SMBH field is $\sim GMH/R^3$ at disc radius R, since $H/R \ll 1$. We can crudely think of the self-gravity as the attraction between clumps of disc gas of linear size $\sim H$, separated by distances H, so $\sim G\rho H^3/H^2 \sim G\rho H$. So the condition for stability against breaking into clumps is

$$\rho \lesssim \frac{M}{R^3}. \tag{4.96}$$

As always in discussions of tidal fields we get a density criterion, here that the local disc density should be smaller than the 'distributed' density of the SMBH – as if its mass were distributed over a sphere of radius R. Multiplying (4.96) by the disc volume $\sim 2\pi R^2 H$ gives an estimate of the mass of a marginally self-gravitating disc as

$$M_{sg} \sim \frac{H}{R}M. \tag{4.97}$$

As almost all of the disc mass is in the gas-dominated region described by (4.91), this gives an estimate

$$M_{sg} \sim 10^{-3}M, \tag{4.98}$$

using (4.94).

The disc mass inside a radius R_d is

$$M(R_d) \simeq 2\pi \int_0^{R_d} \Sigma R dR. \tag{4.99}$$

Using the full disc equations to find the value R_d for which $M(R_d)$ agrees with (4.98) gives (Collin-Souffrin & Dumont, 1990)

$$R_{sg} = 3 \times 10^{16} \alpha_{0.1}^{14/27} \dot{m}^{8/27} M_8^{1/27} \text{ cm}, \tag{4.100}$$

where $\alpha_{0.1} = \alpha/0.1$ and as usual $\dot{m} = \dot{M}/\dot{M}_{Edd}$.

This is another quantity that is remarkably constant. This results since (4.96) implies $\Sigma \sim HM/R^3$ for $R = R_{sg}$, while (4.44) and (4.33) lead to $\Sigma \propto (\dot{M}/\alpha)(R/H)^2(GMR)^{-1/2}$. Equating these two expressions for Σ shows that R_{sg} is mainly sensitive to $(H/R)^2$, which, as we have seen, is itself almost constant. We can also express R_{sg} as

$$R_{sg} \simeq \frac{3}{2}\frac{H}{R}\alpha c_s \frac{M}{\dot{M}}. \tag{4.101}$$

Because gas cooling in the outer disc is rapid, mass at this disc radius is likely to collapse to form stars, or to be expelled by those stars which do form, on a timescale close to dynamical, that is, $t_{dyn} \sim (R_{sg}^3/GM)^{1/2} \sim 2\,\mathrm{yr}$. Evidently, R_{sg} sets an upper limit to the size of any accretion disc around an SMBH of any mass and accretion rate. As an illustration of this, the ring of stars around the centre of our own Galaxy are at radii slightly larger than R_{sg}, suggesting that they took some of the angular momentum of gas originally further in, which accreted to the black hole Sgr A*.

The near-constant value $R_{sg} \simeq 3 \times 10^{16}\,\mathrm{cm}$ is an unusually precise and robust prediction, so offers a real test of disc theory. There seems little prospect of doing this by direct measurement for any galaxy but our own. But there has been extensive work on looking for light echoes from AGN accretion discs responding to increased irradiation following rapid brightening of the central source. Called 'reverberation mapping', this technique measures the time lag Δt between increased X-ray or UV emission from the centre of the disc and an optical or IR echo from larger disc radii. If the echo has an emission line spectrum where the linewidths are $\sim v$, then the assumption that v is of order the Kepler velocity $v_K = (GM/R)^{1/2}$ (or escape velocity $\sqrt{2}v_K$) with $R \simeq c\Delta t$ for low-redshift AGN gives an estimate of the SMBH mass as

$$M \simeq \frac{cv^2}{G}f\Delta t, \tag{4.102}$$

where f is a geometric factor of order unity, depending on the details of the gas kinematics and the irradiation. This is now a widely used method for estimating SMBH masses.

Since the disc cannot be larger than R_{sg}, and observational selection favours AGN discs that are relatively face-on, it follows that in the AGN rest frame the lag cannot be longer than the light travel time

$$\Delta t_{\mathrm{max, rest}} \simeq \frac{R_{sg}}{c} \simeq 12\,\mathrm{days}. \tag{4.103}$$

This should also limit light echoes from the ensemble of stars that may have formed close to R_{sg}, or of emission-line clouds associated with the accretion. For a distant AGN, the longest *observed* lag would be

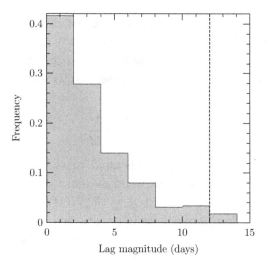

Figure 4.5 The distribution of observed reverberation lags. The rest-frame lags are shorter by the factor $(1 + z)$, where z is the redshift. The distribution cuts off at ~12 days (vertical line), as expected from (4.103) for modest redshifts. Credit: Lobban and King (2022).

$$\Delta t_{\text{max, obs}} \simeq \frac{R_{\text{sg}}}{c}(1 + z) \simeq 12(1 + z)\,\text{days}, \qquad (4.104)$$

where z is the redshift. In principle this allows a redshift measurement independent of that from the line spectrum. The longest lags from quasars at redshift $z \gtrsim 6$ (see Section 5.4) should therefore be $\gtrsim 84$ days.

Figure 4.5 shows that the distribution of observed lags does cut off sharply at 12 days, as predicted by (4.103) for modest redshifts. This confirmation shows that R_{sg} offers a standard 'measuring rod' for AGN. We will see that R_{sg} plays an important role in setting the time- and lengthscales of AGN disc accretion.

4.10 Discs with Super-Eddington Mass Supply

There is good reason to believe that at least some accreting SMBH experience phases where the mass supply $\dot{M}$ rate exceeds $\dot{M}_{\text{Edd}}$. Then from (4.95), the disc structure inside R_{rad} is determined by radiation pressure and electron scattering, and changes radically from (4.91). The Equation (4.90) now give

$$\begin{aligned}
\Sigma &= 10.6\alpha^{-1}\dot{M}_{26}^{-1}M_8^{-1/2}R_{14}^{3/2} \\
H &= 1.6 \times 10^{14}\dot{M}_{26}f.
\end{aligned} \qquad (4.105)$$

The first equation implies that the disc is viscously unstable, since $\dot{M}$ has minima where Σ has maxima. Formally in (4.35) we have $\Sigma \propto \dot{M}^{-1} \propto (\nu\Sigma)^{-1} \propto \mu^{-1}$, so that

$$\frac{\partial \mu}{\partial \Sigma} \propto \frac{\partial \mu}{\partial(1/\mu)} < 0. \tag{4.106}$$

Of course, we have used the α-prescription for viscosity to reach this conclusion, and it is conceivable that the full theory might remove the predicted instability. But we get a clue about what happens when radiation pressure becomes dominant by using the second equation of (4.105), where α does not appear. This shows that the disc semi-thickness H is constant with radius, and

$$\frac{H}{R} \simeq \frac{\dot{M}}{\dot{M}_{\mathrm{Edd}}} \frac{R_{\mathrm{ISCO}}}{R}. \tag{4.107}$$

We see that the disc thickens vertically as the accretion rate approaches the Eddington value, making $H \sim R$ near the ISCO. The thin disc approximation breaks down there.

There have been several attempts to interpret the instability of thin discs dominated by radiation pressure. Because of (4.107), these usually abandon the thin disc assumption $H \ll R$. But because H is constant with radius, the approximations

$$v_z = \frac{\partial v_R}{\partial R} = \frac{\partial v_\phi}{\partial z} = 0, \tag{4.108}$$

which hold if $H \ll R$, are reasonable even when this disc is not thin, that is, for $H \gtrsim R$. Accordingly, one can look for steady solutions of this type. General-relativistic effects for a non-rotating black hole are often mimicked by using a quasi-Newtonian potential (Paczyński & Wiita, 1982)

$$\Psi = -\frac{GM}{R - 2R_g} \tag{4.109}$$

in place of the Newtonian one.

This procedure gives a soluble system for an assumed steady state, defining the so-called *slim disc* solutions (Abramowicz et al., 1988), provided that the solution is matched to a Shakura–Sunyaev solution at an outer boundary R_{ext}, making this an eigenvalue problem which must be solved numerically. The results are shown in Figures 4.6 and 4.7. Slim disc solutions exist for modestly super-Eddington mass supply rates $\dot{M}_{\mathrm{supp}} \lesssim 30\dot{M}_{\mathrm{Edd}}$, and give disc luminosities $\sim 3L_{\mathrm{Edd}}$. They deal with the 'excess' mass supply $\sim \dot{M}_{\mathrm{supp}} - \dot{M}_{\mathrm{Edd}}$ by having v_ϕ low enough and v_R high enough that gas can accrete without radiating significantly before crossing the event horizon, so they are an example of the possibility (a) for dealing with a super-Eddington mass supply discussed in Section 1.2. The 'excess' (super-Eddington) radiation is trapped inside the flow and falls into the black hole, so the emission per unit disc face area goes as

$$D(R) \simeq \frac{3GM\dot{M}(R)}{8\pi R^3} f \tag{4.110}$$

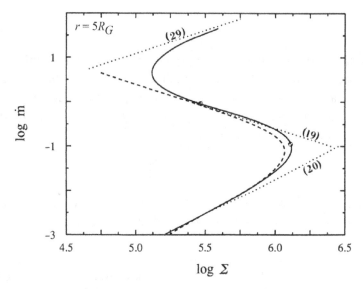

Figure 4.6 The disc $\dot{M}(\Sigma)$ relation for $R = 5R_g$, showing the full transonic solution (solid line), the Shakura–Sunyaev approximation (dashed line), and analytic asymptotic solutions for gas-pressure dominated (dotted line (20)) and radiation-pressure dominated (dotted line (19)), and an approximation for the upper branch slope (dotted line (29)). Credit: Abramowicz et al. (1988).

from (4.45), so the total disc luminosity is

$$
L = \frac{3GM}{2}\left[\int_{R_{\text{sph}}}^{\infty}\frac{f\dot{M}}{R^3}RdR + \int_{R_{\text{in}}}^{R_{\text{sph}}}\frac{f\dot{M}(R)}{R^3}RdR\right] \simeq \frac{GM\dot{M}_{\text{Edd}}}{R_{\text{in}}}\left(1 + \ln\frac{R_{\text{sph}}}{R_{\text{in}}}\right),
$$
(4.111)

that is,

$$
L \simeq L_{\text{Edd}}(1 + \ln\dot{m}),
$$
(4.112)

using (4.115) and $R_{\text{in}} \simeq R_{\text{ISCO}}$.

Because matter has to disappear behind the event horizon without significant radiative emission, slim-disc solutions apply only to black hole accretion: at stellar-mass scales, slim discs cannot describe accretion on to any star with a physical surface, such as a neutron star.

A problem for these solutions is that numerically they are available only for $R_{\text{ext}} \lesssim 50R_g$. At larger disc radii the relevant Shakura–Sunyaev solutions are radiation-pressure dominated, and therefore unstable. So slim disc solutions do not offer a straightforward description of the inner parts of a thin disc with a super-Eddington mass supply at large radius. Some of these difficulties may arise because of the restriction to steady solutions. As we have seen, the ionization instability (Section 4.7) can prevent a disc settling to a steady state for a given mass supply

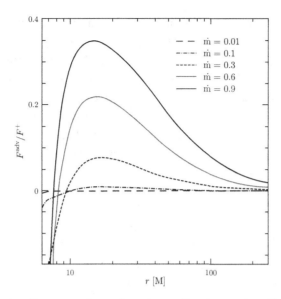

Figure 4.7 Ratio of energy advected to that radiated at each radius for various mass accretion rates, for a non-rotating black hole ($a = 0$). Positive values imply that the heat generated by viscous dissipation is stored in the accreting matter, and negative ones that this heat is radiated. The algebraic sum of these gives the total heat from viscous dissipation. Credit: Sadowski (2009).

rate. This is expected for slim disc solutions too, which may show quasi-periodic outbursts if the supply rate lies in the unstable radiation-pressure dominated regime below the slim disc minimum, that is, $\dot{M}_{Edd} \lesssim \dot{M}_{supp} \lesssim 30\dot{M}_{Edd}$. These would oscillate between gas-pressure dominated low states with accretion rates $\dot{M}$ and luminosities L obeying

$$\dot{M} < \dot{M}_{Edd} < \dot{M}_{supp}, \text{ and } L < L_{Edd}, \tag{4.113}$$

and slim-disc high states with

$$\dot{M} > \dot{M}_{supp} \gtrsim \dot{M}_{Edd}, \text{ and } L_{Edd} \lesssim L \lesssim 3L_{Edd} < \eta c^2 \dot{M}_{supp}. \tag{4.114}$$

But without solving a time-dependent problem with realistic initial conditions at large radius one cannot be sure that any particular steady solution for the flow near the black hole is asymptotically realized at sufficiently large times. These difficulties are generic, and independent of (but compounded by) the separate problem that we do not yet have a fully predictive first-principles theory of disc viscosity (see Section 4.12). They are related to the general problem of setting boundary and initial conditions in astrophysics.

Another type of solution in some ways similar to slim discs is advection-dominated accretion flows (ADAFs). Here again cooling is assumed to be slow compared with infall, potentially allowing very rapid mass gain by the accretor, with rather low radiative output. From the Soltan argument we know that this cannot be the dominant mode of SMBH mass growth, although it may be relevant for very under-luminous AGN, such as Sgr A* in the Galactic Centre. ADAF discs are genuinely thick rather than slim. The radial infall velocity is significant, as thick discs without this property are subject to a global dynamical instability (Papaloizou & Pringle, 1984) and could never be set up in reality. Numerical studies of these (effectively adiabatic) disc flows suggest that much of the matter may actually move slowly *outwards* because of convection, making the net accretion rate very small as inflow and outflow co-exist at each radius. Although these simulations start from well-defined initial conditions and are evolved in time, they still do not have a full physical prescription for the viscosity. Chapter 11 of APIA3 gives an extended discussion of these and other thick disc solutions.

The original paper by Shakura and Sunyaev (1973) suggested a different way out of the difficulties presented by a super-Eddington mass supply. Noting the tendency of the disc to swell vertically, they assumed that it might simply eject the excess mass by radiation pressure when this became critical. Evidently this idea is an example of possibility (b) of Section 1.2 for reacting to a super-Eddington mass supply. The largest radius where ejection is possible is the 'spherization radius' R_{sph} where $H \sim R$, and so from (4.107),

$$R_{\text{sph}} \simeq \frac{\dot{M}_{\text{out}}}{\dot{M}_{\text{Edd}}} R_{\text{ISCO}} = \dot{m} R_{\text{ISCO}}, \tag{4.115}$$

where $\dot{m} = \dot{M}_{\text{out}}/\dot{M}_{\text{Edd}}$ is the Eddington factor of the mass supply rate. The accretion rate must decrease inwards from its value at this radius to keep each disc radius from exceeding its radiation pressure limit. In Newtonian terms the energy release is proportional to $GM\dot{M}/R$. Then the disc must blow gas away so that

$$\dot{M}(R) \simeq \dot{M}_{\text{Edd}} \frac{R}{R_{\text{in}}}. \tag{4.116}$$

Here R_{in} is the innermost disc radius, which we expect to be close to R_{ISCO}. The emission per unit disc face area then goes as

$$D(R) \simeq \frac{3GM\dot{M}(R)}{8\pi R^3} f \tag{4.117}$$

from (4.45). So just as for the slim disc solution, we get the total emitted disc luminosity

$$L \simeq L_{\text{Edd}}(1 + \ln \dot{m}), \tag{4.118}$$

but this time for a different reason – the excess infalling gas is blown away before it can produce radiation, rather than trapping its radiation and dragging it into the hole. We see again that the emitted luminosity increases above L_{Edd} only by the logarithmic factor $\ln \dot{m}$ found for slim discs. Again, in practice this means that even a highly super-Eddington mass supply does not significantly increase the emitted luminosity above L_{Edd}.

But otherwise, the character of this 'ejection' solution clearly distinct from the slim disc case discussed earlier. There the black hole accreted the entire mass supply, so could grow its mass rapidly, whereas here it can accrete mass only at the rate $\dot{M}_{\text{Edd}}$. The ejection solution expels gas at the rate

$$\dot{M}_{\text{out}} = \dot{M}_{\text{supp}} - \dot{M}_{\text{Edd}}, \tag{4.119}$$

since the accretion rate is just $\dot{M}_{\text{Edd}}$ at the inner edge of the disc. An important question is what effects such outflows might have on the surroundings of an SMBH fed at a super-Eddington rate. We shall see in Chapter 6 that these can be very significant.

4.11 When Is the Steady Disc Assumption Valid?

A lot of this chapter has treated cases where the disc structure is steady. The ordering of timescales $t_{\text{dyn}} \lesssim t_{\text{th}} << t_{\text{visc}}$ given in (4.77) means that the structure at any given radius of even an evolving disc passes through a sequence of steady states evolving viscously – $\Sigma(R, t)$ and so $\dot{M}(R, t)$ are given by the diffusion equation (4.31), and the local disc equation (4.91) give all the other disc quantities at R and t.

But this natural-looking (and very convenient) assumption can lead to trouble if misused. The subtlety here is that the steady-state assumption holds locally, but not necessarily globally, that is, for the whole disc simultaneously. A disc is only globally steady when supplied with gas at a rate constant on timescales longer than its longest viscous timescale. This is usually found at the outer radius, where the mass is supplied. If the condition on the mass supply timescale does not hold, assuming a globally steady disc structure (e.g. as a 'sub-grid' recipe in a numerical simulation of repeated black hole feeding) then amounts to assuming that a mass of gas fed into the outer disc instantly causes the same amount of gas to fall into the SMBH in the disc centre. In reality it takes a viscous timescale for the inner disc and the SMBH to 'notice' the new gas. Using the estimate (4.71) at the disc outer edge $R \simeq R_{\text{sg}}$ gives

$$t_{\text{visc,sg}} \sim 1.7 \times 10^7 \, \text{yr} \tag{4.120}$$

almost independently of all parameters. It is possible that most – if not all – SMBH accretion discs are *never* globally steady, but are instead episodic structures that evolve a finite total gas mass as it is drained by central accretion.

4.12 Disc Viscosity

We have arrived at a fairly coherent picture of *local* disc accretion for SMBH. It appears to agree reasonably well with observations, even though we have used the semi-empirical α-prescription to represent the basic driving process of angular momentum transport in discs. This success comes from the fact, noted earlier, that none of the disc properties depend on large powers of α. The highest power is unity, as the viscous timescale t_{visc} is $\propto 1/\alpha$ (see (4.71)), which conveniently allows a simple observational estimate of its value (see (4.81)).

But it is still important to reach a deeper understanding of viscosity, as the α-prescription obviously removes information about the physical dependence and variation of ν. We need this for a full understanding of things such as disc stability, for example. The arbitrariness implicitly imported by using α is also what allows the apparent freedom to consider rather different types of steady disc flows which are nevertheless characterized by the same parameters, such as slim discs and ejection solutions, or steady thin discs and ADAFs. A full deterministic theory specifying ν as an explicit function of physical variables would fix the evolution of all disc quantities from given initial conditions and remove this freedom.

So there have been many attempts to explain the physical origin of the angular momentum transport in accretion discs. Clearly ordinary 'molecular' viscosity (i.e. where the random motions around the main streaming motion of the disc gas are simply ion thermal motions) is far too weak to explain the observed disc timescales, since the observational estimate (4.81) requires mean free paths of order the disc thickness H. The obvious next candidate is hydrodynamic turbulence, but this too fails. The basic reason is that the turbulence has to transport angular momentum outwards, yet the Rayleigh criterion tells us that a Keplerian disc is very stable against axisymmetric hydrodynamic perturbations, as its angular momentum increases outwards. The desired outward transport would have to move angular momentum *up* the angular momentum gradient.

A physical quantity that decreases outwards is the angular *velocity*. This suggests that a possible disc viscosity mechanism could result from some process sensitive to the differences in disc angular velocity rather than angular momentum. The only likely candidate for communicating these differences involves magnetic fields anchored at different disc radii. This is the basis of what is now the favoured possibility, following work by Balbus and Hawley (1991), who rediscovered a weak magnetic field instability originally found by Velikov (1959) and Chandrasekhar (1960). A vertical magnetic field is unstable if the disc angular velocity decreases outwards (the magnetorotational instability or MRI). Given a small radial perturbation a weak vertical field feels two opposing forces. The imbalance between gravity and centrifugal support tries to make the perturbation grow, while the magnetic tension tries to straighten the fieldline and reduce the perturbation. This suppression wins if the original field is too strong or the wavelength of the perturbation is too short, but the perturbation grows linearly if the field is below the value making the longest unstable wavelength $\lesssim H$. Then the radial field is sheared azimuthally and grows. The balance of total (gas plus magnetic) pressure means that the mass density inside flux tubes is lower than in the surroundings, so the flux tubes are buoyant and rise (the Parker instability), regenerating vertical field and reinforcing the original perturbation.

There have been many numerical simulations of the MRI mechanism. These generally agree that it does produce outward transport of angular momentum, but at rates significantly lower than observed. One can calibrate the results, at least for thin discs, by defining an equivalent α using (4.70). This allows a comparison with observations of the variation timescales of disc-accreting systems, which give $\alpha \sim 0.1$–0.4 (see (4.81)). Simulations consistently give α at least 10 times smaller than these values, for reasons that are still unclear (see the discussion in King, Pringle & Livio, 2007). A possible clue here is that simulations that assume that the disc has a net vertical flux *can* produce values of α which are compatible with observation. A permanent net flux of this kind is inherently implausible, but it seems possible that fields with net vertical fluxes might appear and disappear in a transient fashion.[8]

The physical nature of viscosity is important in discussing disc-accreting systems showing significant luminosity variations on long timescales. This is seen in AGN, and in all other accreting systems such as X-ray binaries and protostars. The problem here is that we expect the central regions of the disc to be the most luminous, but these also have inappropriately short viscous timescales if we use the α-prescription with fixed α (see (4.71)). Lyubarskii (1997) showed that variations in the local accretion rate at each radius caused by fluctuations in viscosity could be arranged to overcome this problem, provided that the characteristic timescale was of order the local viscous time t_{visc} itself. Our discussion of the MRI mechanism actually suggests instead that this characteristic timescale is of order the magnetic dynamo timescale, which is probably close to the dynamical time t_{dyn} (see (4.73)), and far shorter than t_{visc}. But these rapid fluctuations in independent neighbouring disc annuli can occasionally by random coincidence produce a poloidal magnetic field which is coherent enough to affect the accretion rate by extracting angular momentum in a wind or jet (King et al., 2004). The sign of the local poloidal field

---

[8] Some numerical simulations assume initial magnetic fields in the form of loops. If these are smaller than the computational volume they formally have zero *net* vertical flux, but can locally produce $\alpha \sim 0.1$, at least initially – once the loops have contracted and cancelled their fluxes the effective α becomes small. This is rather unrealistic, particularly as strong loops must also spread outside the computational volume on a magnetic diffusion timescale.

at each disc annulus of radial extent H must oscillate randomly on the timescale t_{dyn}, so after a time

$$t_{\mathrm{mag}} \sim 2^{R/H} t_{\mathrm{dyn}} \qquad (4.121)$$

the field from a radial width $\sim R$ is coherently poloidal, and can drive significant angular momentum loss and so increase the local accretion rate. The effect is to add a term to the basic disc diffusion equation, which becomes

$$\frac{\partial \Sigma}{\partial t} = \frac{3}{R} \frac{\partial}{\partial R} \left[R^{1/2} \frac{\partial}{\partial R} (\nu \Sigma R^{1/2}) \right] - \frac{1}{R} \frac{\partial}{\partial R} [R \Sigma U_R], \qquad (4.122)$$

where $U_R < 0$ is the radial velocity induced by the extra angular momentum loss in the jet or wind. This form assumes that the jet or wind removes angular momentum but not a significant amount of mass, which is clear from global energy conservation – it is not possible to put all the released accretion energy into driving the accreting mass back to where it came from. From (4.121) we see that the flares recur more frequently if the disc is thicker (larger H/R), so a variety of flaring behaviour is possible (see Figure 4.8).

An important related question is whether a viscous disc can significantly drag in a poloidal magnetic field (i.e. one lying in planes containing the rotation axis of the disc) and so produce a stronger field near its centre. Strong poloidal fields produced like this are the basis for the magnetically arrested disc (MAD) picture (Figure 4.9; see Narayan, Igumenshchev & Abramowicz, 2003). The postulated strong central field concentrated at some disc radius $R_m \gtrsim R_g$ then acts effectively like a hard surface even though we have a black hole accretor. Matter can only penetrate the field and fall into the hole through instabilities allowing it to slip between fieldlines. As $R_m \sim R_g$, the accretion efficiency $\eta = GM/c^2 R_m$ could in principle approach unity. The lifetime of this configuration is unclear, as the postulated field must attempt to expand both outwards and inwards under its own pressure, and fall under gravity, if not held by orbiting and differentially rotating gas. This is investigated via numerical simulations. The initial configuration leading to the formation of the MAD is evidently dynamically unstable, and evolves rapidly before reaching the MAD state, so there remains the question of how the initial condition would come about in reality (see Section 4.13). The same dragging idea appears in discussions of the Blandford–Znajek (BZ) effect (Blandford & Znajek, 1977) often invoked for extracting black hole spin energy to drive jets (but see Section 7.3).

The main difficulty in appealing to this inward dragging is that a realistic disc has a finite diffusivity η (see Section 3.6). Viscosity would pull in the poloidal field *inside* the disc on the viscous timescale $t_\nu \sim R^2/\nu$, but much less above and below it, as its own magnetic pressure resists this. So an initially vertical field B_z develops a significant radial component B_R near the disc mid-plane, which increases as R

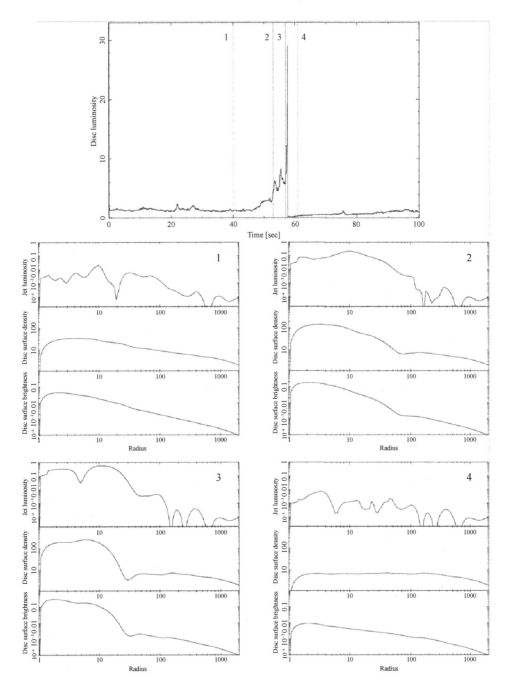

Figure 4.8 A large disc outburst resulting from random local viscosity fluctu-ations sporadically combining to produce a coherent poloidal magnetic field that extracts angular momentum in a wind or jet. Credit: King et al. (2004).

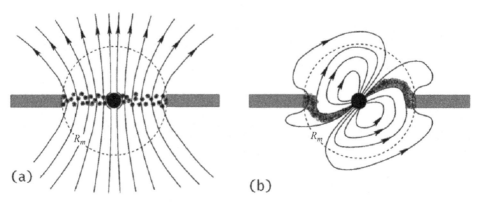

Figure 4.9 (a) A magnetically arrested disc (MAD). An axisymmetric accretion disc is disrupted at radius R_m by a strong poloidal magnetic field that has accumulated at the centre. Inside R_m the gas accretes as magnetically confined blobs which diffuse through the field with a relatively low velocity. Surrounding the blobs is a hot low-density plasma. (b) The accretion flow around a compact magnetic star. An axisymmetric disc is disrupted at the magnetospheric radius R_m by the strong stellar magnetic field. Inside R_m the gas follows the magnetic fieldlines and falls freely on the polecaps of the star. Credit: Narayan, Igumenshchev, and Abramowicz (2003).

decreases (see Figure 4.10). Finite diffusivity η then means that the field in the disc slips outwards with respect to the inward disc flow on a timescale

$$t_\eta \sim \frac{R^2}{\eta} \frac{H}{R} \frac{B_z}{B_R} \tag{4.123}$$

(Lubow, Papaloizou, & Pringle, 1994). At large R this timescale is much longer than t_ν as $B_R/B_z << 1$, so the field is pulled in. But t_η decreases inwards, making inward dragging more difficult. Dragging becomes impossible within the point where $t_\eta \sim t_\nu$. Then we have

$$\frac{B_R}{B_z} \sim \frac{H}{R} \frac{\nu}{\eta}. \tag{4.124}$$

This shows that inward dragging of magnetic fields can happen only if

$$D = \frac{R}{H} \frac{\eta}{\nu} = \frac{R}{H} \frac{1}{\mathcal{P}_m} \lesssim 1. \tag{4.125}$$

If the viscosity ν is itself the result of a turbulent magnetic dynamo process, as suggested in this section, then this is probably also true for the diffusivity η, so it is likely that $\nu \sim \eta$, that is, $\mathcal{P}_m \sim 1$. Then (4.125) shows that

significant dragging-in of poloidal fieldlines is not possible unless the disc is thick, that is, $H/R \gtrsim 1$.

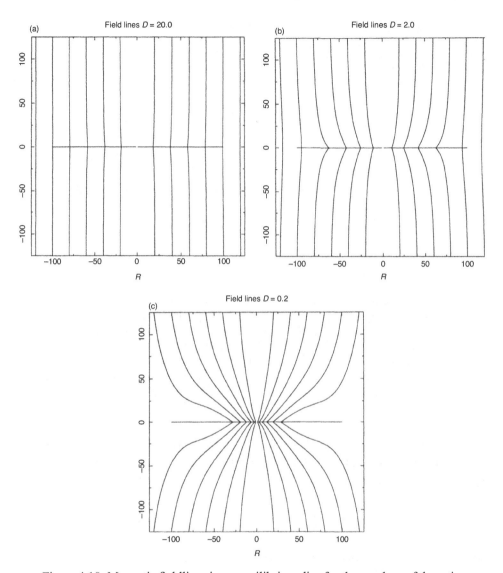

Figure 4.10 Magnetic fieldlines in an equilibrium disc for three values of the ratio $D = R\eta/H\nu$. Significant dragging of the field into the disc centre is possible only if $D \lesssim 1$, which in practice requires a thick disc $H \gtrsim R$, as in general $\eta \sim \nu$ if the diffusivity and viscosity both result from a turbulent dynamo process. Credit: Lubow, Papaloizou, and Pringle (1994).

Fieldlines emerging from the disc make an angle i to the vertical, where

$$\tan i = \frac{B_R}{B_z} \simeq \frac{1.52}{D}. \tag{4.126}$$

Blandford and Payne (1982) (hereafter BP) consider the possibility that a disc could have an embedded magnetic field which could centrifugally expel gas attached to

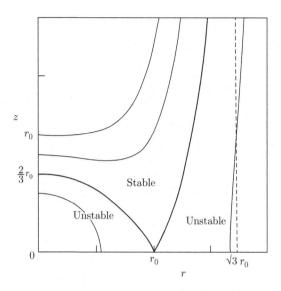

Figure 4.11 Equipotential surfaces for a bead on a wire, co-rotating with the
Keplerian angular velocity $(GM/r_0^3)^{1/2}$ at radius r_0, released at rest from r_0. The
surfaces are separated by equal intervals of $\phi(r, z)$. If the projection of the wire on
the meridional plane makes an angle of less than $60°$ with the equatorial plane, the
equilibrium at $r = r_0$ is unstable. If the angle is greater than $60°$, the equilibrium
is stable. The dashed line at $r = \sqrt{3}r_0$ is the asymptote to the surface of marginal
stability which reaches $z = \infty$. Credit: Blandford and Payne (1982).

rotating fieldlines. They show that the equipotential surfaces for a bead on a wire,
co-rotating with the Keplerian angular velocity $(GM/r_0^3)^{1/2}$ at radius r_0, released at
rest from r_0 have equation

$$\phi(r, z) = -\frac{GM}{r_0}\left[\frac{1}{2}\left(\frac{r}{r_0}\right)^2 + \frac{r_0}{(r_0^2 + z^2)^{1/2}}\right] = \text{constant}, \qquad (4.127)$$

so this kind of magnetically driven wind from the disc requires the fieldlines to
bend outwards sufficiently, that is, $\tan i \gtrsim 1/\sqrt{3} = 0.58$ (see Figure 4.11). From
(4.124) we see that this condition is very similar to (4.125), that is,

a magnetically driven disc wind requires a thick disc ($H/R \gtrsim 1$).

Many numerical simulations of accretion discs assume they are sufficiently thick
to allow at least one of the MAD, BZ, or BP effects to work efficiently. This has the
further attraction of relaxing the stringent requirements on numerical resolution
posed by thin discs. But accretion discs in the best-observed cases (cataclysmic
variables and X-ray binaries) do not appear to be thick, and there is little direct
observational evidence that SMBH discs are thick either. Accordingly, the status

of several widely invoked theoretical ideas about discs is still unclear. The root of these difficulties is evidently the fundamental one of understanding in fully quantitative terms how discs transport angular momentum outwards. This would give a complete theory, and one could hope to follow the evolution of a disc from given initial conditions and so test how realistic ideas such as those mentioned in this section. But currently, although there is a promising idea (the MRI instability), this does not yet explain the observed magnitude of the disc viscosity.

This mismatch between theory and observation leaves the subject of disc accretion in a situation comparable to stellar structure and evolution before astrophysicists began to understand nuclear burning processes in stars in the 1930s. The basic energy source of the system is not yet clearly related to the other physical variables. As we have seen, the internal structure of discs is not very sensitive to α, but this quantity controls the timescale for viscous evolution. So a practical recipe is simply to regard α as a parameter lying in the range 0.1–0.4. This of course leaves us in the unsatisfactory position that there is still no first-principles theory of disc accretion that fits observation.

4.13 Numerical Simulations of Accretion Flows

The complexity of most accretion flows gives a strong motive for numerical simulations, which now make up a large fraction of the research effort in theoretical astrophysics. As all numerical codes have finite resolution, each of them has relative advantages and disadvantages in tackling any given problem, and it is important to be aware of these at the outset. The aim of this section is to indicate where simulations of various kinds are likely to produce results one should take seriously, and not to offer guidance on how to write codes. There is an abundance of published material on numerical methods (e.g. Bowers & Wilson, 1991, the Astrophysics Source Code Library (ascl.net), or the documentation of the Athena++ code (https://www.athena-astro.app)).

Numerical methods for astrophysical gas or plasma flows largely divide into Lagrangian codes, where the gas is treated as a collection of interacting particles, and Eulerian codes, where the gas motions are measured with respect to a coordinate system (sometimes called finite difference hydrodynamics or FDH).

Lagrangian codes – mainly in the form of smoothed particle hydrodynamics (SPH) – deal naturally with local conservation laws (density, momentum, angular momentum). They have resolution determined by the local particle number, so that it follows the mass. So a thin accretion disc is relatively straightforward to simulate, despite its very strong vertical density gradients. Lagrangian codes deal easily with free boundaries, for example where the gas expands into a vacuum. But fixed boundaries, or outlets or inlets (e.g. where a stream of gas feeds a disc) can be hard to handle. Lagrangian codes can be computationally expensive – that is, they require more computer time – if the particle number is low in regions of interest, for example the outer atmosphere of a gas close to hydrostatic equilibrium. It is more difficult to add additional physics such as radiative transfer, and particularly MHD, to an SPH code. As a result, almost all astrophysical codes that treat MHD are Eulerian.

Eulerian codes are widely used and can straightforwardly accommodate additional physics such as radiative transfer and MHD. They deal well with fixed boundaries, and are often cheaper in numerical terms than Lagrangian codes if the mass density does not vary strongly, as in cases close to hydrostatic equilibrium. Global conservation laws are harder to track than in Lagrangian codes, and one may have to choose between conserving, for example, linear or angular momentum to high accuracy. The inevitably finite resolution means that dealing with sharp density contrasts can be problematic. Thin discs pose awkward resolution requirements, as almost all of the matter is confined to a much narrower range of the vertical coordinate than of the horizontal ones.

All numerical codes introduce 'numerical viscosity' – terms involving second-order spatial derivatives of the velocity field $\mathbf{v}$ – and so add diffusive effects to the equation of motion. In Eulerian simulations where the chosen coordinates are anisotropic, such as the spherical or cylindrical polars often adopted in discussing accretion discs, the numerical diffusion itself may be strongly anisotropic.[9] This tends to coerce gas to follow coordinate boundaries, and so, for example, may try to enforce circular motion ($R = $ constant) by suppressing eccentric gas orbits associated with spiral structures in accretion discs.

A classic example of this is the modelling of 'superhumps', which occur in cataclysmic variables (CVs). These are close stellar-mass binary systems where a white dwarf accretes from a low-mass companion star via an accretion disc (see, e.g. APIA3). Short-period systems of this kind are observed to show a photometric modulation with a period slightly longer than the binary's orbital period, and whose amplitude is independent of the inclination of the binary plane to the line of sight. Lagrangian simulations capture this behaviour without problems, but Eulerian codes, which generally use polar coordinates in the disc plane, do not. Early Lagrangian simulations gave insights that allowed an analytic understanding of the physical origin of superhumps. This is a resonance between gas orbits in the disc and the motion of the companion star (see APIA3, Section 5.11, and Lubow, 1991). The resonance tends to make the disc eccentric and precess slightly prograde with

---

[9] By contrast, in SPH simulations the numerical viscosity is usually not strongly anisotropic, so it is actually sometimes convenient to leave it as the only source of viscosity in cases where the absolute value of the viscous timescale is not at issue.

respect to the companion. The separations of the gas streamlines, and so the viscous dissipation, are modulated at this period also, explaining the slightly longer photometric period, and why the modulation does not depend on the inclination of the binary plane. But Eulerian codes using polar coordinates do not easily find superhumps, because they are 'too circular' – the anisotropic numerical viscosity suppresses the tendency of the accretion disc to become eccentric and precess. Given the understanding of the physical origin of superhumps found initially from the SPH simulations, and then from analytic calculations, Euclidean treatments arranged with sufficient care (e.g. coordinate systems that allow for the eccentricities of the gas orbits) now do find superhumps.[10] A global high-resolution Cartesian grid would alleviate the problem, at a significant cost in computing time, but it is not clear that grid-based codes can completely escape it. A similar problem occurs in simulating jets from accretion discs. If the simulation uses polar coordinates, numerical viscosity is likely to suppress any jet not oriented along the polar axis.

A compromise designed to handle complex geometries is adaptive mesh refinement (AMR), where the coordinate grid is made finer in selected regions of the computational volume. This introduces new problems – it is difficult to ensure simultaneous conservation of both linear and angular momentum, since it is often not obvious how the momentum of the cell being divided should be distributed over the new, previously unresolved subcells. Although global conservation of angular momentum remains, this procedure introduces small-scale vorticity, which can give spurious effects.

Most MHD codes are Eulerian, and as well as numerical gas 'viscosity' they generally introduce a significant 'numerical diffusivity' of the magnetic field. The computed evolution then allows the field to diffuse unphysically, for example weakening the frozen-in magnetic field behaviour characteristic of ideal MHD. Then although any given computation may set out to solve the ideal MHD equations, non-ideal effects must appear at some level because of numerical diffusivity. Problems like these are recognized in the solar physics and fluid-MHD literature as significant barriers to progress. The consequences may include apparent changes in the energy of the gas. Because a practicable code must be numerically stable, the change is likely in practice to be negative. If the flow is nevertheless assumed steady, this will give the appearance that some other agency, perhaps gravity or rotation, perhaps mediated by an apparent viscosity, is doing work on the gas to replace the energy it must otherwise be losing.

Other problems can arise because the magnetic field evolution is diffusive rather than wavelike, and as we have noted (footnote 7 in Chapter 3), may imply superluminal numerical propagation speeds. Then even though a boundary condition

---

[10] Similarly, disc breaking (see Section 5.6) first appeared in SPH simulations, and was only found later in unprecedentedly high-resolution Eulerian simulations of thin discs (Liska et al., 2021).

may be set apparently harmlessly inside the event horizon, where it should have no influence on the outer spacetime, by using coordinates (e.g. Kerr–Schild) that remain regular across it, the condition can *numerically* still potentially influence events outside the horizon, in conflict with GR. Similar difficulties associated with diffusivity also appear in codes designed to treat plasma interactions in the force-free electrodynamics (FFE) approximation (see Mahlmann et al., 2021a, 2021b for discussions).

A hybrid type of code useful for plasma simulations is particle-in-cell (PIC). Here individual particles or fluid elements are followed in phase space in Lagrangian fashion, but resulting quantities such as densities and currents are evaluated at stationary Eulerian mesh points. If the particles are assumed to interact only through the average fields the code is usually called particle-mesh (PM). Codes including direct particle-particle interactions are called particle-particle (PP), and those including both types of interaction are PP-PM or P3M. These codes have advantages in particular cases. For example, they can in principle handle electric charge separation, which is important in some astrophysical situations.

Irrespective of the numerical method used, two other pitfalls can produce unreasonable simulation outcomes. The first is assuming initial conditions so inherently unstable that nature would never have allowed a real physical system to reach this state – the fluid equivalent of computing the motion of a pencil initially standing on its point. This can be seductive, as the early evolution of the system may be gratifyingly dramatic. The second – often unavoidable through the demands on computer time – is not running the simulation long enough to remove transient effects.[11]

The list of potential difficulties listed here might seem discouraging. But they illustrate that a major part of the skill in numerical simulations is in choosing a code whose main defects (which are inevitably present) do not have a serious impact on the problem being studied. Numerical simulations in astrophysics are very different in character from computations in many other areas of science, where one can specify the experimental conditions in the laboratory to high accuracy, and often isolate the effect under study. It may then be possible to run a series of simulations of increasing accuracy. But in astrophysics we are never able to specify the full physical situation under study with any accuracy, and are often fairly ignorant of what the competing processes are. Accordingly, numerical simulations in astrophysics are not calculations producing precise results to be compared with experiment, with both being repeated to gain still greater accuracy, but are instead themselves experiments that may reveal something unexpected. The early Lagrangian superhump simulations discussed earlier in this section illustrate this very clearly. The 'experimental' appearance of superhump-like behaviour in them

---

[11] These two characteristics often appear in combination. If so, remedying the second may cure the first.

focused attention on the possible effects of resonances within a disc, and so led to a theoretical understanding of how superhumps arise.

As with any experiment, the significance of an unexpected result from a simulation is unclear until there is a full understanding of the physics producing this outcome. An astrophysical simulation is at best the start of the process of understanding, not the end.[12]

---

[12] This is sometimes expressed as the rather cynical 'never believe an experimental result unless it is supported by theory', usually attributed to Eddington.

5

Supermassive Black Hole Growth

5.1 Introduction

We have seen that there is substantial and growing evidence that almost every galaxy has a supermassive black hole at its dynamical centre. Our own Galaxy is apparently typical of huge numbers of others, and a vast body of very precise observations of various types show that it hosts a central SMBH with mass around $4 \times 10^6 M_\odot$. Observations give masses for many SMBH in other galaxies, and in combination with the Soltan argument and the observed SMBH scaling relations make it hard to avoid the conclusion that effectively all but possibly some of the smallest galaxies have a central SMBH.

It follows that the difference between active and inactive galaxies is simply whether their holes are accreting gas or not – active phases mark the epochs when the SMBH is growing through accretion of gas. Our Galaxy is typical in this sense: observations of the Galactic Centre suggest that an accretion event happened some six million years ago, and the gamma-ray instrument *Fermi*-LAT found bubbles of emission located symmetrically at about 5 kpc on each side of the Galactic plane (Figures 1.6 and 1.7). One can interpret these as remnants of a relatively minor AGN episode at the same epoch (e.g. Zubovas et al., 2013), confirming that our Galaxy is active from time to time. The bipolar nature of the *Fermi* bubbles follows if there was a roughly spherical outflow from the Galactic Centre, as the higher density of the Galactic plane prevents this propagating and so focuses the outflow into roughly symmetric bubbles on each side of this plane. (The alternative idea that the central SMBH produces jets aligned with the Galaxy disc axis is improbable, as observation shows that the SMBH accretion geometry is always randomly oriented with respect to the host galaxy structure – see Chapter 1.) Together with the Soltan relation (Soltan, 1982), this makes it likely that the central SMBH of every galaxy accretes gas from time to time.

This property is fundamental to understanding SMBH evolution. Accretion theory first gained widespread acceptance in the context of close binary star systems with compact accretors – a white dwarf, neutron star, or stellar-mass black hole – for two main reasons. First, these systems have very convenient timescales for observation (often hours to days, down to seconds or less), giving fairly precise constraints on theory. Second, the source of the accreting matter is clear: the companion star transfers mass to the accretor either because the binary separation shrinks as a result of orbital angular momentum losses, for example, via gravitational radiation or because the companion itself expands for stellar-evolutionary reasons. This links accretion processes closely to the overall evolution of the binary system.

Neither of these fortunate circumstances applies to SMBH accretion. Although some variation timescales are easily observable, most are not – AGN lifetimes are clearly too long for direct estimates. Knowing the fraction of active galaxies among all galaxies at any cosmic epoch would tell us the duty cycle of these growth phases, but estimates of the AGN fraction at low redshift vary widely, partly because the concept of 'active' is not easy to define uniquely. So we know only that active phases must be short. Then by the Soltan argument, SMBH accretion rates in these rare phases must be high to explain why the holes are able to grow to the masses we measure.

Equally important, how and why a supermassive black hole manages to capture and accrete gas at all remains far less clear than in the stellar-mass binary case. The fact that gas accretes on to SMBH at all implies that the central region of their host galaxies cannot be in complete dynamical equilibrium, so SMBH accretion must be related to how galaxies form. Numerical simulations of cosmological structure growth strongly suggest that galaxies grow through hierarchical merging with other galaxies. So it is plausible that galaxy mergers are what ultimately prevents galaxy spheroids reaching complete equilibrium, promoting accretion on to their central SMBH. There is, for example, general agreement that SMBH in the Universe collectively grew at their highest rates at redshift $z \sim 2$, that is, about 3.5 billion years after the Big Bang, simultaneously with the peak in star formation rate.

This picture has an attractive economy in placing SMBH as a link between the growth of galaxies and their subsequent evolution. But the vast difference in length-scales – from $\lesssim 3 \times 10^{16}$ cm $\sim 10^{-2}$ pc for an SMBH accretion disc (see (4.100)) to $\gtrsim 10$ kpc for the galaxy spheroids – together with an even wider range of baryonic mass scales, that is, $M_{\text{disc}} < (H/R)M \sim 10^{-3}M \sim 10^{-6}M_b$ (see (4.98), (1.19)), mean that numerical simulations cannot yet follow such possible consequences of mergers. Attempts to verify the connection between galaxy mergers and AGN observationally are not yet conclusive and in any case would probably not adequately specify the resulting accretion rates on to the SMBH, the key quantity governing the interaction between SMBH and their host galaxies.

5.2 Feeding the Hole

Many theoretical treatments and numerical simulations sidestep these difficulties by simply assuming that gas has somehow rid itself of angular momentum and flows inwards via spherically symmetric accretion on to the SMBH from an external medium. This is assumed to be at rest, with uniform density ρ_∞ and pressure P_∞. These two variables define an adiabatic sound speed c_∞ through the relation $c_\infty^2 = \gamma P_\infty/\rho_\infty$, where γ is the specific heat ratio (=5/3 for a monatomic gas). Bondi (1952) was the first to formulate an accretion problem in this form, in the context of stars gaining mass from the interstellar medium, and the result is usually called *Bondi accretion*. Bondi gave a full solution (see Pringle & King (2007)), but simple physical arguments give the main results to order of magnitude. We expect that the external matter will not feel the gravity of the SMBH outside a typical distance

$$R_B \sim \frac{GM}{c_\infty^2},\qquad(5.1)$$

called the Bondi radius. (This is the same as the radius of influence (eqn (1.15)) with the local sound speed c_∞ in place of the velocity dispersion σ.) Gas of density ρ_∞ must flow inwards through the sphere of radius R_B at something like the sound speed, so the Bondi accretion rate is

$$\dot{M}_B \sim 4\pi R_B^2 \rho_\infty c_\infty.\qquad(5.2)$$

The full calculation gives this answer multiplied by $[2/(5-3\gamma)]^{(5\gamma-3)/2(\gamma-1)}$, which is unity in the limit $\gamma = 5/3$ and reaches $e^{3/2} \simeq 4.5$ for $\gamma = 1$ (see AF for details).

Given the resolution difficulties we mentioned at the end of the previous section, the Bondi formula, or variants of it, is very widely used as a subgrid recipe in cosmological simulations of galaxy formation that try to follow SMBH growth. But it is hard to see it as a realistic representation of SMBH accretion, for at least two reasons.

The first is that the Bondi formula assumes that gas falls towards the SMBH because of the destabilizing influence of its gravity. But the hole mass is so small (see (1.19)) compared to that of even a small part of the galaxy bulge that this is clearly unrealistic. The implicit assumption is that the gas would be in equilibrium in the potential generated by the stars and is drifting inwards because of the perturbation by the central mass's gravity. But the mass now making up the hole was almost certainly inside the Bondi radius, and by spherical symmetry exerting the same gravitation pull on the gas further out, even before the hole formed. In reality the problem must be time-dependent.

A second difficulty is that all the gas in a galaxy bulge has significant angular momentum and so cannot fall in radially as assumed in Bondi accretion. Angular momentum is the main barrier to accretion, and we have seen (eqn (4.4)) that its importance grows as matter falls towards the accretor. Even a very small amount of angular momentum makes the flow markedly non-spherical at small radii. Neglecting angular momentum is a singular limit.

So we must take account of the large velocity dispersion observed in the host spheroid. Although this is generally measured for stars, the gas must be at least partially entrained by them and have a similar velocity dispersion. A measure of the difficulties this introduces is that the maximum size of any gaseous accretion disc around the SMBH is limited by self-gravity to $R_{sg} \simeq 3 \times 10^{16}$ cm $= 10^{-2}$ pc (see (4.100)), while the gas that supplies the hole mass must originally have come from a region of size $\gtrsim f_g^{-1} R_{inf}$, where $R_{inf} \sim$ few pc (see (1.16)) is the radius of the sphere of influence and $f_g < 1$ is the gas fraction relative to stars (the cosmological average value is $f_g \simeq 0.16$). This means that the gas that grows the hole must lose almost all its original specific angular momentum

$$J_{inf} \gtrsim f_g^{-1} R_{inf} \sigma \sim 10^{28} M_8 \sigma_{200}^{-1} \, \text{cm}^2 \, \text{s}^{-1} \tag{5.3}$$

to reach the value

$$J_{sg} \sim (GMR_{sg})^{1/2} \sim 10^{25} M_8^{1/2} \, \text{cm}^2 \, \text{s}^{-1} \tag{5.4}$$

needed to form a disc within R_{sg}. This is the condition to get any gas to accrete through a disc on to the SMBH at all.[1] But we know from observation that many AGN accrete at rates close to $\dot{M}_{Edd}$, and the condition for this is tighter still. If a significant gas mass m is added to the disc at its self-gravity radius R_{sg}, the resulting accretion follows essentially from (4.38). Then after one viscous time, that is, at $\tau \simeq 1$, the central accretion rate is

$$\dot{M}_{SMBH} \simeq 3\pi \nu \Sigma(0, 1) \sim \frac{m}{t_{visc,sg}}. \tag{5.5}$$

Comparing with the Salpeter timescale $t_{Sal} = M_{SMBH}/\dot{M}_{Edd} \simeq 5 \times 10^7 \eta_{0.1}$ yr (from (1.9)), we see that

$$t_{visc,sg} \sim \frac{0.35}{\eta_{0.1}} t_{Sal}, \tag{5.6}$$

---

[1] For SMBH masses $M \lesssim 10^5 M_\odot$ there is an even tighter constraint from the viscous timescale (4.71), which gives t_{visc} larger than the Hubble time t_H unless the disc forms within a radius $R_H \simeq R_{sg} M_5^{1/3}$, where $M_5 = M/10^5 M_\odot$.

so (5.5) implies

$$\dot{M}_{\text{SMBH}} \sim \frac{3m}{t_{\text{Sal}}} \eta_{0.1} \sim \frac{3m}{M} \dot{M}_{\text{Edd}}. \tag{5.7}$$

This shows that it is difficult to increase the SMBH accretion rate rapidly by adding gas at the outer disc edge, where we expect $m < M$ for self-consistency. Yet we observe SMBH accreting at rates close to $\dot{M}_{\text{Edd}}$. Postulating that mass is added at much smaller disc radii, where the viscous time is shorter, simply shifts the problem back to why the gas had so little angular momentum and was so well aimed at the SMBH.

The basic cause of these difficulties is the gas angular momentum, and the argument of the previous paragraph, strongly suggests that the observed high accretion rates must result from something more violent than pure disc accretion – something must destroy the gas angular momentum rather than gently exporting it outwards. The clue here is that the bulge stars and gas are generally not in net global rotation, so their total angular momentum about the central SMBH is low. This suggests that the basic driver of SMBH accretion involves processes that disturb the gas motions, causing opposing bulge gas angular momenta to cancel through collisions.

We noted in discussing (5.1) that the SMBH mass M is too small to affect the dynamics of the bulge gas directly. So this is not a good candidate for disturbing the orbits of gas moving in the gravitational field of the galaxy bulge. We have already remarked in Chapter 1 that the one property of the SMBH that does have global significance for the host galaxy spheroid is the gravitational binding energy $\eta M c^2$ the hole must release as it accretes. Since this is evidently the only way that the spheroid notices the presence of the SMBH, it is worth asking if feedback from the hole may somehow cause further accretion on to it. In this case we would conclude that feedback causes feeding: the SMBH somehow hunts for its own food, in a process sometimes called black hole foraging. Clearly, a process like this can only start at all if triggered by some event that is not itself connected with feedback, but puts enough gas close enough to the hole (i.e. within $\sim 10^{-3}$ pc). A galaxy merger is a reasonable candidate for this. This would imply that the beginning of SMBH growth would be a fairly random process, but more likely in perturbed or merging galaxies.

To see how triggering of black hole feeding by feedback might work, we note that any kind of feedback from the SMBH to the bulge must produce a strongly radial transmission of energy and momentum, as the SMBH and its accretion disc are so small in size and mass compared with the bulge. This implies that feedback cannot directly affect the angular momentum of the gas orbiting the galaxy centre. In Chapter 6 we will find the likely explicit form of this feedback. In particular we will see that it is quasi-spherical and episodic. It sweeps the bulge gas outwards but, except at particular epochs, does not unbind it. This kind of SMBH feedback increases (i.e. makes less negative) the specific orbital energies E of cold gas clouds orbiting the galaxy centre, but leaves their specific angular momenta J fixed.

When a feedback episode ends, the thrust supporting the bulge gas is removed, so this gas falls inwards on ballistic orbits from the radius R_{shell} where it was pushed. For simplicity we initially consider how this would affect a cloud orbit of semi-major axis a and eccentricity $0 \leq e < 1$ about a fixed central mass M_c. This has specific energy and angular momentum

$$E = -\frac{GM_c}{2a}, \quad J = (GM_c a)^{1/2}(1 - e^2)^{1/2}. \tag{5.8}$$

Feedback adds energy $\Delta E > 0$, so

$$\Delta E = \frac{GM_c}{2a^2}\Delta a, \tag{5.9}$$

implying $\Delta a > 0$. Since J remains fixed, the second equation of (5.8) requires that e increases. Then using this equation again, the pericentre distance

$$a(1 - e) \propto \frac{J^2}{1 + e} \tag{5.10}$$

must *decrease*. We will see later in this section that this effect is even larger for a more realistic potential. We get a collection of very eccentric cloud orbits, where the clouds themselves are tidally stretched into streams. These crowd together at pericentre, allowing the possibility of partial cancellation of angular momenta.

In reality the clouds do not orbit a constant central mass M_c, but move under the combined gravitational potential $\Phi = -GMR^{-1} + \Phi_{bulge}(R)$, where $\Phi_{bulge}(R)$ is the bulge potential. The cold gas in the bulge is probably not in large-scale coherent rotation, but instead the ensemble of gas clouds or density maxima has a distribution of (partly) opposing angular momenta, and so only a small net total angular momentum.

So we consider a population of clouds with the same peri- and apocentric radii, with specific orbital energies and angular momenta

$$E = \frac{R_+^2 \Phi_+ - R_-^2 \Phi_-}{R_+^2 - R_-^2}, \quad J^2 = \frac{2R_-^2 R_+^2 (\Phi_+ - \Phi_-)}{R_+^2 - R_+^2}, \tag{5.11}$$

where $\Phi_\pm = \Phi(R_\pm)$, but with orbital axes and phases randomly arranged. For clouds or streams that do not collide and have no internal shocks, Liouville's theorem (see, e.g., Binney & Tremaine, 2008) implies that the phase-space density of the clouds is conserved and is just the product of delta functions of E and J^2. Integrating over velocities gives the spatial mass density

$$\rho(R) = \frac{M_g C}{R \sqrt{2R^2(E - \Phi(R)) - J^2}},$$

(5.12)

where

$$C^{-1} = 4\pi \int_{R_-}^{R_+} \frac{R \, dR}{\sqrt{2R^2(E - \Phi(R)) - J^2}},$$

(5.13)

gives the normalization constant and M_g is the total mass of gas. The apocentres of the gas orbits are at the radius of the initial swept-up shell R_{shell}, so we can numerically evaluate C and the mass

$$M_g(R) = 4\pi \int_{R_-}^{R} \rho(R) R^2 \, dR$$

(5.14)

inside radius R. The corresponding ρ and $M_g(R)$ are shown in Figure 5.1 for several pericentres: the apocentre is fixed at $R_+ = 2R_{\text{inf}}$. Eccentric orbits spend a long time near apocentre, so there is a density maximum there. Most of the gas is now further from the SMBH than before, in a thick shell near R_{shell}, where the clouds are near their initial positions, with modest relative velocities. But the infalling gas creates a second maximum near pericentre, where the clouds or streams can potentially collide with significant relative velocities and a frequency $\propto \rho^2$, causing significant cancellation of angular momentum. The colliding gas shocks and loses orbital energy by cooling and so cascades through a sequence of smaller and more circular orbits until it forms a disc. If more gas falls into the disc it is destroyed and replaced by a smaller one. We will see in Section 5.6 that still more angular momentum cancellation is likely as the disc becomes sensitive to the direction of the black hole spin.

This process is clearly chaotic. It is evidently able to cancel most of the original angular momentum of at least a small fraction of the gas orbiting in the galaxy bulge after each feedback event. This transfers mass from a radius $R_{\text{gas}} \sim 10$–100 pc to an accretion disc at $R_{\text{disc}} \sim 10^{-3}$–$10^{-2}$ pc.

The resulting mass infall rate must be close to dynamical – a mass $\sim f_g R_{\text{gas}} \sigma^2 / G$ of orbiting gas at radius R_{gas} is robbed of centrifugal support and falls inwards on the dynamical timescale $t_{\text{dyn}} \sim R_{\text{gas}} / \sigma$. This feeds the SMBH accretion disc at a maximum rate

$$\dot{M}_{\text{dyn}} \sim \frac{m}{t_{\text{dyn}}} \sim \frac{f_g \sigma^3}{G} \sim 280 \sigma_{200}^3 \, M_\odot \, \text{yr}^{-1}$$

(5.15)

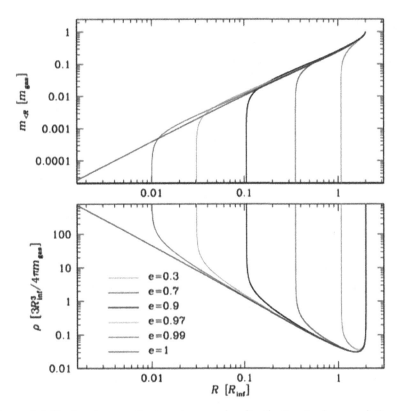

Figure 5.1 Enclosed mass (top) and mean density (bottom) of a population of clouds/streams orbiting a black hole with the same apocentric radius $R_+ = 2R_{\rm inf}$ (corresponding to $M \simeq M_\sigma/2$: for other choices the picture is very similar) but different eccentricities $e = (R_+ - R_-)/(R_+ + R_-)$, decreasing from the right as indicated. The bulge was modelled as an isothermal sphere. Credit: Dehnen and King (2013).

for short durations, at times which are stochastically distributed. As we have seen, there is in any case an upper limit of $\sim (H/R)M \sim 10^{-3}M$ to the mass of any accretion disc around the SMBH, and any excess is likely to form stars.

For SMBH close to the M–σ relation (6.2) we have $\dot{M}_{\rm Edd} \sim 4.4 M_\odot$ yr^{-1} and so an Eddington ratio

$$\dot{m} = \frac{\dot{M}_{\rm dyn}}{\dot{M}_{\rm Edd}(M_\sigma)} \simeq \frac{64}{\sigma_{200}} \simeq \frac{54}{M_8^{1/4}}. \tag{5.16}$$

This implies SMBH accretion at rates $\gtrsim \dot{M}_{\rm Edd}$ and, as we shall see in Chapter 6, outflows with momentum scalars $\dot{M}_{\rm out} v \simeq L_{\rm Edd}/c$. This self-consistently produces the feedback required for further infall to feed an SMBH accretion disc.

The controlling timescales of the feedback/feeding cycle are the infall time

$$t_{\text{infall}} \sim \frac{R_+}{\sigma} \sim \frac{R_{\text{inf}}}{f_g \sigma} \sim \frac{GM}{f_g \sigma^3} \sim \frac{M}{\dot{M}_{\text{dyn}}} \sim \frac{t_{\text{Sal}}}{\dot{m}} \sim 1 \times 10^6 M_8^{1/4} \eta_{0.1} \text{ yr} \qquad (5.17)$$

and the viscous timescale of the accretion disc, which from (4.120) is

$$t_{\text{visc}} \sim 1.7 \times 10^7 \left(\frac{R_{\text{disc}}}{R_{\text{sg}}} \right)^{3/2} \text{ yr}. \qquad (5.18)$$

The ratio of infall to viscous times can be > 1 or < 1 depending on the parameters $R_+/R_{\text{inf}}, R_{\text{disc}}/R_{\text{sg}}, \eta$, and α, so a variety of behaviours are possible.

Although this picture was invented to explain how gas feeds AGN, it has several other desirable consequences.

1. The geometry suggested here, with a density enhancement at a few pc from the SMBH and orientation similar to the accretion disc within it, is very similar to the properties inferred for the obscuring AGN torus (Antonucci & Miller, 1985; Antonucci 1993) to account for the existence of unobscured (Type I) and obscured (Type II) AGN. It naturally removes the main problem in trying to model the torus physically, that is, it must consist of cool material, but has to have a large scaleheight. The large solid angle is as expected if the obscuring matter has not yet settled into a disc, but is instead still falling on orbits with a wide range of inclinations. Setting the bulge potential term $\Phi_{\text{bulge}} = 0$ in (5.12) gives the column density

$$\Sigma = \int \rho(R) dR \sim \frac{m}{2\pi (R_- + R_+) \sqrt{R_- R_+}}. \qquad (5.19)$$

This becomes large for small pericentric radii R_-, so the SMBH must be obscured along many lines of sight and for extended times. The obscuring matter is not necessarily arranged simply in a torus, but is a collection of cool clouds and streams on eccentric orbits about the SMBH so that the observer's view of the AGN structure depends on the precise line of sight. The obscuration must have some slow time dependence, in line with observations suggesting occasional changes between Seyfert types (see Alloin et al., 1985; Shappee et al., 2013).

The infalling gas was impacted by the outflow from the black hole before falling, so is likely to form stars at some rate. Since angular momentum was cancelled in making the inflow, these newly formed stars must have near-parabolic orbits about the SMBH. This gives three further consequences.

2. Stars falling close to the SMBH may be tidally disrupted (see Sections 1.5 and 5.8).

3. Similarly, tidal disruption of close binaries formed in this flow ejects the lower-mass star of the binary as a hypervelocity star (Hills, 1988), while the captured component spirals in towards the SMBH. If the SMBH has a mass $\gtrsim 10^7$–$10^8 M_\odot$ it swallows the captured star whole, but for lower-mass SMBH the captured star eventually fills its tidal lobe at pericentre and produces quasi-periodic eruptions (QPEs: see Section 5.9), which are observable in X-rays if the star is a white dwarf.

4. Massive stars in this flow inject metal-enriched gas into their surroundings. This is likely to remain near to the hole and may form part of the feedback outflow later on. Gas may be recycled repeatedly in this way and so account for the high metallicities seen in AGN spectra (see Sections 1.5 and 6.14).

5.3 Chaotic and Misaligned Accretion

In the 'foraging' picture described in the previous Section, both the value of the mass feeding rate and its duty cycle are fixed by processes that are essentially stochastic. This makes it difficult to go further than the simple estimates made there either analytically or numerically. But this randomness agrees with many aspects of what we know about SMBH accretion, such as the complete lack of correlation between the jet–torus structure of AGN and the geometry of their host galaxies. Randomness is natural in any picture where the gas feeding the SMBH is originally quite far from it, so its large-scale motions are not strongly determined by the hole. Foraging is one example, as the feedback from the hole only perturbs gas orbits set by the gravitational potential of the bulge. This generic idea is often called chaotic accretion.[2]

If accretion on to the SMBH is chaotic, we must abandon some assumptions that are often made largely for simplicity. For example, we have so far considered only discs that are completely axisymmetric, with the central disc plane orthogonal to the black hole spin vector. This was a natural assumption in the first studies of discs in stellar-mass binary systems, where the binary orbital plane gives a preferred orientation for the disc. But even there, there is no guarantee that the disc is orthogonal to the black hole spin (see King & Nixon, 2016). For example, if the black hole was born in a supernova, observations strongly suggest that this may well have been asymmetric. Some misalignment of an accretion disc with respect to the spin of the accretor appears to be generic on all mass scales, making axisymmetry a singular limit – the assumption of complete alignment rules out several effects which are important even when the degree of misalignment is very

---

[2] This description – suggested by Piero Madau at a conference in the Papal Observatory, Castel Gandolfo, Italy in 2005 – is sometimes challenged on the grounds that mathematically the accretion process is stochastic rather than chaotic. But the original name has stuck.

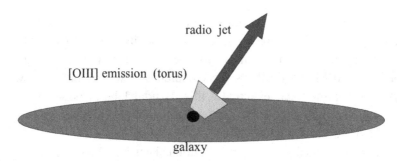

Figure 5.2 Schematic figure of the orientations of the radio jet and the AGN torus in a galaxy (The torus is observed in the forbidden emission line radiation of doubly ionized oxygen (i.e. [OIII]). Note that there is no relation between these components and the large-scale structure of the galaxy.

Figure 5.3 Schematic figure of the chaotic surroundings of a supermassive black hole in a galactic nucleus.

small. The most obvious is Lense–Thirring precession, which causes vertical shear between disc orbits and can have major effects, as we discuss after (5.20).

An active galactic nucleus has even less symmetry than an accreting binary – there is clearly no preferred plane for accretion. Instead it seems very likely that gas getting close enough to the SMBH to feed it must come from a variety of directions at various epochs. The most obvious characteristic of chaotic accretion is that accretion occurs in disconnected events, and in particular, the plane of each accretion disc episode must be randomly oriented with respect to the previous one and with respect to the SMBH spin. We can immediately see that there is a potential connection here (see Figures 5.2, 5.3) to the observation that the jet–torus structure in AGN is randomly aligned with respect to the large-scale structure of the host galaxy (e.g. Nagar & Wilson, 1999; Kinney et al., 2000). To discuss these possibilities we need first to understand how a misaligned disc evolves.

Misaligned accretion on to a black hole gives another example of Hawking's (1972) result that a stationary black hole is either static (i.e. not spinning) or axisymmetric. A spinning black hole in a non-axisymmetric environment cannot remain stationary, but must experience torques trying to remove the

Lense–Thirring effect

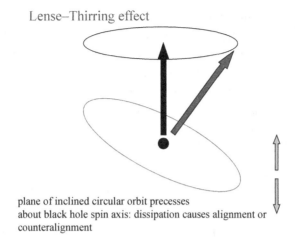

plane of inclined circular orbit precesses
about black hole spin axis: dissipation causes alignment or
counteralignment

Figure 5.4 The Lense–Thirring effect: a circular orbit with angular momentum (inclined arrow) misaligned from the black hole spin (vertical arrow) precesses differentially, with a frequency depending on its radius.

non-axisymmetry (or conceivably, but less probably, to reduce the hole spin to zero). Here the symmetrizing torques arise naturally from Lense–Thirring (LT) precessions (see Figure 5.4). Lense–Thirring precessions are differential, that is, the local frequency changes with radius (as $\omega \sim R^{-3}$). The disc is no longer plane but warped, as its local plane changes with radius. The disc rings shear against neighbouring rings and produce viscous torques, but as well as transferring rotational angular momentum outwards these also try to dissipate the local up–down motions. The forced LT precession produces a resonant radial pressure gradient (see Figures 5.11, 5.12) and so a 'vertical' viscosity coefficient ν_2 in addition to the usual 'horizontal' coefficient $\nu = \nu_1$ (see (5.53)).

We discuss these effects further in Section 5.6, but for the moment consider how the hole–disc system evolves. The 'vertical' torques must tend to make the hole–disc system axisymmetric around its the total angular momentum, so in equilibrium the disc axis must end up fully aligned or fully counteraligned with respect to the spin.[3]

The outcome here has important implications. The black hole spin fixes the accretion efficiency, and so – via the Eddington limit – just how fast the SMBH can grow and how its spin (and so accretion efficiency) evolves. In the aligned case accretion is prograde. The small ISCO radius means high accretion efficiency and

---

[3] Another example occurs if a spinning black hole is immersed in a misaligned uniform magnetic field. A torque acts to try to align the spin and the field; King and Lasota (1977); King and Nixon (2018). See Sections 2.3 and 7.3 for discussions of the cases where the sources of the field, respectively, have far higher, and far lower, inertia than the hole.

little spin-up, so the Eddington limit allows only relatively slow mass growth. In the counteraligned case, accretion is retrograde, and the large ISCO radius implies low accretion efficiency and rapid mass growth as well as a bigger effect on spinning the hole *down* because of its greater lever arm.

The hole and disc influence each other only through precessions, so the torque between them cannot have a component along J_h, the SMBH spin angular momentum, or J_d, the angular momentum of the warped part of the disc. Then it must have the general form (King et al., 2005)

$$\frac{dJ_h}{dt} = -K_1[J_h \times J_d] - K_2[J_h \times (J_h \times J_d)], \tag{5.20}$$

where K_1, K_2 are functions of the disc structure, which we can regard as effectively constant during any alignment process. The first term on the right-hand side of (5.20) gives the torque inducing precession, and the second one describes alignment or counteralignment. This last process must be dissipative since it involves viscosity, and we will show after (5.27) that this requires $K_2 > 0$.

The scalar product of (5.20) with J_h gives $dJ_h^2/dt = 0$. The magnitude of the spin remains constant during precession and alignment or counteralignment, that is, the tip of the vector J_h moves on a sphere. The total angular momentum $J_t = J_h + J_d$ is a constant vector in magnitude and direction. Using this and $J_h.dJ_h/dt = 0$, we get

$$\frac{d}{dt}(J_h.J_t) = J_t.\frac{dJ_h}{dt} = J_d.\frac{dJ_h}{dt}. \tag{5.21}$$

Equation (5.20) now shows that if K_2 is positive,

$$\frac{d}{dt}(J_h.J_t) = K_2[J_d^2 J_h^2 - (J_d.J_h)^2] = K_2 J_d^2 J_h^2 \sin^2 \theta \equiv A \geqslant 0. \tag{5.22}$$

Both J_h and J_t are constant, so (5.22) shows that if K_2 is positive,

$$\frac{d}{dt}(\cos \theta_h) \geqslant 0, \tag{5.23}$$

where θ_h is the angle between the spin J_h and the total angular momentum J_t. Then θ_h always decreases, that is, the spin vector moves to align itself along the fixed total angular momentum vector J_h. Although the total angular momentum of the system (hole plus disc) is conserved, alignment occurs because of dissipation. The magnitude of the spin of the hole remains unchanged, so dissipation must reduce the magnitude of the disc angular momentum. We can see this (and justify the assumption $K_2 > 0$ used to derive (5.22), (5.23)) by considering the quantity A defined in (5.22). From the definition of $\boldsymbol{J}_t$ and the fact that $J_t = $ constant, we have

$$A = \frac{\mathrm{d}}{\mathrm{d}t}(\boldsymbol{J}_h \boldsymbol{.} \boldsymbol{J}_t) = \frac{\mathrm{d}}{\mathrm{d}t}(J_h^2 + \boldsymbol{J}_h \boldsymbol{.} \boldsymbol{J}_d) = \frac{\mathrm{d}}{\mathrm{d}t}(\boldsymbol{J}_h \boldsymbol{.} \boldsymbol{J}_d). \tag{5.24}$$

Then

$$\frac{\mathrm{d}}{\mathrm{d}t}(\boldsymbol{J}_h \boldsymbol{.} \boldsymbol{J}_d) \equiv A \geqslant 0. \tag{5.25}$$

We also have

$$0 = \frac{\mathrm{d}}{\mathrm{d}t}J_t^2 = \frac{\mathrm{d}}{\mathrm{d}t}(J_h^2 + 2\boldsymbol{J}_h \boldsymbol{.} \boldsymbol{J}_d + J_d^2), \tag{5.26}$$

so

$$\frac{\mathrm{d}}{\mathrm{d}t}J_d^2 = 2\boldsymbol{J}_d \boldsymbol{.} \frac{\mathrm{d}\boldsymbol{J}_h}{\mathrm{d}t} = -2A \leqslant 0. \tag{5.27}$$

This tells us that the magnitude of the disc angular momentum J_d^2 decreases as J_h aligns with J_t if and only if $K_2 > 0$, justifying the assumption made after (5.20) that K_2 is positive, as a negative value of K_2 would require the alignment process to feed energy into the disc. As we are considering timescales that are short compared with that for accretion, this is not possible.

This gives us the criterion for co- or counteralignment. The three vectors $\boldsymbol{J}_t, \boldsymbol{J}_h$ and $\boldsymbol{J}_d$ make a triangle. During co- or counteralignment the first vector is completely constant, the second moves with constant length, while the third shortens. Once $\boldsymbol{J}_h$ lines up with $\boldsymbol{J}_t$, it is counteraligned with $\boldsymbol{J}_d$ if and only if

$$J_h^2 > J_t^2, \tag{5.28}$$

so counteralignment ($\theta \to \pi$) occurs if and only if the hole spin vector is *longer* than the total angular momentum vector. Then the disc angular momentum has to be counteraligned to the spin. Given an initial angle θ between $\boldsymbol{J}_h$ and $\boldsymbol{J}_d$, we get from the cosine theorem that

$$J_t^2 = J_h^2 + J_d^2 + 2J_h J_d \cos\theta. \tag{5.29}$$

So counteralignment ($\theta \to \pi$) occurs if and only if the initial angle has

$$\cos\theta < -\frac{J_d}{2J_h}, \tag{5.30}$$

which requires both

$$\theta > \pi/2 \quad \text{and} \quad J_d < 2J_h \tag{5.31}$$

– see Figure 5.5.

There is a cone of directions for the initial disc angular momentum for which counteralignment is the outcome. The process is not necessarily monotonic, as we

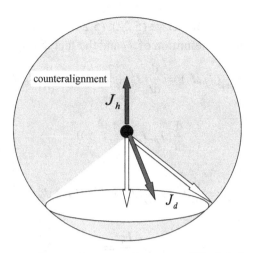

Figure 5.5 Geometry of co- and counteralignment of black hole and disc angular momenta $\mathbf{J}_h$, $\mathbf{J}_d$. The ratio $J_d/2J_h$ defines a cone of directions in which counteralignment occurs. The evolution to the final aligned or counteraligned states is not necessarily monotonic.

have seen – the spin and disc angular momenta can initially move towards one outcome, but in the end reach the opposite one.

The first published discussion of alignment (Scheuer & Feiler, 1996) actually concluded that counteralignment of disc and spin was impossible in direct disagreement with the results above. This incorrect result followed from assuming, simply for convenience, a Newtonian rest frame in which the vector $\mathbf{J}_d$ had a fixed direction in space. In reality, only the *total* angular momentum $\mathbf{J}_t = \mathbf{J}_d + \mathbf{J}_h$ is fixed in a Newtonian sense, so the assumption that $\mathbf{J}_d$ is fixed is only compatible with the conservation of angular momentum if the magnitude J_d is infinite. Equation (5.30) shows that counteralignment could never occur, as it would require $\cos\theta \to -\infty$. This subtle error was not discovered for almost a decade and had seriously misleading consequences for the subject, as we will see in Section 5.4.

In reality, with $J_d < 2J_h$ and $\mathbf{J}_h$ and $\mathbf{J}_d$ in random directions, counteralignment occurs in a fraction

$$f = \frac{1}{2}\left(1 - \frac{J_d}{2J_h}\right) \tag{5.32}$$

of all cases. The disc spin is related to J_h, J_t, and θ by the relevant root of (5.29), so

$$J_d = -J_h\cos\theta \pm (J_t^2 - J_h^2\sin^2\theta)^{1/2} \tag{5.33}$$

with the upper/lower sign for alignment/counteralignment. In both cases J_d decreases monotonically with time, reaching the final values $J_t - J_h > 0$ and $J_h - J_t < 0$, respectively.

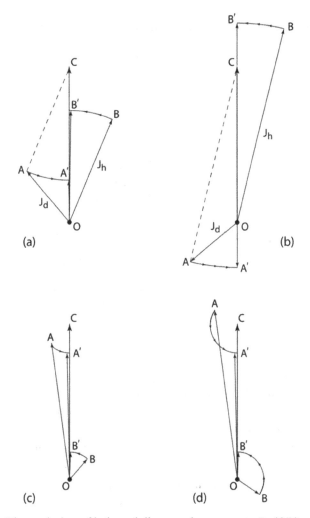

Figure 5.6 The evolution of hole and disc angular momenta $\mathbf{J}_h$ (OB) and $\mathbf{J}_d$ (OA) under the alignment torque viewed in the plane they define. This plane precesses around the fixed total angular momentum vector $\mathbf{J}_t$ (OC). (a) A case where the initial angle θ between $\mathbf{J}_h$ and $\mathbf{J}_d$ satisfies $\cos\theta > -\mathbf{J}_d/2\mathbf{J}_h$: the two angular momenta align. (b) A case where $\cos\theta < -\mathbf{J}_d/2\mathbf{J}_h$: the angular momenta counteralign. (c,d) Two cases where $\mathbf{J}_d \gg \mathbf{J}_h$, as assumed by Scheuer and Feiler (1996), which alignment always occurs. Credit: King et al. (2005).

We can derive the equation for the change of θ over time. From (5.25) we get

$$A = J_h \frac{\mathrm{d}}{\mathrm{d}t}(J_d \cos\theta) = J_h J_d \frac{\mathrm{d}}{\mathrm{d}t}(\cos\theta) - A \frac{J_h}{J_d}\cos\theta, \qquad (5.34)$$

where we have used (5.27) to write $\mathrm{d}J_d/\mathrm{d}t = -A/J_d$. Then using $A = K_2 J_h^2 J_d^2$, we get

$$\frac{d}{dt}(\cos\theta) = K_2 J_h \sin^2\theta (J_d + J_h \cos\theta), \tag{5.35}$$

and so from (5.33) we have finally

$$\frac{d}{dt}(\cos\theta) = \pm K_2 J_h \sin^2\theta (J_t^2 - J_h^2 \sin^2\theta)^{1/2} \tag{5.36}$$

in terms of the constant quantities J_t, J_h. This implies

$$\pm K_2 J_h t = \int \frac{d\theta}{\Delta \sin\theta}, \tag{5.37}$$

where $\Delta = [1 - (J_h/J_t)^2 \sin^2\theta]^{1/2}$, which integrates to

$$\pm K_2 J_h t = -\frac{1}{2J_t} \ln \frac{\Delta + \cos\theta}{\Delta - \cos\theta} + C, \tag{5.38}$$

where C is an integration constant. Then the connection between t and θ is

$$A_\pm e^{\pm K_2 J_t J_h t} = \frac{[1 - (J_h/J_t)^2 \sin^2\theta]^{1/2} + \cos\theta}{[1 - (J_h/J_t)^2 \sin^2\theta]^{1/2} - \cos\theta}, \tag{5.39}$$

where the $A_\pm$ are constants. For a positive sign in the exponential the left-hand side tends to ∞ as $t \to \infty$, so $\theta \to 0$, that is, the black hole spin aligns. For a negative sign the left-hand side $\to 0$ as $t \to \infty$, so $\theta \to \pi$, and we get counteralignment.

Expanding equation (5.35) about $\theta = 0, \pi$ shows that these two equilibria are both stable. There is no contradiction here between the global counteralignment criterion (5.30) and the local equation (5.36), as θ does not always decrease monotonically for alignment or increase monotonically for counteralignment (see Figure 5.7).

Figure 5.7 illustrates the evolution of misalignment angle θ for a particular evolution rate, namely for $K_2 = $ constant in (5.35). The general behaviour of the solutions is independent of the detailed form of K_2, provided that K_2 is positive (i.e. $dJ_d/dt < 0$.) The highest and lowest curves in Figure 5.7 show a monotonic approach to misalignment and alignment, respectively. The middle two curves are close to the bifurcation between the two end states. From (5.35) it follows that non-monotonic behaviour in time can occur in a misaligned disc that begins an approach towards alignment, but in the end becomes misaligned (as seen in the second-highest curve of Figure 5.7).

This situation is realized for an initial state with $J_d < -2J_h \cos\theta < 2J_d$. In such cases the disc never gets close to alignment; that is, θ never drops below $\pi/2$. A bifurcation between end states of alignment and misalignment occurs for cases where initially $J_d = -2J_h \cos\theta$. A misaligned disc can change the sign of $\dot\theta$ from negative (heading towards alignment) to positive (heading towards misalignment) once J_d has been sufficiently reduced. The required reduction is larger for initial

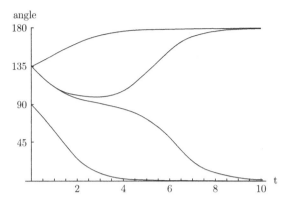

Figure 5.7 The time evolution of the disc–black hole misalignment angle θ (in degrees) as a function of dimensionless time, normalized by τ, the disc spin-down timescale for $\theta = 0°$. The evolution is determined by equation (5.35) with an assumed constant value $K_2 = 1/2\tau J_h$. Initial misalignment angles are $\theta = 135°$ for the uppermost three curves and $90°$ for the lowest curve. The initial angular momentum ratios from the highest to lowest curves are $J_d/J_h = 0.5, 1.40, 1.42$, and 0.5, respectively. The curves, from top down, correspond to the cases (a), (b), (d), and (c), respectively in Figure 5.6. For an initial misalignment angle $135°$, (5.35) predicts that the transition between long-term alignment and counteralignment occurs when initially $J_d/J_h = \sqrt{2} \simeq 1.414$, as displayed in the middle two curves, which are on opposite sides of the transition. Note that the second-highest curve also shows non-monotonic time behaviour. Credit: King et al. (2005).

values of J_d/J_h closer to the bifurcation value $-2\cos\theta$, where misalignment is achieved at a time when $J_d(t)$ is zero.

Significantly, the criterion (5.30) for counteralignment itself depends on the quantity J_d associated with the disc. But is it obvious that since the physics of the alignment process involves disc viscosity, this quantity cannot refer to the total angular momentum of a large disc, but only that part of it in viscous communication with the hole, that is, inside the disc warp. Then a hole whose spin is initially counteraligned with respect to a disc at first only interacts with a relatively small central region, for which the relevant value of J_d is small compared with J_h. According to the criterion (5.30), the hole will force this central region to align with its spin, and so be counteraligned with respect to the parts of the disc further out. The opposing sense of the gas angular momenta across the warp must cause rapid cancellation of gas angular momentum, and so we expect a significant increase in the local mass accretion rate there.

But as time proceeds the larger angular momentum of the disc must eventually overwhelm this central counteraligned hole–disc combination and force it in its turn to align with the outer disc. Figures 5.8 and 5.9 show a numerical calculation of a case like this, where the initial condition has a hole whose spin is counteraligned

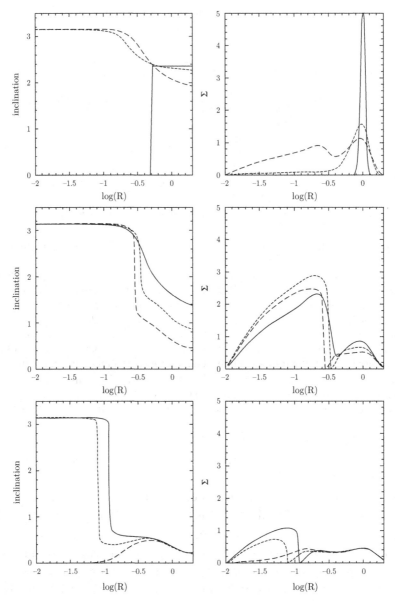

Figure 5.8 Numerical calculation of alignment evolution in a case where the black hole spin $\mathbf{J}_h$ is initially strongly counteraligned with respect to the disc angular momentum. The central region of the disc has angular momentum $< J_h$, so it initially aligns with the hole, counter to the rest of the disc. The disc warp spreads outwards on a viscous timescale to include disc regions with angular momentum larger than that of the counteraligned hole and central disc. This entire combination is itself ultimately aligned by the disc. Left-hand panels show inclination versus radius at a succession of epochs, measured in units of the viscous time of the gas ring situated at $R = 1$. Top panel: at $t = 0$ (solid line), $t = 0.01$ (short-dashed line), and $t = 0.02$ (long-dashed line). Middle panel: at $t = 0.03$ (solid line), $t = 0.04$ (short-dashed line), and $t = 0.05$ (long-dashed line). The right-hand panels show the evolution of surface density. Credit: Lodato and Pringle (2006).

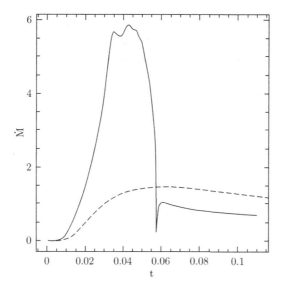

Figure 5.9 Time evolution of the mass accretion rate on to the black hole for the case shown in Figure 5.8 where initially almost anti-parallel angular momenta eventually align. The dashed line shows the corresponding evolution for a flat spreading ring. The accretion rate is greatly enhanced during the accretion of the counteraligned inner disc as the gas angular momenta are opposed across the disc warp. Credit: Lodato and Pringle (2006).

with repect to the surrounding disc. In reality an accretion event opposed in angular momentum to the hole spin would itself accumulate near the hole on a viscous timescale, but the figure qualitatively reproduces all the features we expect from the discussion in this section.

5.4 Chaotic Accretion and SMBH Growth

We have seen that chaotic accretion leads to initially misaligned disc formation around the SMBH, and the possibility of accretion episodically co- or counter-aligned with respect to the current black hole spin, tending to reduce its magnitude in the second case. Black hole spin controls the efficiency η of the energy release and so has potential implications for the mass growth rate, as we expect that the hole can at most accrete at the rate $\dot{M}_{Edd} = L_{Edd}/\eta c^2$, giving the Eddington luminosity. Then a low spin means low η and so a larger $\dot{M}_{Edd}$ for a given black hole mass.

This makes understanding the spin evolution essential in discussing constraints on models of black hole growth. The spectra of high-redshift quasars give SMBH mass estimates from the velocity widths of their rest-frame UV emission lines, and luminosity estimates from the νF_ν method (see (4.60)). These suggest that all of these systems are radiating at close to their Eddington luminosities L_{Edd}. The observed redshifts give the lookback times to the Big Bang, setting limits on the times that the SMBH can have been growing their masses by gas accretion. The results are striking: SMBH with masses $M > 5 \times 10^9 M_\odot$ are found at redshift $z = 6$, only 10^9 yr after the Big Bang (e.g. Willott, McLure & Jarvis 2003).[4]

Unless these SMBH somehow evaded the Eddington limit, their accretion rates are limited to

$$\dot{M}_{\mathrm{acc}} \leqslant \frac{L_{\mathrm{Edd}}}{\eta c^2}. \tag{5.40}$$

Not all of this infalling mass grows the black hole mass M since an amount L_{Edd}/c^2 is lost to radiation. So the SMBH mass grows at the rate

$$\dot{M} = \frac{L_{\mathrm{Edd}}}{\eta c^2} - \frac{L_{\mathrm{Edd}}}{c^2}. \tag{5.41}$$

Rearranging and using the expression (1.6) for L_{Edd}, we get

$$\dot{M} < \frac{1 - \eta}{\eta} \frac{M}{t_{\mathrm{Edd}}}, \tag{5.42}$$

where

$$t_{\mathrm{Edd}} = \frac{\kappa c}{4\pi G} = 4.5 \times 10^8 \text{ yr.} \tag{5.43}$$

Integrating (5.42), we find

$$\frac{M}{M_0} < \exp\left[\left(\frac{1}{\eta} - 1\right)\frac{t}{t_{\mathrm{Edd}}}\right], \tag{5.44}$$

where M_0 is the initial mass at $z \gg 1$. At redshift $z = 6$ the lookback time is $t = 10^9$ yr, so $t/t_{\mathrm{Edd}} \simeq 2.2$. We subtract from this a timescale ~ 100 Myr to allow for the formation of the first stars to see if growth from stellar-mass black holes can explain the observed masses, so $t/t_{\mathrm{Edd}} \simeq 2$, and

$$\frac{M}{M_0} < \exp\left[2\left(\frac{1}{\eta} - 1\right)\right]. \tag{5.45}$$

The crucial point is that the right-hand side of this equation is extremely sensitive to η and so to the spin parameter a. A maximally spinning hole ($a = 1$) has $\eta = 0.42$, so

$$\frac{M}{M_0} < e^{2.76} = 15.8. \tag{5.46}$$

---

[4] A potential caveat here comes from suggestions that the true luminosities – and so the SMBH masses – might be overestimated because the quasars could be gravitationally lensed.

The 'standard' efficiency $\eta = 0.1$ ($a = 0.85$) changes this to

$$\frac{M}{M_0} < e^{18} = 6.67 \times 10^7, \tag{5.47}$$

while for $a = 0$ ($\eta = 0.057$) we get

$$\frac{M}{M_0} < e^{33.1} = 2.34 \times 10^{14}. \tag{5.48}$$

The minimum initial masses M_0 required to grow to the observed $M > 5 \times 10^9 M_\odot$ at $z = 6$ therefore differ dramatically, depending on the value of the spin parameter a: we have

$$M_0 > 3.16 \times 10^8 M_\odot \text{ for a} = 1, \tag{5.49}$$

but only

$$M_0 > 7.5 M_\odot \text{ for a} < 0.85, \tag{5.50}$$

with limits far below a stellar mass for the non-spinning case $a = 0$ and any retrograde spin $a < 0$. This means that except for cases with rapid spins ($a \gtrsim 0.9$) and almost permanently prograde accretion, black holes can grow to the masses measured at $z = 6$ from very modest (stellar) initial masses.

These estimates assume that the holes have accreted for most of the lookback time. This is not unreasonable since masses $\sim 5 \times 10^9 M_\odot$ at $z = 6$ are extreme – there is a strong selection effect in that the most massive SMBH have the highest values of L_{Edd}, and we would expect them to correspond to cases where accretion was unusually prolonged. In any case, for spin $a = 0$, accretion for only 61 per cent of the lookback time is needed if $M_0 = 10 M_\odot$, and 56 per cent of this time if $M_0 = 50 M_\odot$, a mass found by the LIGO–Virgo gravitational wave experiments for a black hole descending from stellar evolution.[5]

These estimates used values of the efficiency η corresponding to a given spin parameter a. To check that they represent realistic evolutions we use (4.8) – (4.12) to evolve (M, a) numerically as a black hole accretes from a succession of randomly oriented discs. Since we know that the timescale for co- or counteralignment is short compared with disc accretion, we can assume that accretion is always aligned either fully prograde or fully retrograde, depending on the criterion (5.30). We assume that accretion is at the Eddington rate given by the current spin value a. We limit the disc mass to the self-gravity value $(H/R)M \sim 10^{-3} M$, which shows that the condition $J_d < 2J_h$ (see (5.30), (5.31)) is in general easily satisfied, allowing either co- or counteralignment depending on the initial disc orientation.

---

[5] The stellar evolution preceding the formation of a $50 M_\odot$ black hole takes $\lesssim 1$ per cent of the lookback time at $z \simeq 6$.

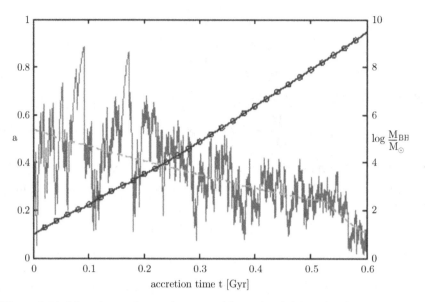

Figure 5.10 Mass (curve increasing smoothly to the right) and spin parameter s (strongly varying curve) of a supermassive black hole accreting from a sequence of randomly oriented discs limited by self-gravity. The spin decreases because retrograde accretion has a larger lever arm than prograde. Note that the hole mass grows faster at epochs when the spin is relatively low, as the radiative efficiency is then lower, so that the limiting Eddington luminosity corresponds to a higher accretion rate. The time axis gives only the time when the hole is accreting gas, showing that black hole growth from stellar masses to observed values $M \sim 5 \times 10^9 M_\odot$ is possible in the lookback time $\simeq 10^9$ yr at redshift $z = 6$ without violating the Eddington limit. Coalescences with other SMBH would take the hole discontinuously off both curves, but they are attractors, so the hole reverts to both after roughly doubling its mass by gas accretion. The dashed line shows the average spin evolution predicted by King, Pringle and Hofmann (2008).

Figure 5.10 gives the hole mass and spin parameter a as functions of the time actually spent accreting. It shows that the spin decreases stochastically over time, as expected since retrograde accretion has a bigger lever arm than prograde. (The same also holds for mergers with other SMBH, for the same lever arm reason – see Hughes & Blandford, 2003.) We conclude that unless accretion is somehow arranged to produce consistent spin-up, the large SMBH masses observed at red-shift $z = 6$ could have grown from accretion on to what were originally stellar-mass black holes.

The figure describes only the effects of gas accretion. It is possible for an SMBH to merge with another black hole in the course of a galaxy merger. This would discontinuously alter its spin and increase its mass, so move it off the relevant curves in the figure. But these are attractors, so the SMBH reverts to them after roughly further doubling its mass by gas accretion. An exception to this can occur for the very largest SMBH, such as those in the nuclei of giant ellipticals. A suitably oriented merger might give the merged hole an unusually large spin $a \sim 1$, and there would be too little gas available to double the SMBH mass and reduce this.

Most SMBH instead tend asymptotically to a state of small but not quite zero spin. The condition $J_d \lesssim J_h$ requires

$$M_d \lesssim Ma \left(\frac{R_g}{R_d} \right)^{1/2}, \tag{5.51}$$

so it is difficult to decrease the prograde spin episodes below

$$a \simeq \frac{M_d}{M} \left(\frac{R_d}{R_g} \right)^{1/2}. \tag{5.52}$$

The spin then oscillates between this value and an even smaller (closer to zero) negative value, as spin-down is more efficient than spin-up (see King, Pringle & Hofmann, 2008 for quantitative details).

Since $J_d \gtrsim J_h$, accretion in this initially counteraligned case quickly spins the hole up to an aligned state with $J_h \sim J_d$. So at late times – low redshift – the hole has spin angular momentum comparable with that of the last disc accretion event. In all cases the system quickly reaches axisymmetry, with the AGN jet parallel to the axis of the accretion torus, in agreement with what is observed.

Since the self-gravity constraint limits the disc mass to $M_d \sim (H/R)M \sim 10^{-5}M$, it follows (King & Nixon, 2015) that the typical variation timescale for AGN accretion is

$$t_{\text{AGN}} \sim 10^{-3} t_{\text{Sal}} \dot{m}^{-1} \sim 10^5 \dot{m}^{-1} \text{ yr}, \tag{5.53}$$

where t_{Sal} is the Salpeter timescale (1.9) and $\dot{m} \lesssim 1$ is the Eddington ratio. This agrees with the variation timescale ('AGN flickering') deduced from observations (Schawinski et al., 2015).

Now we can see the effect of the earlier incorrect treatment of disc-spin alignment discussed after 5.31. This implied that any retrograde disc would almost instantly flip the hole spin to make accretion prograde. Then from (4.12) it is clear that after increasing the original hole mass M_0 by a factor no larger than $\sqrt{6} = 2.45$, its spin parameter would reach the maximal value $a = 1$. From (5.46) it follows that the hole mass could grow by no more than a further factor 15.8 by redshift 6, making the hole mass at redshift 6 no larger than $39M_0$, so requiring $M_0 \gtrsim 10^8 M_\odot$ to explain the observations of SMBH masses $5 \times 10^9 M_\odot$ at $z = 6$. These massive 'seed' black holes would need to have masses larger than many typical SMBH at low redshift, which must instead have grown their masses significantly by gas accretion (the Soltan relation) since $z = 6$.

5.5 Massive Seed Black Holes

The reasoning of the last paragraph of Section 5.4 motivated a large body of work on forming seed black holes with masses much larger than stellar values. These could then overcome the apparent problem in growing to the values observed at redshift 6 if black holes were forced to spin up. Even though clarifying the spin evolution means that there is now no absolute necessity for these, it is still interesting to consider whether such seeds could exist.

The first black holes were probably the remnants of the first generation of stars, called population III. These must have had almost pure H and He composition, as the Big Bang made negligible amounts of other elements. They probably formed inside small dark matter (DM) halos, with virial temperatures T_{vir} of order 10^3–10^4 K, and gas cooling though emission from molecular H. For masses $\gtrsim 250 M_\odot$ these explode as pair-instability supernovae and produce black holes of about half their mass. The small DM halos merge and grow, and the seed black holes sink to the centres of the halos as they grow further by accreting gas. If, as we suggested in Sections 5.3 and 5.4, chaotic accretion keeps the spin and accretion efficiency low, there is no need for more complicated seeding. But this conclusion must change if the infall is somehow more orderly.

One idea is that dense stellar clusters form, and the frequent stellar collisions within them directly form more massive black holes. Dark matter self-annihilation might heat the early stars so that they retard the onset of H-burning and let the stars grow to larger masses. An entirely different possibility is that massive ($M > 10^5 M_\odot$ clouds of primordial gas in DM halos with virial temperatures $T_{vir} > 10^4$ K are unable to cool if H_2 formation is suppressed by suitable ionizing sources. Then atomic H emission cools the gas to $\sim 10^4$ K, and simulations predict the formation of thick, stable pressure-supported discs. The cores of these discs can then collapse to separate objects with masses $\sim 10^4 M_\odot$. One way of ensuring the suppression of H_2 may be that Lyman alpha radiation is trapped in the collapsing gas and keeps the temperature above 10^4 K (Spaans & Silk, 2006).

Efficient collapse of the central objects of the primordial discs requires a way of removing angular momentum. One idea is that for haloes with $T_{vir}/T_{gas} < 1.8$, angular momentum is transported by local gravitational instabilities, keeping the disc in a state of marginal stability. If instead the clouds are unstable to bar formation, this could remove angular momentum and allow dynamical collapse. If gas cooling remains effective this process can repeat, as 'bars within bars' (Shlosman, Frank & Begelman, 1989). If the gas in the centre of a DM halo is massive enough, it can form a supermassive star. Then general-relativistic instabilities may cause it to collapse a significant fraction of its mass to a black hole by various routes.

An intriguing idea is that if the mass infall is sufficiently rapid the star cannot reach thermal equilibrium (Begelman, 2010). Then the core collapses when H-burning stops, leaving a stellar-mass black hole inside a massive H envelope like a greatly enlarged version of a red giant (Begelman, Volonteri & Rees, 2006), called a quasi-star. This has the virtue that the accretion rate on to the black hole is limited only by the Eddington rate of the entire object, which is initially much larger than that of the black hole alone. The excess energy is carried away by convection.

The discussion here is almost entirely theoretical, as the constraints on seed formation are currently very indirect. This may change with the advent of the James Webb Space Telescope (JWST) and the Large Interferometer Space Antenna (LISA). The presence and nature of central black holes in dwarf galaxies, where their masses cannot be much larger than the putative seed masses,[6] offers one of the few avenues to checking any of these ideas at present.

5.6 Disc Warping, Breaking, and Tearing

We have seen that the inner part of a misaligned viscous disc where Lense–Thirring precession is rapid aligns fastest with the spin plane, and so a disc *warp* travels outwards. In general, warps can propagate in either of two ways. If the disc viscosity is significant, that is, $\alpha > H/R$, as implicitly assumed in the discussion to this point, warps propagate through the disc by viscous diffusion. This is usually considered the relevant regime for black hole discs, but there are cases where it may not hold.

With instead $\alpha < H/R$, pressure forces dominate over viscous ones and instead propagate warping disturbances through distances $> R$ as waves. The small value of α means viscosity is too small to damp the wave locally (i.e. scales of order H) and allow it to propagate. A second, more subtle, condition must also be satisfied for wave propagation to be efficient – the disc must be close to Keplerian, that is, $|1 - \kappa^2/\Omega^2| \lesssim H/R$, where κ is the epicyclic frequency (see, e.g., section 5.1 of Wijers & Pringle, 1999). This makes the forcing resonant in a warp, as Keplerian motion implies that the orbital, vertical, and epicyclic frequencies are all the same, that is, $\Omega = \Omega_z = \kappa$. The first treatment of the behaviour of warped discs in this wavelike regime is by Papaloizou and Lin (1995), who show that these waves, driven by pressure gradients, propagate at approximately half the local sound speed. In this section we will mainly discuss the diffusive warp propagation regime (i.e. assume $\alpha > H/R$), but we also describe some cases where wavelike propagation may occur.

In many cases, if the disc is steadily fed matter misaligned with respect to the spin, the diffusive assumption $\alpha > H/R$ results in a smooth steady warp, a result

---

[6] And are occasionally at risk of being smaller.

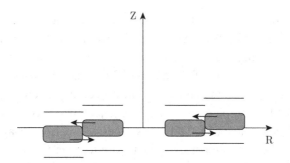

Figure 5.11 A schematic view of a warped disc viewed orthogonal to its axis. The shaded areas show regions of higher pressure. The arrows show the directions of the horizontal pressure gradients induced by the warp. A fluid element orbiting the centre feels an oscillating radial pressure gradient, whose amplitude is a linear function of height z. Credit: Lodato and Pringle (2007).

first found by Bardeen and Petterson (1975). For a time there was a general belief that this was the only possible outcome. But the behaviour of the viscosity in a warped disc is more complex than pictured in that paper, which simply assumed that there is a second independent viscosity coefficient α_2 relating to 'vertical' shearing of disc annuli.

If one instead makes the reasonable assumption that *dissipation* by viscosity is isotropic (see Papaloizou & Pringle, 1983), this enforces a relation of the form

$$\frac{\nu_2}{\nu_1} = \frac{\alpha_2}{\alpha_1} \simeq \frac{1}{\alpha^2}, \tag{5.54}$$

where $\alpha_1 \simeq \alpha$ is the 'standard' Shakura–Sunyaev coefficient defining the isotropic dissipation. This implies the apparently surprising result that

$$\alpha_2 \simeq \frac{1}{\alpha}, \tag{5.55}$$

so that the warp diffusion scales inversely with the dissipation.

The key to understanding this is that in a warped disc, neighbouring annuli exert resonant torques on each other (see Figures 5.11, 5.12), with the result that the effects of dissipative damping work in a way that at first appears counterintuitive. The presence of a warp in a thin disc, whose internal density and pressure are stratified on the scaleheight H, causes a horizontal pressure gradient with the same frequency Ω as each fluid element in the disc. This is in resonance with the epicyclic motions of these elements, which therefore try to grow. The induced pressure gradient is

$$\frac{\partial P}{\partial R} \sim \frac{\partial P}{\partial z}\psi \sim \frac{P\psi}{H}, \tag{5.56}$$

Figure 5.12 A projection of a 3D model disc (here actually around a protostar) after it has been warped. The annotations show the radial pressure gradients induced by the warp at different azimuths as a fluid element orbits the disc. The curved arrows show the rotation of the disc. The shaded boxes represent the pressure between adjacent gas rings. At $\phi = 0$ and $\phi = \pi$ the rings are aligned, so there is no pressure gradient. At all other azimuths, the heights of the annuli are offset, resulting in an oscillating radial pressure gradient as the gas follows an orbit. The direction of the resulting pressure gradient is shown by the horizontal black arrows. Lighter regions of the disc have higher surface density. Credit: Rowther, Nealon and Farzana (2021).

where

$$\psi = R \left| \frac{\partial \mathbf{l}(R)}{\partial R} \right| \tag{5.57}$$

is the dimensionless warp amplitude, where

$$\mathbf{l}(R) = \frac{\mathbf{J}(R)}{J(R)} \tag{5.58}$$

is the unit vector specifying the local direction of the angular momentum of a fluid element. This gives a force term

$$\frac{1}{\rho} \frac{\partial P}{\partial R} \sim \frac{c_s^2 \psi}{H} \sim H \Omega^2 \psi \tag{5.59}$$

in the horizontal momentum equation.

This forcing varies linearly with disc height z and so excites a response in the radial velocity field of the form

$$v_R = v_\perp \left(\frac{z}{H}\right) \cos \phi, \tag{5.60}$$

where ϕ is the usual disc azimuthal coordinate. Here $v_\perp$ is the induced horizontal velocity at the surfaces $z = \pm H$ close to the disc surface. If $x(t)$ is the radial Lagrangian displacement of a disc surface particle with respect to an unperturbed particle on a circular orbit, we have

$$x(t) = \left(\frac{v_\perp}{\Omega}\right) \sin \Omega t. \tag{5.61}$$

The particle feels three forces: a restoring force producing the epicyclic frequency Ω, a damping force caused by viscous dissipation, and a pressure force caused by the warp, given by (5.59). So $x(t)$ has the equation of motion

$$\ddot{x} + \frac{\dot{x}}{\tau} + \Omega^2 x = R\Omega^2 \left(\frac{\psi H}{R}\right) \cos \Omega t. \tag{5.62}$$

Here τ is the damping timescale. Assuming this is simply caused by a kinematic viscosity v_z, we have

$$\tau \sim \frac{H^2}{v_z}, \tag{5.63}$$

since the typical lengthscale of the horizontal shear is H. Assuming that viscous dissipation of the vertical and horizontal shear occurs at the same average rates, then $v_z = v$, which gives

$$\tau \sim \frac{H^2}{v_z} = \frac{1}{\alpha \Omega}. \tag{5.64}$$

In view of (5.61) we seek a solution of (5.62) of the form

$$x(t) = a \sin \Omega t, \tag{5.65}$$

where a is the amplitude of the horizontal motions at the disc surface. Substituting into (5.62) gives

$$a = \tau \Omega \psi H, \tag{5.66}$$

or a horizontal velocity

$$v_\perp = a\Omega = \Omega H \frac{\psi}{\alpha} \sim c_s \frac{\psi}{\alpha}, \tag{5.67}$$

since $H = c_s/\Omega$. (For $\alpha < H/R$ warps propagate as wavelike disturbances through the disc, and this estimate does not hold.)

This then tells us that the amplitude of the vertical shear is

$$S = \frac{dv_R}{dz} = \frac{\Omega a}{H} = \frac{\psi}{\alpha}\Omega. \tag{5.68}$$

The 'vertical' viscosity coefficient v_2 acts on the radial derivative of the vertical motion, as this is the effect that dissipates the warp. But as we have seen, this process itself happens because of the damping of the resonant horizontal shear propagating the warp radially. The definition of v_2 actually embodies the assumption we have made that these two processes dissipate energy at the same rate, that is,

$$v_2 \left\langle \frac{dv_z}{dR} \right\rangle^2 = v_z \left\langle \frac{dv_R}{dz} \right\rangle^2, \tag{5.69}$$

where the angle brackets imply suitable averaging over radius and azimuth. Since we have $v_z = \psi R\Omega$ and $dv_z/dR = \psi\Omega$, (5.69) gives

$$v_2 = v_z(\Omega t)^2 = \Omega H^2 (\Omega \tau), \tag{5.70}$$

using (5.63) in the second step. Then with $v_2 = \alpha_2 c_s H$ we get

$$\alpha_2 = \Omega \tau = \frac{1}{\alpha}, \tag{5.71}$$

assuming that the damping comes from isotropic viscous dissipation. Then since $v_1 = \alpha c_s H$, we get finally that

$$\frac{v_2}{v_1} \sim \frac{1}{\alpha^2}. \tag{5.72}$$

This relation implies the initially surprising result that

warp damping is faster if the viscous dissipation is smaller.

This happens because the viscosity also determines the resonant velocity amplitude, and a smaller isotropic viscosity α produces a larger resonant velocity $v_R \propto 1/\alpha$. The dissipation rate then goes as $vv_R^2 \propto 1/\alpha$ and so increases as α decreases, making warp damping faster. (Obviously this scaling fails at some point since it would predict arbitrarily high dissipation as $\alpha \to 0$. In reality this is prevented because, as we noted at the beginning of this section, for $\alpha < H/R$, warps propagate in a wavelike fashion.) If the warp is small in the sense that $|\psi| \ll H/R$, then for discs evolving diffusively (i.e. $H/R \lesssim \alpha \ll 1$) Papaloizou and Pringle (1983) show that

$$\alpha_2 = \frac{1}{2\alpha}. \tag{5.73}$$

Ogilvie (1999) extends these results to larger α, ψ by expanding in powers of H/R, but retaining the assumption that viscous dissipation is isotropic. Then for a small warp ($|\psi| \ll 1$) this gives

$$\frac{\nu_2}{\nu_1} = \frac{1}{2\alpha^2} \frac{4(1 + 7\alpha^2)}{4 + \alpha^2}. \tag{5.74}$$

We will discuss the significant effects warping can have on discs later in this section, but first we consider rather more closely the ways in which a disc can warp. The warp amplitude ψ defined in (5.57) uses the unit tilt vector $\mathbf{l}$. We can write this as

$$\mathbf{l} = (\cos \gamma \sin \beta, \sin \gamma \sin \beta, \cos \beta), \tag{5.75}$$

where $\beta(R, t)$ is the disc *tilt* and $\gamma(R, t)$ is the disc *twist*. So discs can be warped, that is, have $\psi > 0$, in different ways. One way is to have β varying with radius R, and another distinct mode instead has γ varying while β is non-zero – see Figure 5.13. A steep variation in either β or γ can cause a significant warp. In a general warped disc both of β, γ vary with R. On scales $\lesssim H$ there is no distinction between tilt and twist in specifying the relative orientation of neighbouring disc annuli, so local stability is specified simply by ψ rather than either β or γ. But the global disc behaviour does discriminate between changes in tilt or twist.

The increase of warp damping as viscosity decreases has an important consequence. As we can see from (5.54) and (5.72), smaller α makes α_2 increase strongly so that annuli are pulled more strongly in the vertical direction than they are along the disc plane. In effect, the forces trying to hold the disc together actually become weaker as the disc warps. For a sufficiently large warp amplitude it is possible for the disc to *break* – undergo an almost discontinuous change of plane. Matter accretes from the outer to the inner disc through a sequence of orbits whose inclination changes very sharply over a very narrow range of radii. This effect was first explicitly found in SPH simulations by Lodato and Price (2010) (see Figure 5.14), who assumed an initial disc which was already strongly warped.[7] Specifically, the components of the unit disc normal vector $\mathbf{l}$ are initially

$$l_x = \begin{cases} 0 & \text{for } R < R_1, \\ \frac{A}{2}\left[1 + \sin\left(\pi \frac{R - R_0}{R_2 - R_1}\right)\right] & \text{for } R_1 < R < R_2, \\ A & \text{for } R > R_2, \end{cases}$$

---

[7] Although there had been earlier indications in other simulations which were not fully investigated (see Nixon & King, 2012 for a discussion).

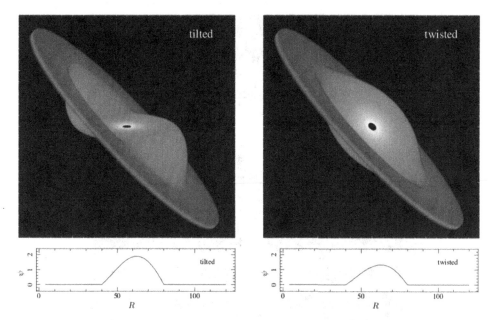

Figure 5.13 Examples of how a disc can be warped. Left: a three-dimensional rendering of a warp where only the tilt angle β varies with radius. Right: only the disc twist angle γ varies. The inner disc radius is $R = 4$ and the radius is $R = 120$ in both cases. The tilted case has $\gamma = 0$ for all R with $\beta = 0$ for $R < 40$ and $= 45°$ for $R > 80$, varying smoothly between these two radii as a cosine. This gives the warp amplitude $\psi(R) = R|\partial\beta/\partial R|$ shown below the image. For the twisted case, $\beta = 0$ at all radii, and $\gamma = 0$ for $R < 40$ and $\gamma = 45°$ for $R > 80$, varying smoothly between $0°$ and $45°$ as a cosine. This gives the warp amplitude $\psi = R|\sin\beta\partial\gamma/\partial R|$ shown below the image. The orientation of the outer disc is the same in the two cases. Credit: Raj, Nixon and Doğan (2021).

where A is a constant, and $R_0 = (R_1+R_2)/2$, together with $l_y = 0$ and $l_z = \sqrt{1 - l_x^2}$. The warp amplitude is then

$$\psi = R|\frac{\partial \mathbf{l}}{\partial R}| = \frac{R}{l_x}\frac{\partial l_x}{\partial R}, \tag{5.76}$$

which reaches a maximum at $R \simeq R_0$, with value

$$\psi_{\max} \simeq \frac{\pi R_0}{2(R_1 - R_2)}\frac{A}{(1 - (A/2)^2)^{1/2}}. \tag{5.77}$$

One might argue that the resulting break is somehow imprinted into the disc by the strongly warped initial state. But Nixon and King (2012) showed that disc breaking can result self-consistently from the viscous evolution of an initially plane but misaligned disc at an initial angle θ_0 to the spin of the central black hole. Nixon et al. (2012) made a simple comparison of precession and viscous torques to show

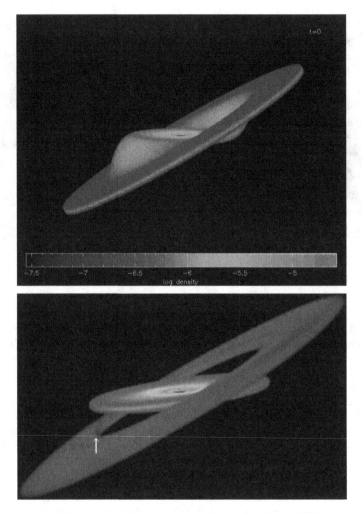

Figure 5.14 Breaking found in SPH simulations of a disc initially placed at a fixed inclination to the spin of the central black hole (top figure). After viscous evolution the disc attains the 'broken' configuration of the lower figure. Here the disc plane changes sharply between the original inclination, maintained in the outer disc, and the plane orthogonal to the black hole spin vector in the inner disc because the forced Lense–Thirring precession rate grows strongly near the hole. The tenuous gas between these two regions follows a series of orbits whose inclinations change very rapidly with radius. Part of this gas is just visible above the white arrow. Credit: Lodato and Price (2010).

that breaking was likely in binary systems where the accretion disc was misaligned. Their result can be written as

$$\sin 2\theta \gtrsim 0.06 \left(\frac{\alpha}{0.1} \right) \left(\frac{H/R}{0.01} \right) \qquad (5.78)$$

(see Problem 5.2 at the end of the book). This work used a ring code – that is, one where each disc annulus obeys conservation of mass and angular momentum while interacting viscously with its neighbours. Pringle (1992) and Ogilvie (1999) show that the disc's angular momentum density $\mathbf{L}(R, t)$ obeys the equation

$$\frac{\partial \mathbf{L}}{\partial t} = \frac{1}{R}\frac{\partial}{\partial R}\left\{ \frac{(\partial/\partial R)[\nu_1 \Sigma R^3(-\Omega')]}{\Sigma(\partial/\partial R)(R^2\Omega)}\mathbf{L} \right\}$$
$$+ \frac{1}{R}\frac{\partial}{\partial R}\left[\frac{1}{2}\nu_2 R|\mathbf{L}|\frac{\partial \mathbf{l}}{\partial R} \right]$$
$$+ \frac{1}{R}\frac{\partial}{\partial R}\left\{ \left[\frac{\frac{1}{2}\nu_2 R^3\Omega|\partial \mathbf{l}/\partial R|^2}{(\partial/\partial R)(R^2\Omega)} + \nu_1\left(\frac{R\Omega'}{\Omega}\right) \right]\mathbf{L} \right\} \qquad (5.79)$$
$$+ \frac{1}{R}\frac{\partial}{\partial R}\left(\nu_3 R|\mathbf{L}|\mathbf{l} \times \frac{\partial \mathbf{l}}{\partial R} \right)$$
$$+ \mathbf{\Omega}_p \times \mathbf{L}.$$

Here ν_1, ν_2, ν_3 are the effective viscosities, $\Omega(R)$ is the local azimuthal angular velocity, $\Sigma(R, t)$ is the disc surface density, $\mathbf{l}(R, t)$ is the unit angular momentum vector, $\mathbf{\Omega}_p$ is the precession frequency induced in the disc by the LT effect, $\mathbf{L} = \Sigma R^2 \Omega \mathbf{l}$, and the prime stands for differentiation with respect to R.

Equation (5.79) is useful for understanding the physics of disc warping. We see that five independent torques act on the rings of disc gas. The first four terms on the right-hand side are the internal disc torques communicating angular momentum between the rings, and the last represents the LT torque.

From the presence of the azimuthal shear viscosity ν_1 we can see that the first term on the right-hand side of (5.79) describes the usual viscous diffusion of mass, which of course affects the angular momentum budget of the disc. The second term is also diffusive, and the presence of the vertical viscosity ν_2 and the radial tilt derivative $\partial \mathbf{l}/\partial R$ shows that it transports angular momentum given by the disc tilt $\mathbf{l}$, that is, misaligned from the main disc axis parallel to the black hole spin. The third term directly advects the disc angular momentum density $\mathbf{L}(R, t)$ as disc mass moves.

These three torques are just as expected from conservation of mass and angular momentum (Pringle, 1992). The vector product in the fourth term shows that it is a precessional torque, found by Ogilvie (1999), causing disc rings to precess around the vector $\mathbf{l} \times \partial \mathbf{l}/\partial R$. This evidently only acts if the disc is warped, causing dispersive wavelike warp propagation. The direction of the precession depends on the disc viscosity α and the warp amplitude ψ (see Ogilvie, 1999, figure 5). Its form is the same as the second term of (5.79), as is seen by making the transformations $\nu_2 \rightarrow 2\nu_3$ and $\partial \mathbf{l}/\partial R \rightarrow \mathbf{l} \times \partial \mathbf{l}/\partial R$. This term involves the

Lense–Thirring precession we have already encountered several times – the disc ring angular momentum vectors precess around the black hole spin vector with frequency

$$\Omega_p = \frac{2G\mathbf{J}_h}{cR^3},$$

(5.80)

where $\mathbf{J}_h$ is the hole's angular momentum vector, with magnitude $J_h = GM^2 a/c$. The torque this exerts on the disc is balanced by the reaction on the hole, which gives the precessional torque

$$\frac{d\mathbf{J}_h}{dt} = -2\pi \int \Omega_p \times \mathbf{L} R dR$$

(5.81)

familiar from Section 5.3.

The treatment by Ogilvie (1999) shows that the disc's viscous response to a warp is modified by the presence of the warp itself. Assuming a locally isothermal disc, the effective viscosity coefficients must be consistent with the internal fluid dynamics of the disc if we retain the assumption of a locally isotropic physical viscosity.[8]

Local conservation of mass and angular momentum fixes the non-linear coefficients Q_1, Q_2, Q_3 relating the effective viscosity component to Shakura–Sunyaev forms (Ogilvie, 1999, eqns 10, 11 and 12). All three of these coefficients decrease in magnitude in a strong warp, as the lack of perfect contact reduces the rings' ability to communicate angular momentum. Communication evidently requires the gas particles to deviate from completely circular orbits, and this becomes rarer as the relative inclination of neighbouring disc rings increases. Beyond some critical relative inclination this only happens where the rings cross. The effective viscosities are given by

$$\nu_1 = \frac{Q_1 I \Omega^2}{\Sigma R d\Omega/dR},$$

(5.82)

$$\nu_2 = \frac{2Q_2 I \Omega}{\Sigma}$$

(5.83)

and

$$\nu_3 = \frac{Q_3 I \Omega}{\Sigma},$$

(5.84)

where

$$I = \frac{1}{2\pi} \int_0^{2\pi} \int_{-\infty}^{\infty} \rho z^2 dz d\phi.$$

(5.85)

---

[8] The isothermal assumption here amounts to neglecting pressure work. This can lead to significant differences compared with cases where this assumption is not adopted – see, for example, Rowther, Nealon and Farzana (2021), which studies whether this can remove or reduce the tendency of a protoplanetary disc to show self-gravitational effects such as spiral arms.

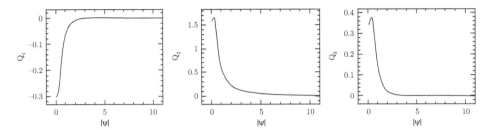

Figure 5.15 The effective viscosity coefficients Q_1, Q_2, and Q_3 plotted against warp amplitude $|\psi|$ for a case with $\alpha = 0.2$. The coefficients change significantly for $0 < |\psi| \lesssim 5$, after which they are approximately constant at a small but non-zero value. As expected, these coefficients all decrease in magnitude in the presence of a strong warp, where the disc rings are highly inclined to each other. Credit: Nixon and King (2012).

With the assumption that the disc is locally isothermal, and Σ is constant in azimuth, this gives $I \propto \Sigma c_s^2 / \Omega^2$. Further, with $\Omega = (GM/R^3)^{1/2}$ for a Keplerian disc around a point mass, we find

$$v_1 = -\frac{2}{3} Q_1 [(H/R)^2 R^2 \Omega], \tag{5.86}$$

$$v_2 = 2 Q_2 [(H/R)^2 R^2 \Omega] \tag{5.87}$$

and

$$v_3 = Q_3 [(H/R)^2 R^2 \Omega]. \tag{5.88}$$

(For an unwarped disc, $v_1 = \alpha c_s H = (H/R)^2 R^2 \Omega$, and $Q_1(|\psi| = 0) = -3\alpha/2$ (Ogilvie, 1999), so v_1 takes exactly the usual Shakura–Sunyaev form $v = \alpha c_s H$.)

The quantity $x = [(H/R)^2 R^2 \Omega]$ would be constant for a disc sound speed dependence $c_s \propto R^{-3/4}$, while in a steady-state Keplerian disc we have $c_s \propto R^{-3/8}$. So taking x as constant is a reasonable approximation, with the helpful property of making all the viscosities independent of radius.

From Ogilvie (1999) we get the coefficients Q_i shown in Figure 5.15. Figures 5.16 and 5.17 show how the disc tilt behaves in a case where the warp propogates, and in one where it breaks. We see that, as expected, the reduction of the effective viscosities in warped regions promotes the breaking of the disc into two distinct planes.

Doğan et al. (2018) give a systematic answer to the question when discs with $\alpha > H/R$ (and so diffusive warp propagation) break, and the nature of the instability, by finding the dispersion relation for warp propagation. This is a cubic, and instability occurs if any of its three roots has a positive real part. The growth rate of the instability is always close to dynamical. The physics of the instability involves a

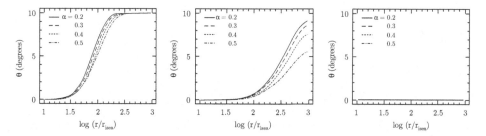

Figure 5.16 Disc structure specified by plotting the tilt angle θ between the disc and the hole against the logarithm of the radius. The initial misalignment of the disc is $\theta_0 = 10°$. The various curves correspond to differing initial α for the effective viscosities used in the simulation. From left to right the panels correspond to $t = 0.01, 0.1$, and 1 viscous times after the start of the calculation. Note that in the latter case the warp has moved to the edge of the grid and the disc becomes axisymmetric (aligns or counteraligns). Credit: Nixon and King (2012).

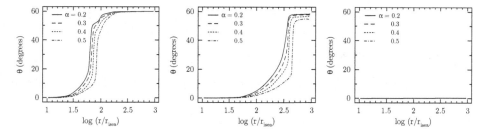

Figure 5.17 Disc structure specified by plotting the tilt angle between the disc and the hole against the logarithm of the radius. The initial misalignment of the disc is $\theta_0 = 60°$. The various curves correspond to differing initial α for the effective viscosities used in the simulation. From left to right the panels correspond to $t = 0.01, 0.1$, and 1 viscous times after the start of the calculation. We see that initially the disc profile is significantly steepened, and the disc breaks, for all values of α. The break propagates outwards in the disc, reaching its outer edge in about one viscous timescale for all values of α, whereupon the disc becomes fully aligned or counteraligned with the black hole spin. The comparison with Figure 5.16 suggests that breaking is more likely if the initial inclination with respect to the black hole spin is high, and that the non-linear effect of the disc viscosity also promotes breaking. Credit: Nixon and King (2012).

combination of a term which would in a plane disc give the standard (Lightman–Eardley; see 4.78)) viscous instability

$$\frac{\partial(\nu\Sigma)}{\partial\Sigma} < 0 \tag{5.89}$$

(see Section 4.7) but is here modified by the warp to the form

$$\frac{\partial(\nu_1|\psi|)}{\partial|\psi|} < 0, \tag{5.90}$$

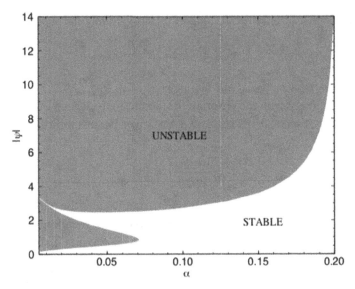

Figure 5.18 Stable (white) and unstable (shaded) regions for disc breaking, in the
$(\alpha, |\psi|)$ plane for discs with $\alpha > H/R$ (and so with diffusive warp propagation).
The figure gives the critical warp amplitudes $|\psi|$ required for breaking to occur
in discs with a given value of α. A flat disc is always stable against breaking if
we assume an equation of state suppressing Lightman–Eardley-like instabilities,
where $\partial(\nu\Sigma)/\partial\Sigma < 0$ – see equation (4.78). With $0.01 < \alpha < 0.2$ an almost flat
disc with $|\psi| \lesssim 0.1$ is also always stable. For a given $\alpha \leq 0.2$ there is always a
minimum value of the warp parameter $|\psi|$ which causes breaking. Credit: Doğan
et al. (2018).

and a similar condition acting on the diffusion of the warp amplitude given in
simplified form by

$$\frac{\partial(\nu_2|\psi|)}{\partial|\psi|} < 0. \tag{5.91}$$

These two criteria map out the regions of instability on the $(\alpha - |\psi|)$ plane, as
shown in Figure 5.18.

A simple physical picture of the instability uses the fact that if $\alpha \ll 1$, warps
are smoothed faster than mass is transported radially by viscosity (Papaloizou &
Pringle, 1983). So the most likely destabilizing torque is the radial diffusion of
misaligned angular momentum caused by ν_2. Retaining only these terms in the
angular momentum conservation equation of Ogilvie (1999) gives

$$\frac{\partial}{\partial t}(\Sigma R^2 \Omega \mathbf{l}) \sim \frac{1}{R}\frac{\partial}{\partial R}\left(Q_2 \Sigma c_s^2 R^3 \frac{\partial \mathbf{l}}{\partial R}\right). \tag{5.92}$$

This is a diffusion equation for the flux of misaligned angular momentum,

$$\mathbf{F}_{\mathrm{mis}} = Q_2 \Sigma c_s^2 R^2 \frac{\partial \mathbf{l}}{\partial R}, \tag{5.93}$$

with a rate of transfer $\sim |\mathbf{F}_{\mathrm{mis}}/\Sigma R^2 \Omega|$.

By a direct analogy with the criterion for the thermal-viscous Lightman–Eardley type of instability, that is, $\partial(\nu\Sigma)/\partial\Sigma < 0$, this gives the instability criterion for disc breaking as

$$\frac{\partial(\nu_2|\psi|)}{\partial|\psi|} < 0. \tag{5.94}$$

We can interpret this as saying that if $\nu_2(|\psi|)$ arranges that the maximum diffusion rate of misaligned angular momentum is not at the same location as the maximum warp amplitude, then there is nothing to stop the local maxima in this amplitude $|\psi|$ growing, so the disc must eventually break. The large warp amplitude means large torques so that mass is transferred rapidly out of this region and the surface density Σ drops rapidly. This is very similar to the Lightman–Eardley (4.78) instability for a flat disc, except that here the warp amplitude $|\psi|$ replaces the surface density Σ. The instability happens only because the effective viscosities themselves depend on the warp amplitude.

It is clear from the 'whale's face' Figure 5.18 that disc breaking is potentially very common in SMBH discs, where it is likely that successive accretion episodes have random inclinations with respect to the hole spin.[9] This can make a major change to the observational appearance of a disc compared with the original Bardeen–Petterson smooth warp. But it can produce even stronger effects. Even after a break occurs, the forced Lense–Thirring precession still operates, and its rate depends strongly on disc radius (frequency $\omega \propto R^{-3}$). Then for sufficiently high inclination, rings of gas can break off from the misaligned outer disc, and through differential precession neighbouring rings can end up in partial *counter-rotation*. As they spread viscously they can cancel angular momentum and cause gas to fall to smaller radii. Then the whole process can repeat itself as the infalling matter circularizes.

The net result is that disc matter has 'borrowed' angular momentum from the black hole via the Lense–Thirring effect in order to cancel its own (see Figures 5.19 and 5.20), so is somehow complicit in feeding itself more rapidly than simple viscous evolution would allow. This process is called *disc tearing* (Nixon et al., 2012) (see Figures 5.20 and 5.21) and can sometimes make accretion far more violent

---

[9] Breaking can also appear in accreting stellar-mass black hole binaries if the spin of the black hole or neutron star is misaligned with respect to the binary orbit. (The neutron star's mass multipole moments also imply precessions of accretion disc gas orbits.) Misalignments are actually observed and are probably an effect of the earlier supernova explosion forming the accretor. Nixon and Salvesen (2014) appeal to breaking to give a physical basis for the X-ray state changes observed in stellar-mass X-ray binaries.

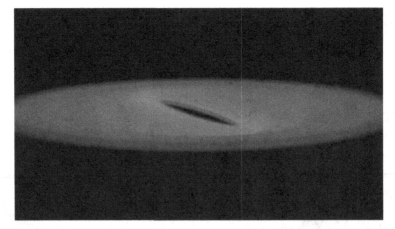

Figure 5.19 Full three-dimensional surface rendering of a small inclination simulation. The whole disc is initially inclined at 10° to the hole, with no warp. This snapshot is after approximately 500 dynamical times at the inner edge of the disk ($50R_g$). The disc does not break or tear. Credit: Nixon et al. (2012).

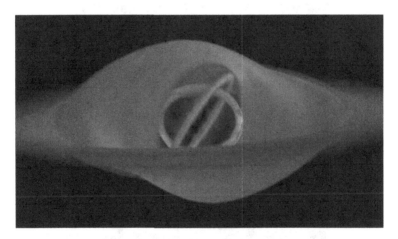

Figure 5.20 Full three-dimensional surface rendering of a simulation of a disc at high inclination with respect to the black hole spin. Here the disc *tears*, into rings which precess independently. Neighbouring rings can precess into relative counterrotatation and through viscous spreading cancel angular momentum and cause rapid infall. Here the whole disc was initially inclined at 60° to the hole with no warp. This snapshot is after approximately 500 dynamical times at the inner edge of the disk ($50R_g$). Credit: Nixon et al. (2012).

than usual (see Figure 5.22). It is easiest to find in SPH simulations because of their Lagrangian character, but it is now clearly verified in grid-based simulations (e.g. Liska et al., 2019, which includes both GR and MHD: see Figure 5.23).

Figure 5.21 Column density projection of a tearing disc structure. Credit: Nixon and Salvesen (2014).

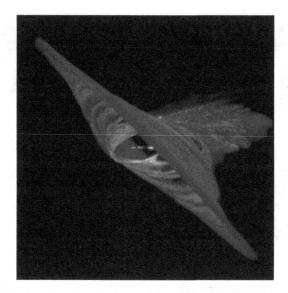

Figure 5.22 Disc tearing (here for accretion on to an SMBH binary; see Section 5.7) producing violent outflow. Credit: Nixon, King and Price (2013).

It is clear that disc tearing can alter the instantaneous accretion rate on to the central accretor (see Figure 5.22), but not its long-term average, which is set by the viscous timescale of the outer disc. A possible application (e.g. Raj & Nixon, 2021) is to 'changing-look' AGN, where the observational appearance of some AGN changes on timescales shorter than given by straightforward estimates of the standard local viscous timescale.

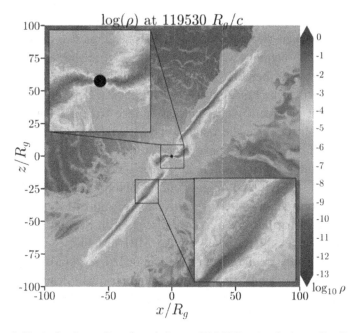

Figure 5.23 A broken disc found in a GRMHD simulation. Credit: Liska et al. (2019).

We have so far said little about wavelike warp propagation, which occurs for $\alpha < H/R$, that is, either when the viscosity is weak, or when the disc is relatively thick. Most of the work in this area is confined to the linear wavelike regime. Nixon and King (2016) show that in this case (with α formally set to zero) the tilt vector **l** obeys

$$\frac{\partial^2 \mathbf{l}}{\partial t^2} = \frac{1}{\Sigma R^3 \Omega} \frac{\partial}{\partial R} \left(\Sigma R^2 \frac{c_s^2}{4} \frac{\partial \mathbf{l}}{\partial R} \right), \tag{5.95}$$

where the wavelike character is clear.

As a warp wave propagates it can lead to either local or global bending of the disc. The disc's response to a warping event depends on its wavelength λ, which in turns depends on the scale of the disturbance causing the warp. At any disc radius R, the disc responds by warping if $H \ll \lambda < R$, but by bending as a whole, while remaining plane, if instead $\lambda > R$. In Nixon and Pringle (2010) a perturber passing the edge of a disc excites a warp of wavelength about $1/3$ of the outer radius. This produces a warped outer region, but as the wave moves inwards, the inner disc with $R < \lambda$ tilts as a whole. The outer region precesses globally, and the wave travel time across the entire disc is short enough that it is able to communicate the precession to the central bodily tilting region, making the entire disc process.

Many simulations show similar results (e.g. Larwood & Papaloizou, 1997; Fragile et al., 2007; Fragner & Nelson 2010). This requires two conditions: (a) the time ($\sim R/c_s$) for the bending waves to cross the disc is shorter than the precession period and (b) the bending waves are not damped locally, which requires $\alpha \ll H/R$. Numerical treatments may inadvertently satisfy both conditions and incorrectly predict persistent global precession (see Nixon & King, 2016). Condition (a) may hold because the outer computational boundary has been made artificially small for numerical convenience so that bending waves reflect there, instead of propagating outwards for far longer times and leaving a steady disc shape (see Lubow, Ogilvie & Pringle, 2002). Condition (b) reflects the fact that bending wave propagation is very sensitive to the ratio of α to H/R, and this may be distorted in various ways. As we discussed in Section 4.12, attempts to model the viscosity explicitly by using the MRI instability give values of the effective α significantly smaller than is likely in reality, so simulations that incorporate this viscosity mechanism may perhaps unrealistically leave bending waves undamped. Also, discs may be taken as thicker than realistic in order to increase vertical resolution, artificially making $H/R \gg \alpha$, and again suppressing wave damping.

We saw at the beginning of this section that the usual approach to implementing a viscosity for warped discs is the α parametrization, where the local stress is assumed proportional to the pressure. Then the horizontal and vertical components of the shear are damped by viscous dissipation at the same average rates. This is the origin of the term 'isotropic viscosity', but we saw that this did not imply that the effective viscosities (torque coefficients ν_1, ν_2, ν_3) are isotropic. Instead, the internal structure of a warp means that these torque coefficients take very different values that depend on the disc shear, on α, and on the warp amplitude. Angular momentum transport is mainly by Reynolds rather than viscous stresses, but this process is well described by a viscosity. Modelling turbulent fluids by means of effective viscosities has long been a common procedure in both astrophysics and fluid dynamics.

But it is not obvious that this local isotropy holds for a viscosity driven by MHD effects, typically due to the magnetorotational instability (MRI), and this question is unresolved. Pringle (1992) points out that the azimuthal shear is secular, but that the vertical shear is oscillatory. Gas parcels displaced by a small radial displacement ΔR drift increasingly further apart, but those at the same radius but displaced vertically by Δz simply oscillate. In a simplistic MRI picture it then seems likely that more energy is dissipated in the azimuthal viscosity, leading to a larger ν_1 torque. But the reduced dissipation in the vertical direction could allow a stronger resonant response and so a larger ν_2 torque. These opposing effects leave it unclear what the effect of including MHD in this picture would be.

If we instead assume that the MRI drives turbulence, and that the velocity field is uncorrelated on scales $\ll H$ (as seems likely from numerical simulations – see figure 14 in Simon, Beckwith, and Armitage, 2012), then the action of the turbulent viscosity seems likely to be the same whatever the direction of the shear on which it acts. This in turn suggests that assuming isotropic α dissipation would be correct. In particular, MRI-driven viscosity is likely to confirm the weakness of the vertical viscosity relative to horizontal, which is the basis of disc breaking and tearing.

Torkelsson et al. (2000) and Ogilvie (2003) investigate whether α dissipation is isotropic, by numerical and analytic techniques, respectively. Both conclude that the isotropic α assumption is valid for an MRI-turbulent disc. King et al. (2013) use observations of warped disc dynamics to constrain the problem, finding support for the isotropic picture.

The viscosity assumptions in the diffusive and wavelike cases are effectively the same. Nixon and King (2016) show that the term describing α damping of propagating waves (in their eqn 34) is exactly the ν_2 torque in the diffusive case. The physics of both torques is the same – for a Keplerian disc, α damping controls the resonant response to the radial pressure gradients forcing the gas at the local orbital frequency. When $\alpha < H/R$ (i.e. the diffusive case) the disc responds by propagating a wave that damps non-locally through α. For wavelike propagation with $\alpha > H/R$ we can instead think of this as the wave being damped locally, so acting diffusively.

5.7 SMBH Binaries and the Last Parsec Problem

As we have noted several times, mergers of galaxies are integral to the modern view of how structure forms in the Universe. An obvious question then is what happens to the central SMBHs when two galaxies merge. We saw in Section 1.4 that dynamical friction against the stars in a galaxy's central bulge is efficient in bringing supermassive black holes into the centre of a galaxy, and it is reasonable to think that this will hold when two galaxies merge. We would expect that the central bulges merge first, as they are the most massive entities in the problem, and the resulting configuration then drags the two SMBH into the merged bulge. In line with this, there is observational evidence that a small number of galaxies (e.g. NGC 6240; Komossa et al., 2003; Kollatschny et al., 2020) have double nuclei, suggesting the presence of two (or even three) SMBH. The scarcity of double nuclei suggests that pairs of SMBH in the nucleus of a merged galaxy themselves merge relatively quickly.

But dynamical friction works because massive bodies (here the two SMBH) give up angular momentum to stars. This makes the stars move in wider orbits, where they become less effective in absorbing angular momentum. Theoretical estimates (e.g. Milosavljević & Merritt 2003; Merritt 2013) suggest that without special assumptions, the two black holes in a typical galaxy merger eject so many stars from their close vicinity that dynamical friction is seriously weakened (this effect is called *loss cone depletion*). As the holes spiral inwards the number of stars remaining eventually becomes too low for dynamical friction to drive them to a small enough separation that gravitational radiation completes the merger process in a short time. The typical separation between the two SMBH when this occurs is $\sim$1 pc, so this apparent paradox is often called the *last parsec problem*.[10]

There have been several suggestions of how to solve this problem. Since the nub of it is to extract energy from the SMBH binary, these generally introduce extra matter close to this pair. One idea is to invoke a second merger, introducing a third SMBH. The orbits of the triple black hole system generally become chaotic, with the frequent outcome that one of the holes (usually the one of lowest mass, as for comparable kinetic energies it moves fastest) is ejected. This extracts angular momentum from the orbit of the remaining pair, which becomes tighter and can decay rapidly by emitting gravitational radiation.

Although this double merger idea may well work in some mergers, it is inherently improbable that all galaxy mergers are rapidly followed by a second one. This then leaves only the introduction of gas as a way to engineer the required orbital shrinking. This is plausible, as we know that SMBH do accrete most of their mass from gas, even if we do not yet fully understand how this process works (see Section 5.2). By the usual arguments leading to disc formation in any gas accretion event we expect the gas to orbit the SMBH binary as a disc, whose plane is set by its angular momentum.

In general this disc is not coplanar with the SMBH binary, but by analogy with the similar problem of a misaligned disc around a spinning black hole (see Section 5.3), we can expect that its viscous evolution must bring this about.

To see how this works, we initially consider for simplicity an SMBH binary with masses M_1 and $M_2 \ll M_1$. (We will treat the more complicated case of eccentric binaries later in this section.) We take a coordinate system where the binary angular momentum points along the z-axis of cylindrical polar coordinates (R, ϕ, z). Since $M_2 \ll M_1$, we place M_1 at the origin, with M_2 orbiting at radius a_b in the (R, ϕ) plane. The orbit has angular velocity

$$\Omega_b = \left(\frac{G(M_1 + M_2)}{a_b} \right)^{1/2} . \tag{5.96}$$

A disc particle orbiting the binary at radius $R \gg a_b$ (but not in its plane) would have a circular orbit with angular velocity $(GM_1/R)^{1/2}$ if both M_2/M_1 and a/R were zero,

---

[10] Or *final parsec problem*.

instead of simply being small. The non-zero values of these quantities introduces small perturbations from the circle. Some of the perturbations have inertial-frame frequencies $2\Omega_b$ and multiples of it and are oscillatory, giving no long-term cumulative ('secular') effects. These come instead from the azimuthally symmetric and time-independent term in the potential of the binary SMBH system. This term is given by replacing the orbiting mass M_2 by the same mass spread uniformly over its orbit, giving a ring of mass M_2 and radius a_b in the (R, ϕ) plane. Adding this to the fixed point mass M_1 gives the effective potential felt by a disc particle as

$$\Phi(R, z) = -\frac{GM_1}{(R^2 + z^2)^{1/2}} - \frac{GM_2}{2\pi} \int_0^{2\pi} \frac{d\phi}{r}, \tag{5.97}$$

where r is the distance between the particle position and a general point on the ring of mass so that

$$r^2 = R^2 + a^2 + z^2 - 2Ra_b \cos\phi. \tag{5.98}$$

Expanding (5.97) in powers up to second order of a_b/R and z/R gives

$$\Phi(R, z) = -\frac{G(M_1 + M_2)}{R} + \frac{GM_2 a_b^2}{4R^3} + \frac{G(M_1 + M_2)z^2}{2R^3} - \frac{9GM_2 a_b^2 z^2}{8R^5} + \dots . \tag{5.99}$$

The orbital and vertical oscillation frequencies of a particle in this potential are given by

$$\Omega^2 = \frac{1}{R}\frac{\partial\Phi}{\partial R} \tag{5.100}$$

and

$$\nu = \frac{\partial^2\Phi}{\partial\phi^2}, \tag{5.101}$$

both evaluated at $z = 0$. The nodal precession frequency is $\Omega_p = \Omega - \nu$, giving

$$\Omega_p = \frac{3}{4}\left[\frac{G(M_1 + M_2)}{R^3}\right]^{1/2} \frac{M_2}{M_1 + M_2}\frac{a_b^2}{R^2}. \tag{5.102}$$

This is very similar to that for Lense–Thirring precession around a spinning black hole, which goes as R^{-3}, rather than $R^{-7/2}$ as here. If the disc and binary axis are misaligned by an angle θ with $0 \le \theta \le \pi/2$, the precession frequency is simply multiplied by $\cos\theta$. The opposite case, with the disc partially counteraligned (i.e. $\theta > \pi/2$), is equivalent to the $\theta < \pi/2$ case with the binary angular momentum reversed. But this reversal leaves the precession frequency unchanged since we are dealing only with the $m = 0$ part of the potential. Then for all values of θ with $0 < \theta < \pi$ the precession frequency is $\Omega_p(\theta) = \Omega_p|\cos\theta|$. This result differs from the Lense–Thirring case, where the factor $\cos\theta$ appears without modulus signs.

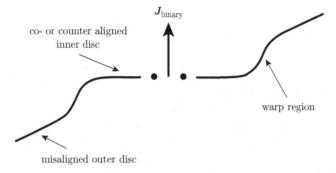

Figure 5.24 The warped disc shape expected after the inner disc co- or counteraligns with the binary plane, but the outer disc stays misaligned. Eventually the entire disc will co- or counteralign with the binary plane, depending on the global criterion (eqn (5.105)). In practice the precession makes the warp non-axisymmetric. Credit: Nixon, King and Pringle (2011).

The argument here shows that the binary potential makes the disc orbits precess. The precession is strongly differential, as rings of gas closer to the binary precess faster. This causes a dissipative torque between adjacent rings of gas, which makes θ tend to 0 or π so that the precession ultimately vanishes.

The precession timescale in the disc increases with radius (see (5.102)). The torque acts faster at small radii either co- or counteraligning the disc orbits with the binary plane. This creates a warp in the disc. The disc inside this is co- or counteraligned, but it is not yet aligned outside it (see Figure 5.24).

Now we can argue in a very similar way to Section 5.3, which discussed the alignment or counteralignment of a disc misaligned with respect to the spin of the black hole rather than with respect to the orbital angular momentum of a black hole binary. As the hole feels only precessions, angular momentum conservation is expressed by the analogue of (5.20), that is,

$$\frac{\mathrm{d}\boldsymbol{J}_b}{\mathrm{d}t} = -K_1[\boldsymbol{J}_b \times \boldsymbol{J}_d] - K_2[\boldsymbol{J}_b \times (\boldsymbol{J}_b \times \boldsymbol{J}_d)]. \tag{5.103}$$

Then just as for the case of LT alignment of a spinning black hole, the magnitude J_b of the binary angular momentum remains constant, while the direction of $\mathbf{J}_b$ aligns with the total angular momentum $\mathbf{J}_t = \mathbf{J}_b + \mathbf{J}_d$, which is of course a constant vector. During this process the magnitude of J_2 decreases because of dissipation in the disc. Counteralignment ($\theta \rightarrow \pi$) occurs if and only if $J_t^2 > J_b^2$. By the cosine theorem

$$J_t^2 = J_b^2 + J_d^2 - 2J_bJ_t \cos(\pi - \theta), \tag{5.104}$$

this requires

$$\cos\theta < -\frac{J_d}{2J_b}. \tag{5.105}$$

So counteralignment of a binary and an external disc is possible and requires

$$\theta > \pi/2, \; J_d < 2J_b. \tag{5.106}$$

Just as in the corresponding discussion of discs misaligned with respect to black hole spin (Section 5.3), we have to think of J_d as the angular momentum within the disc warp radius. In the 'mixed' case where we still have $\theta > \pi/2$ as in (5.106), but now $J_d > 2J_b$, we get the same phenomenon of discs apparently heading towards counteralignment but then ultimately aligning as the warp travels further out.

The typical timescale for co- or counteralignment for an SMBH binary is

$$t_{\text{binary}} \simeq \frac{J_b}{J_d(R_w)} \frac{R_w^2}{v_2}, \tag{5.107}$$

where R_w is the warp radius, $J_d(R_w)$ is the disc angular momentum within R_w, and v_2 is the vertical disc viscosity. This is identical to the formal expression for LT alignment of a spinning black hole if we replace the spin angular momentum J_h with J_b (see Scheuer & Feiler, 1996). The warp radius is given by equating the precession time $1/\Omega_p(R)$ to the vertical viscous time R^2/v_2. Inside this radius the precession timescale is short and the disc dissipates and co- or counteraligns with the binary plane. Outside this radius the disc is not dominated by the precession and so maintains its misaligned plane. The connecting region therefore takes on a warped shape, as sketched in Figure 5.24, where the precession makes the warp non-axisymmetric. The warp propagates outwards and eventually co- or counter-aligns the entire disc with the plane of the binary. Approximating the disc angular momentum as

$$J_d(Rw) \sim \pi R_w^2 \Sigma (GMR_w)^{1/2}, \tag{5.108}$$

with Σ the disc surface density and $M = M_1 + M_2$, and using the steady-state disc relation $\dot{M} = 3\pi v \Sigma$, we find

$$t_{\text{binary}} \sim 3\frac{M_2}{M_1}\left(\frac{a_b}{R_w}\right)^{1/2}\frac{v_1}{v_2}\frac{M}{\dot{M}}, \tag{5.109}$$

since

$$J_b = M_1 M_2 \left(\frac{Ga_b}{M}\right)^{1/2}. \tag{5.110}$$

Since $v_1 < v_2, a_b \ll R_w$ and $M_2 < M_1$, we see that alignment takes place on a timescale shorter than the mass growth of the central accretor(s). The timescale (5.109) is directly analogous to the expression

$$t_{LT} \sim 3a \frac{M_2}{M_1} \left(\frac{R_g}{R_w} \right)^{1/2} \frac{v_1}{v_2} \frac{M}{\dot{M}} \qquad (5.111)$$

for spin alignment of a disc via LT precession, where $a < 1$ is the Kerr spin parameter and R_g the gravitational radius of the spinning hole. Evaluating R_w in the two cases, we find

$$\frac{t_{LT}}{t_{binary}} \sim \frac{3^{1/2}}{2} \left(\frac{aM_1}{M_2} \right)^{1/2} \left(\frac{a_b}{R_g} \right)^{1/4}. \qquad (5.112)$$

Then in general, provided we assume that the ratio v_1/v_2 is similar in the two cases and that the hole spin is not small enough to make $a < (R_g/a_b)^{1/2}(M_2/M_1)$, the binary-disc alignment is rather faster than the corresponding process for aligning a disc with the black hole spin.

This result has significant consequences for SMBH binaries. For random orientations, (5.106) shows that initial disc angles leading to alignment occur significantly more frequently than those giving counteralignment only if $J_d > 2J_b$. In the LT case this fact leads to a slow spin-down of the hole because retrograde accretion has a larger effect on the spin, as we saw in Section 5.4 – see Figure (5.10).

We will see shortly (see the discussion after (5.113) that prograde external discs are rather inefficient in shrinking SMBH binaries and so solving the last parsec problem. This is because of resonances within the disc. In contrast, the slightly rarer retrograde events have a much stronger effect on the binary. They rapidly produce a counterrotating but coplanar accretion disc outside the binary, which has no resonances. We will see that the binary gradually increases its eccentricity as it captures negative angular momentum from the disc, ultimately coalescing once this cancels its own. A non-zero binary eccentricity changes the detailed form of the perturbing potential from that in (5.99), but cannot change the precessional character leading to the torque equation (5.107). The results remain unchanged, particularly the counteralignment condition (5.106), apart from minor modifications of the timescale t_{binary} (cf. (5.109)).

So in a random sequence of accretion events producing circumbinary discs, the prograde events have little effect, and the retrograde ones shrink the binary. In particular, a sequence of minor retrograde events with $J_d < J_b$ has a cumulative effect and must ultimately cause the binary to coalesce once the total accreted retrograde angular momentum exceeds J_b. This is important since the disc mass is limited by self-gravity to $M_d \sim (H/R)M_1$ (eqn (4.97)). The binary must be close to coalescence once the retrograde discs have supplied a total mass M_2, that is, once a total number $n \sim (M_2/M_1)(R/H)$ of retrograde discs have accreted. For minor mergers this requires at most a few randomly oriented accretion disc events, rising to

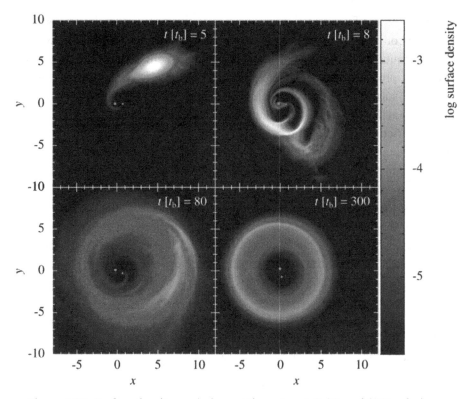

Figure 5.25 Surface density renderings at times $t = 5, 8, 80$, and $300t_b$, during a prograde SPH simulation showing the formation and evolution of a circumbinary disc around an SMBH binary with period $t_b = 3 \times 10^4$ yr. The x and y coordinates are in units of the initial binary separation $a_b = 1$ pc, and surface density is in units Mb_a^{-2}, where M_b is the total binary mass $10^7 M_\odot$. The top left-hand panel shows the cloud on its initial approach to the binary. The top right-hand panel shows the cloud as it interacts with the binary, with tidal streams forming during the cloud's passage. The bottom left-hand panel shows the formation of the disc, as the gas shocks and begins to follow more regular orbits around the binary, and the bottom right-hand panel shows the fully formed circumbinary disc. Credit: Dunhill et al. (2014).

a few hundred for major mergers ($M_2/M_1 > 0.1$). This agrees with the conclusion of Dunhill et al. (2014) (see Figure 5.25), who considered circumbinary discs of various thicknesses and concluded that the longer lifetime of low-mass retrograde discs limited by self-gravity allows them to drive significant binary evolution towards coalescence. (We should note that if there is significant mass lost from the circumbinary discs in forming stars, these themselves contribute to angular momentum loss from the SMBH binary through dynamical friction.)

The effects of prograde and retrograde circumbinary discs are very different. We have seen that the timescale to achieve full co- or counteralignment is short compared with the full viscous evolution, so we compare the two cases, assuming full co- or counteralignment has already occurred and the system is now plane. Viscous torques try to share angular momentum in both prograde and retrograde cases. For a prograde disc the interaction shrinks the binary, but moves the inner edge of the surrounding disc outwards, reducing the torque trying to shrink the binary. This is the gas analogue of the loss-cone depletion which eventually makes dynamical friction ineffective in coalescing SMBH binaries.

A prograde circumbinary disc rapidly becomes a *decretion* disc, simply transporting angular momentum outwards with little net mass transport (Lodato et al., 2009). A retrograde circumbinary disc instead remains an accretion disc, transporting angular momentum outwards and mass inwards. The long-term evolution of the disc-binary system is radically different in the two cases.

First, the disc-binary torque is quite different in the two cases. A prograde disc interacts tidally mainly through resonances, which occur when the condition

$$\Omega^2 = m^2(\Omega - \Omega_b)^2 \tag{5.113}$$

holds. Here Ω_b is the binary orbital frequency, $\Omega(R)$ is the disc angular velocity, and $m = 1, 2, \ldots$ is the wave mode number (see, e.g., Papaloizou & Pringle, 1977). It follows that resonances occur at radii where

$$\Omega(R) = \frac{m\Omega_b}{m \pm 1}, \tag{5.114}$$

where Ω and Ω_b have the same sign. Resonances outside the binary orbit must have $\Omega < \Omega_b$ and correspond to the positive sign in the denominator of (5.114). Then these appear at radii where

$$\frac{\Omega(R)}{\Omega_b} = \frac{1}{2}, \frac{2}{3}, \frac{3}{4}, \ldots \tag{5.115}$$

and the dominant interaction between the binary and the disc comes from the 1:2 resonance. The exterior disc gains angular momentum from the binary and so tends to spread outwards, particularly at its inner edge, while the binary shrinks only marginally. Just as in the case of tidal friction, this process falls victim to its own success in that ultimately the inner edge of the circumbinary disc retreats to radii where there is little possibility of shrinking the binary further. The tidal interaction has effectively dammed up the inflow of the disc gas.

In a retrograde disc Ω and Ω_b have *opposite* signs, so from (5.113) there are no resonances in a circumbinary (i.e. $|\Omega| < |\Omega_b|$) retrograde disc. The disc-binary interaction is direct and occurs once the inner edge of the disc gets close enough to the secondary black hole to feel a significant effect from its gravitation. So a retrograde circumbinary disc remains an accretion disc, but one whose material and angular momentum are gravitationally captured by one of the black holes in the binary. Because this gas was rotating in the opposite sense to the binary, it reduces the binary's angular momentum. This is inherently more promising for shrinking the binary towards coalescence than the prograde case, where the binary dams up the disc.

An important point here is that 'capture' means only that the gas orbits one of the binary holes and so has added its angular momentum – which is negative – to the binary orbit. It is not necessary for the relevant hole to actually accrete this gas (although it may). After this capture into a bound orbit about one of the holes, we can regard that hole plus its orbiting gas as a single body. This is still true if some or all of the captured gas is expelled in some way, for example, by radiation pressure, provided that the process is isotropic in the frame of the orbiting hole. Then there is no change of its orbital specific angular momentum and no effect on the orbital dynamics and eventual coalescence of the binary (see (5.133)).

A retrograde circumbinary disc can have detailed effects depending on how the captured mass is distributed between the two black holes, as this changes the binary mass ratio. So we need to look at the reaction of test particles to the binary. For simplicity we first consider a circular binary with a low mass ratio, that is, $M_2/M_1 \lesssim 0.1$. Then the primary is effectively fixed and the secondary in a circular orbit of radius a around it with velocity $V = (GM_1/a)^{1/2}$. The circumbinary disc gas has circular orbits with velocity $\simeq (GM_1/R)^{1/2}$ at radius R. The disc spreads slowly inwards through viscous evolution, and its counterrotation means there are no resonant effects. Fluid effects do not begin to appear until orbits begin to cross, so we can treat the gas as ballistic. The holes have effective radii R_1, R_2 for capturing gas particles, for example, into a disc around one or other of them. The captured gas then has the same net specific angular momentum as the hole it orbits. Clearly R_1, R_2 cannot in practice be larger than the individual Roche lobe radii for each hole, where centrifugal terms begin to overcome gravity.

Disc particles orbiting close to M_2 initially interact hyperbolically, and so we can use the impulse approximation. Since the relative velocity is approximately $2V$, a particle at the circumbinary disc edge at radius $R = a + b$, where $b \ll a$, acquires an inward radial velocity

$$U_R = \left(\frac{GM_2}{b^2}\right) \times \frac{2b}{2V} = \frac{GM_2}{bV}. \tag{5.116}$$

This shows that the inner edge of the circumbinary disc is not significantly perturbed if its distance b from the outer (secondary) black hole's orbit is large enough that $U_R \lesssim V$, that is, $V^2 \gtrsim GM_2/b$, or equivalently $b \gtrsim (M_2/M_1)a$. This tells us that the secondary black hole pulls gas from the inner edge of the disc at b if $R_2 \gtrsim b$, or

$$\frac{R_2}{a} \gtrsim \frac{M_2}{M_1}. \tag{5.117}$$

The Roche lobe radius for the secondary black hole is

$$\frac{R_2}{a} \lesssim 0.4f \left(\frac{M_2}{M_1}\right)^{1/3}, \tag{5.118}$$

where $f \lesssim 1$ is a dimensionless factor. So the secondary hole captures most of the gas, provided that the mass ratio q satisfies

$$q = \frac{M_2}{M_1} \lesssim q_{\text{crit}} = 0.25f^{3/2} \simeq 0.25. \tag{5.119}$$

For mass ratios larger than this the gas at the inner disc edge feels the gravity of both holes. The flow becomes more complex and we can expect some gas to fall towards the binary centre of mass and be captured by the primary. Numerical simulations show that even in this case most of the gas is captured by the secondary. Evidently the primary only captures a non-negligible amount from a retrograde circumbinary disc in a major merger with $q \gtrsim q_{\text{crit}} \sim 0.25$.

A retrograde circumbinary disc can reduce both the energy and angular momentum of the SMBH binary, which must change its eccentricity e. It is straightforward to understand this process. The binary orbit is shrinking, so it must always have at least a slight eccentricity. For simplicity we assume a mass ratio sufficiently extreme that the primary black hole is effectively fixed at the binary centre of mass and only the distant secondary interacts with the disc. Capture of any of the retrograde disc gas always reduces the secondary's orbital velocity through momentum conservation. This is trivially true for a direct collision and also obvious in the more common case that the secondary hole captures gas into a bound disc around itself. The secondary's mass does not decrease in these interactions, so its specific orbital energy always becomes more negative, and the binary semi-major axis a must decrease. Since the mass capture involves a point interaction, the new orbit must pass through the same point, again with zero radial velocity, so the quantity $a(1+e)$ must remain fixed. Given the decrease in a, this shows that the eccentricity e increases as

$$e \sim \frac{\Delta M}{M_2}, \tag{5.120}$$

where ΔM is the captured mass. Similarly, for mass capture near pericentre, $a(1-e)$ stays fixed despite a further drop in a. Then the eccentricity must *decrease* here by about the same amount ($\Delta M/M_2$) that it increased at apocentre.

The secondary SMBH is always closer to the disc than the primary, so captures disc gas at all points between apo- and pericentre. This means that the effects on the eccentricity are opposed for significant times. If the eccentricity is initially small these times are nearly equal. Then e stays small as the orbit shrinks, provided that $\Delta M/M_2$ is small enough.

If instead the orbit is initially quite eccentric, or the mass grows significantly in one orbit, the pericentre may be so small that there is no interaction with the disc. Then e grows, while keeping $a(1 + e) = a_0 \simeq$ constant. The pericentre distance goes as

$$p = a(1 - e) = 2a - a_0. \tag{5.121}$$

Then the binary coalesces once the original semi-major axis has halved. We will see (eqn (5.133)) that this happens once the secondary has gravitationally captured total disc gas with mass comparable to its own. If the inward gas flow rate through the circumbinary disc is $\dot{M}$ then the binary must be close to coalescence after a time

$$t_{\text{co}} \simeq \frac{M_2}{\dot{M}}. \tag{5.122}$$

Clearly, there must be a critical eccentricity separating cases where the binary remains almost circular from those where the eccentricity grows. This depends on the surface density distribution of the circumbinary disc. At the edge of a disc with scaleheight H, and so aspect ratio H/R, the surface density drops off radially over a lengthscale $H(a) \sim (H/R)a$. The critical eccentricity dividing the near-circular and growing-eccentricity cases is evidently

$$e_{\text{crit}} \sim \frac{H}{R}. \tag{5.123}$$

A black hole binary with $e > e_{\text{crit}}$, or reaching it by capturing a large mass (comparable to the secondary's) in one orbit, must become strongly eccentric. Even a preceding episode of *prograde* accretion can leave the binary with an eccentricity exceeding e_{crit} (e.g. Cuadra et al., 2009) so that growth to high eccentricity is very likely if accretion is chaotic.

This makes it easy to find analytic estimates of the orbital evolution of a binary with an external disc, by considering again the simple case $q = M_2/M_1 \ll 1$. The secondary has specific orbital energy and angular momentum E, J^2, with

$$E = -\frac{GM_1}{2a} = \frac{1}{2}v(R)^2 - \frac{GM_1}{R} \tag{5.124}$$

and

$$J^2 = GM_1 a(1 - e^2). \tag{5.125}$$

We have seen that binary eccentricity often grows quite strongly. In the limiting case the secondary interacts with the disc only very near apocentre $r = a(1 + e)$.

Its velocity v_{ap} is purely azimuthal here, with

$$[a(1 + e)v_{ap}]^2 = J^2. \tag{5.126}$$

Using (5.125) gives

$$v_{ap}^2 = \frac{GM_1}{a} \frac{1 - e}{1 + e}, \tag{5.127}$$

Near apocentre the secondary hole interacts with disc material moving with azimuthal velocity $v_{disc} < 0$, where

$$v_{disc}^2 = \frac{GM_1}{R} = \frac{GM_1}{a(1 + e)}. \tag{5.128}$$

If a mass ΔM of disc matter is captured into orbit about the secondary near apocentre, all of its orbital angular momentum is transferred to the secondary, but there may be mass loss from the subsequent accretion process on to the secondary hole (if, e.g., this is super-Eddington, or mass interacts gravitationally with the secondary but ultimately all accretes to the primary). We can allow for this by assuming that the effective mass of the hole plus disc becomes $M_2 + \alpha \Delta M$ with $0 \leq \alpha \leq 1$. Conservation of linear momentum then implies

$$M_2 v_{ap} - \Delta M v_{disc} = (M_2 + \alpha \Delta M)u, \tag{5.129}$$

where u is the new apocentre velocity of the secondary plus its captured gas disc. The changes $\Delta E, \Delta a$ in orbital specific energy and semi-major axis follow from

$$\frac{GM_1}{2a^2} \Delta a = \Delta E = \frac{1}{2}u^2 - \frac{1}{2}v_{ap}^2. \tag{5.130}$$

Combining (5.127), (5.128), (5.129), (5.130) gives

$$\frac{\Delta a}{a^2} = \frac{-2}{a(1 + e)} \frac{\Delta M}{M_2}[(1 - e)^{1/2} + \alpha(1 - e)] \tag{5.131}$$

to lowest order in ΔM. If the interaction is very close to apocentre, the latter remains fixed, with $a_0 = a(1 + e)$ remaining constant. Then $1 + e = a_0/a$ and $1 - e = 2 - a_0/a$ so that the left-hand term of (5.131) is simply $\Delta(1 - e)/a_0$, and (5.131) becomes

$$\Delta(1 - e) = -\frac{2\Delta M}{M_2}[(1 - e)^{1/2} + \alpha(1 - e)]. \tag{5.132}$$

Using $M_2 = M_{20} + \alpha M$, where M_{20} was the mass of the secondary hole when e was zero, and M is the total mass since transferred from the disc, this integrates to give

$$1 - e = \left(\frac{M_{20} - M}{M_{20} + \alpha M}\right)^2. \tag{5.133}$$

So in this approximation the binary coalesces (i.e. $1 - e = 0$) once the disc has transferred a mass equal to the secondary black hole's original mass, that is, after a time t_{co} (see (5.122)), and this is *independent* of the fraction α. Loss of any of the transferred mass has no effect in slowing the inspiral. We see from (5.131) that the energy dissipated in the shrinkage of an eccentric binary is

$$- M_2 \Delta E \simeq \frac{GM_1 \Delta M}{2a} \qquad (5.134)$$

per binary orbit. This is less than produced by viscous dissipation in the disc, which pulls inwards a mass larger than ΔM on each binary orbit.

All other cases give similar coalescence timescales $\sim t_{\mathrm{co}}$. If, for example, the orbit stays circular we can set $e = 0$ in (5.131) and find

$$\frac{\Delta a}{a} = -2(1 + \alpha)\frac{\Delta M}{M_2}, \qquad (5.135)$$

so that $a \propto M_2^{-2(1+\alpha)}$. If the secondary gains the transferred mass we have $\alpha = 1, a \propto M_2^{-4}$, while if the transferred mass (but not its angular momentum) ends up on the primary we have $\alpha = 0, a \propto M_2^{-2}$. Shrinking the binary from $a = 1$ pc to $a = 10^{-2}$ pc, where gravitational radiation rapidly coalesces it, requires the transfer of between two and nine times the original mass of the secondary in the two cases. This is larger than in the eccentric case because the torque on the binary now decreases with a. In fact the required mass would be formally infinite without gravitational radiation. Figures 5.26 and 5.27 show SPH simulations of external gas discs of various configurations.

We have so far considered for simplicity only initially circular binary SMBH orbits. But dynamical studies (e.g. Aarseth, 2003; Khan, Andreas & Merritt, 2011; Wang et al., 2014) of galaxy mergers suggest that high eccentricities are likely in many cases, and we have already seen that accretion from a retrograde circumbinary disc makes the binary eccentric. This makes the time-averaged binary potential triaxial rather than axisymmetric, and it is known that misaligned discs can precess about both the major and minor axes in these cases (Steiman-Cameron & Durisen, 1984; Thomas, Vine, & Pearce, 1994).

To treat an eccentric SMBH binary we expand its potential to quadrupole order and consider a circular circumbinary gas disc. These approximations are reasonable if the disc is far enough from the binary. We average in time over the binary orbit, which is justified if resonances are unimportant (as we have seen, this is certainly true for retrograde discs) and assume that dissipation is negligible. We write $\mathbf{R} = \mathbf{x}_1 - \mathbf{x}_2$ as the instantaneous separation vector, where $\mathbf{x}_{1,2}$ are the position vectors of the two masses $M_1, M_2 = q < M_1$, with $M_1 + M_2 = M$, the binary-specific angular momentum vector is

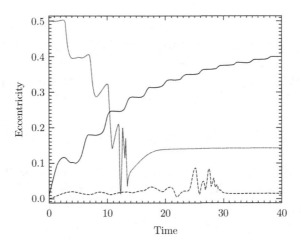

Figure 5.26 Eccentricity growth and decay in SPH simulations of SMBH binaries with external gas discs under various conditions. The solid line describes a case with an initially circular (separation a) black hole binary with mass ratio $q = M_2/M_1 = 10^{-3}$, where the low-mass hole is initially just inside the inner edge of a retrograde circumbinary disc of mass $M_d = 10^{-2}M_1$ which spreads over radii $0.8a$ to $1.5a$. The dashed curve is the same, except that the initial inner disc radius is $0.1a$. The dotted curve describes a simulation which starts with a retrograde disc interior to the binary, that is, a circumprimary disc extending from $0.1a$ to $1.0a$. This starts with initial binary eccentricity 0.5 and the binary components at apocentre. So the secondary begins outside the circumprimary disc but plunges into it before it reaches pericentre. All of these simulations have capture radii $0.1a, 0.01a$ for the primary and secondary black hole, respectively. The first simulation (solid curve) captures mass at apocentre but not at pericentre. The second simulation (dashed curve) captures at both pericentre and apocentre and the third simulation (dotted curve) captures at pericentre but not at apocentre) Credit: Nixon et al. (2011).

$$\mathbf{h} = \frac{q}{(1+q)^2}\mathbf{R} \times \dot{\mathbf{R}}, \tag{5.136}$$

and the eccentricity vector is

$$\mathbf{e} = \frac{\dot{\mathbf{R}} \times (\mathbf{R} \times \dot{\mathbf{R}})}{GM} - \hat{\mathbf{R}}, \tag{5.137}$$

where $\hat{\mathbf{R}}$ denotes the unit vector. The vector $\mathbf{e}$ points from the centre of mass to pericentre and is orthogonal to $\mathbf{h}$. Its magnitude is the eccentricity e, and Kepler's laws give the connection

$$h^2(1+q)^4 = q^2 GMa(1 - e^2), \tag{5.138}$$

where a is the semi-major axis.

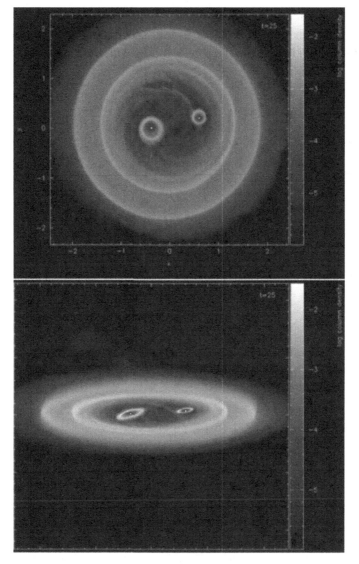

Figure 5.27 Gas capture from a circumbinary disc in a case where the initial binary masses are in the ratio $q = 0.5$, so that both black holes gain gas. The units of length and time are the initial binary separation, and the initial binary period is 2π. The upper panel shows the system face on, and the lower panel shows it viewed at $15°$ to the orbital plane. The discs are not flat and are significantly tilted with respect to the binary plane. This is evidently because mass is accreted in finite amounts and this 'particle noise' removes the symmetry about the orbital axis. Globally mass and angular momentum are conserved but locally the streams supplying the discs need not be plane. Credit: Nixon et al. (2011).

To treat the evolution of the disc-binary system, we again idealize the disc as a gaseous ring of mass m and radius r, with specific angular momentum vector $\mathbf{l} = [G(M + m)r]^{1/2}\hat{\mathbf{l}}$. To avoid the ring colliding with the binary, it must obey

$$r > \frac{a(1 + e)}{1 + q}, \tag{5.139}$$

which is also required for the quadrupole approximation to hold. The tilt angle of the ring to the binary is θ, where $\cos\theta = \hat{\mathbf{l}} \cdot \hat{\mathbf{h}}$.

The quadrupole interaction energy between binary and ring, averaged over both the binary orbit and the ring, is

$$\langle E_{\mathrm{br}} \rangle = -\frac{m\omega^2 a^2 q}{8(1 + q)^2}[6e^2 - 1 - 15e^2(\hat{\mathbf{l}} \cdot \hat{\mathbf{e}})^2 + 3(1 - e^2)(\hat{\mathbf{l}} \cdot \hat{\mathbf{h}})^2], \tag{5.140}$$

where $\omega = [G(M + m)/r^3]^{1/2}$ is the ring's orbital frequency. The time-averaged torque of the binary on the ring is

$$\dot{\mathbf{l}} = \Theta \times \mathbf{l}, \tag{5.141}$$

with

$$\Theta = \frac{\omega q}{4(1 + q)^2}\frac{a^2}{r^2}[5e^2(\hat{\mathbf{l}} \cdot \hat{\mathbf{e}})\hat{\mathbf{e}} - (1 - e^2)(\hat{\mathbf{l}} \cdot \hat{\mathbf{h}})\hat{\mathbf{h}}]. \tag{5.142}$$

From (5.141) we have $\mathbf{l} \cdot \dot{\mathbf{l}} = 0$, which means $\dot{l} = \dot{r} = 0$, so the ring simply precesses. Since $\Theta \propto \partial\langle E_{\mathrm{br}} \rangle / \partial\hat{\mathbf{l}}$, (5.140) implies

$$\frac{d\hat{\mathbf{l}}}{dt}\frac{\partial\langle E_{\mathrm{br}} \rangle}{\partial\hat{\mathbf{l}}} = 0, \tag{5.143}$$

so only angular momentum, and no energy, is exchanged between the binary and the ring.

The torque of the ring back on the binary, averaged over an orbit, is

$$\dot{\mathbf{h}} = -\frac{m}{M}\Theta \times \mathbf{l}, \tag{5.144}$$

so the total angular momentum of the ring–binary system is conserved. For the earlier case of a circular binary we have Θ parallel to $\mathbf{h}$ so that $\dot{\mathbf{h}}.\mathbf{h} = 0$, so $h = |\mathbf{h}|$ is conserved, and the binary simply precesses, as we saw, with an amplitude smaller than the disc by the ratio of their masses m/M.

Instead, for an eccentric binary $\mathbf{h}$ does not simply precess, and h changes over time. We find

$$\dot{h} = -\frac{15}{4}\frac{\omega^2 m}{\Omega M}\frac{e^2 h}{(1 - e^2)^{1/2}}(\hat{\mathbf{l}} \cdot \hat{\mathbf{e}})(\hat{\mathbf{l}} \cdot \hat{\mathbf{k}}), \tag{5.145}$$

where $\hat{\mathbf{k}} = \hat{\mathbf{h}} \times \hat{\mathbf{e}}$, and $\Omega = (GM/a^3)^{1/2}$ is the binary orbital frequency. So h is constant only if $e = 0$ (i.e. a circular binary), or if $\hat{\mathbf{l}}$ is orthogonal to either $\hat{\mathbf{e}}$ or $\hat{\mathbf{k}}$ so that one of $\hat{\mathbf{e}}, \hat{\mathbf{k}}$ is in the plane of the ring. In all other cases, h oscillates because $\hat{\mathbf{l}} \cdot \hat{\mathbf{k}}$ oscillates around zero as the ring precesses. The eccentricity vector changes as

$$\dot{\mathbf{e}} = \frac{3}{4} \frac{\omega^2 m}{\Omega M} e(1-e^2)^{1/2} \{[2-(\hat{\mathbf{l}} \cdot \hat{\mathbf{h}})^2 - 5(\hat{\mathbf{l}} \cdot \hat{\mathbf{e}})^2]\hat{\mathbf{k}} + (\hat{\mathbf{l}} \cdot \hat{\mathbf{k}})(\hat{\mathbf{l}} \cdot \hat{\mathbf{h}})\hat{\mathbf{h}} + 5(\hat{\mathbf{l}} \cdot \hat{\mathbf{e}})(\hat{\mathbf{l}} \cdot \hat{\mathbf{k}})\hat{\mathbf{e}}\}, \quad (5.146)$$

while the eccentricity itself changes as

$$\dot{e} = \frac{15}{4} \frac{\omega^2 m}{\Omega M} e(1-e^2)^{1/2}(\hat{\mathbf{l}} \cdot \hat{\mathbf{e}})(\hat{\mathbf{l}} \cdot \hat{\mathbf{k}})^2 \quad (5.147)$$

and the binary undergoes apsidal precession at the rate

$$\dot{\hat{\mathbf{e}}} \cdot \hat{\mathbf{k}} = \frac{3}{4} \frac{\omega^2 m}{\Omega M} e(1-e^2)^{1/2}[2 - (\hat{\mathbf{l}} \cdot \hat{\mathbf{h}})\hat{\mathbf{h}} - 5(\hat{\mathbf{l}} \cdot \hat{\mathbf{e}})^2]. \quad (5.148)$$

This is prograde for near-plane disc orientations ($|\hat{\mathbf{l}} \cdot \hat{\mathbf{h}}| \sim 1$), but retrograde for near-polar discs ($|\hat{\mathbf{l}} \cdot \hat{\mathbf{e}}| \sim 1$). For $m \ll M$ the binary orientation barely changes, or does so much more slowly than the ring.

Dissipation in viscous disc accretion damps all the precessions, so our analysis shows that a disc around an eccentric SMBH binary is likely to settle into one of four orientations. The disc may be aligned or counteraligned with respect to the binary or in a polar orbit around the binary eccentricity vector with either sense of rotation. As we would expect from the earlier analysis of circular SMBH binaries, the aligned/counteralignment orientation is favoured at low binary eccentricities. The analytic expression for the fraction of polar orbits is

$$f_{\text{polar}} = \frac{1}{\pi} \arccos \frac{1 - 6e^2}{1 + 4e^2} \quad (5.149)$$

(see Figure 5.32), so for higher eccentricities polar alignment is likely.

SPH simulations agree with these conclusions (see Figures 5.29–5.33) and show that the interaction of gas with the binary potential can be very violent. Disc breaking and tearing cause large amounts of gas to lose most of its angular momentum and fall rapidly inwards. Accretion from a polar disc does not promote SMBH binary coalescence and by itself solve the final parsec problem. But this infalling low-angular momentum gas is likely to be violently ejected via gravitational sling-shot (see Figure 5.22) and extract energy from the binary, reducing its separation. Moreover, this process acts on dynamical rather than viscous timescales and so is much faster. It is probably fair to conclude that infall of gas on to SMBH binaries tends to lead to a sufficiently messy and complex situation that in most cases the binary is likely to coalesce fairly quickly.

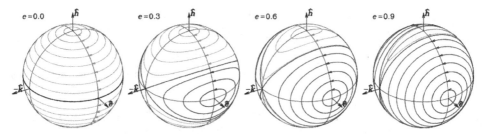

Figure 5.28 Precession paths for the direction $\hat{\mathbf{l}}$ of the angular momentum of a dissipationless circular ring of negligible mass orbiting a binary with eccentricity e as indicated. The binary orbits counterclockwise in the plane perpendicular to the ring's specific angular momentum $\mathbf{h}$ with periapse in the direction $\hat{\mathbf{e}}$. For a circular binary ($e = 0$, left), $\hat{\mathbf{l}}$ always precesses around $\hat{\mathbf{e}}$ in a retrograde sense. For eccentric binaries, prograde polar precession around $\hat{\mathbf{e}}$, the long axis of the time-averaged binary potential, is also possible. The regions of polar and azimuthal precession are separated by two great circles. The four ring orientations and $\hat{\mathbf{l}} = \pm\hat{\mathbf{e}}$ are stable (non-precessing), while the orientations $\hat{\mathbf{l}} = \pm\hat{\mathbf{k}}$ are unstable. Dissipation would damp the precession and eventually align the ring with one of the four stable orientations. In the case of a massive ring, the binary orbit evolves too: the vectors $\mathbf{h}$ and $\mathbf{e}$ oscillate and precess, and $\mathbf{e}$ and $\hat{\mathbf{k}}$ rotate around $\mathbf{h}$. Credit: Aly et al. (2015).

We invoked the presence of a circumbinary disc only indirectly from observation, from the need to get the SMBH involved in galaxy mergers to coalesce rapidly themselves, as dual galaxy nuclei are rare. Direct observation of the gravitational-wave (GW) emission from the mergers themselves is the major target for the long-planned LISA experiment. But if a merging SMBH binary does have a circumbinary gas disc, this potentially offers the possibility of electromagnetic signals associated with the merger (see de Mink & King, 2017 for stellar-mass black hole mergers). In the GW event, a significant fraction of the merging SMBH mass must be lost and this instantaneous decrease must have major effects on a surrounding gas disc. The gas orbits are suddenly considerably less bound to the merged hole than they were to the SMBH binary. Their energies and angular momenta are now no longer those of circular orbits around the new smaller central mass – they become eccentric and collide. Further, the GW reaction imparts a kick velocity to the merged hole, and if this is close to the pre-merger orbital plane it is clear that the dissipative effects are far higher (see Figures 5.35–5.37).

To quantify this we define $R_V = GM/V^2$ as the disc radius where the magnitude of the GW kick velocity V is equal to the Kepler value. Early estimates assumed on dimensional grounds that the electromagnetic energy release from the kick should be $\sim (1/2)V^2 \Sigma_R R_V^2$, where Σ_V is the disc surface density at R_V.

But as we can see from Figure 5.34, this is a severe underestimate for all kicks at angles $\theta < 50°$ from the disc plane.

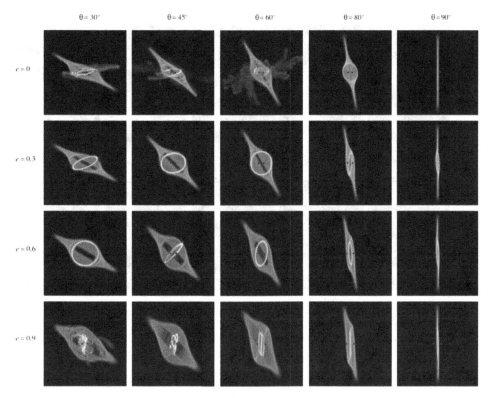

Figure 5.29 Density rendering of SPH simulations of a gas disc around a binary system after about 95 binary orbits, for different binary eccentricities and initial misalignment angles, as indicated. The projections are along the intermediate binary axis **k** with the angular momentum vector **h** pointing upwards and the eccentricity vector **e** to the right. Credit: Aly et al. (2015).

5.8 Tidal Disruption Events

The idea that the strong tidal gravity field of a supermassive black hole might disrupt a passing star has a long history; see Frank and Rees (1976). Carter and Luminet (1982) gave the first theoretical treatment. Rees (1988) (see also Phinney, 1989) showed that the electromagnetic flare produced as the stellar debris is swallowed by the hole had a characteristic time dependence and might be used as a tracer of SMBHs in nearby galaxies which were not otherwise active. Komossa and Bade (1999) identified the first likely candidates in the form of giant outbursts from the galaxies NGC 5905 and IC 3599. Since then observations of likely tidal disruption events (TDEs) have accumulated rapidly.

A TDE occurs when a star of mass and radius M_2, R_2, initially at a large distance from a supermassive black hole, falls on a near-parabolic orbit about the black hole whose pericentric distance R_p is close to its tidal radius

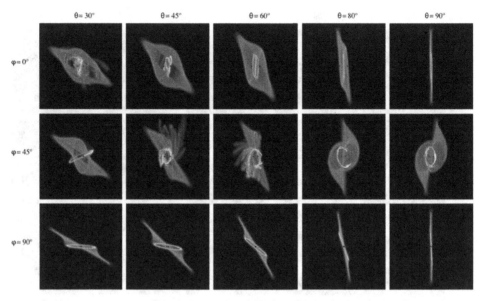

Figure 5.30 Density rendering of gas disc simulations (see Figure 5.28) after about 95 orbits of a binary with eccentricity $e = 0.9$, projected on the x, z plane with different values for ϕ and θ, as indicated in the figure. Credit: Aly et al. (2015).

$$R_T \simeq \left(\frac{M}{M_2} \right)^{1/3} R_2, \tag{5.150}$$

where the gravity of the SMBH overcomes the star's own self-gravity. This condition implies a maximum black hole mass M for a given stellar structure M_2, R_2. The tidal radius R_T must be larger than the ISCO; otherwise the star is simply swallowed without disruption by falling into the black hole from the ISCO before it fills its tidal lobe. This requires $R_T \gtrsim \lambda R_g$, so we must have

$$M \lesssim \left[\frac{c^6 R_2^3}{G^3 M_2 \lambda^3} \right]^{1/2}, \tag{5.151}$$

with $\lambda \sim 1$ depending on the spin parameter a. For a given mean-stellar mass density $\propto M_2/R_2^3$ this limits M. For main-sequence stars the limit is $\sim 10^7 M_\odot$ for a black hole with low spin ($a \ll 1$), but can increase to $\sim 7 \times 10^8 M_\odot$ for high spin rates $a \sim 1$ (Kesden, 2012). Evidently, for more compact stars such as white dwarfs the mass limit allowing tidal disruption is much lower (see Figure 5.38).

To derive the characteristic time dependence after the star is disrupted we note that the star's pressure forces and internal self-gravity are almost completely in equilibrium, so the gas making up the star moves on essentially Keplerian orbits around the SMBH. Each orbit has its own eccentricity, initially very close to the

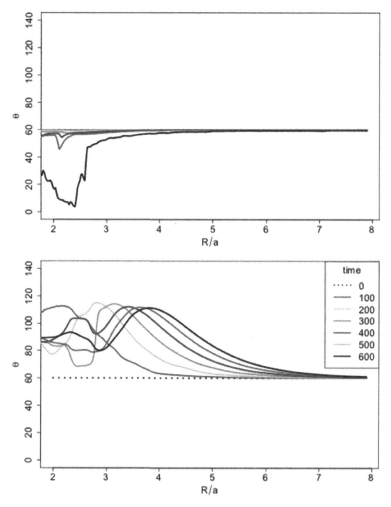

Figure 5.31 Evolution of tilt profiles for $e = 0$ (top) and $e = 0.9$ (bottom) binaries with initial dissipationless circumbinary ring tilt $\theta = 60°$. Times in orbital periods. Credit: Aly et al. (2015).

near-parabola of the star's centre of mass. The distribution of specific mechanical energy in the star is peaked very narrowly around the energy of the centre of mass. As the star moves closer to the hole, these Kepler orbits are squeezed, disturbing the hydrostatic balance. Pressure forces respond by redistributing energy inside the star and widening the specific energy distribution. At pericentre the tides from the SMBH disrupt the star, and its gas has a wide distribution of internal energies, some of it negative, so bound to the hole, and some positive, so remaining unbound. This energy distribution fixes the characteristic light curve of the event. After the encounter the gas elements again move in Keplerian orbits, but with their new

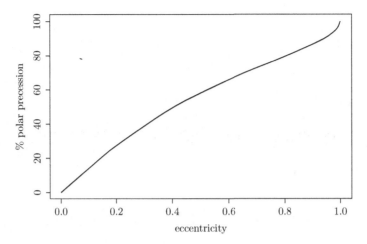

Figure 5.32 Percentage of dissipationless circumbinary ring orientations undergoing polar precession as a function of binary eccentricity. Credit: Aly et al. (2015).

energies. The bound elements return close to pericentre after a Keplerian period T given by their (negative) specific energy E by

$$E = -\left(\frac{\pi GM}{\sqrt{2}T}\right)^{2/3},$$ (5.152)

where M is the SMBH mass since $E = -GM/2a$ and $T = 2\pi(a^3/GM)^{1/2}$. The mass distribution with specific energy $\mathrm{d}M/\mathrm{d}E$ gives a mass distribution of return times $\mathrm{d}M/\mathrm{d}T$. The characteristic flare shape comes from making the simplest possible assumption, that as bound material returns to pericentre it loses its energy and angular momentum on a timescale $\ll T$, accreting instantaneously on to the SMBH and producing the flare. The distribution of mass with return time is then effectively the mass accretion rate of the black hole during the event, and we can easily compute the luminosity. We have

$$\frac{\mathrm{d}M}{\mathrm{d}T} = \frac{\mathrm{d}M}{\mathrm{d}E}\frac{\mathrm{d}E}{\mathrm{d}T} = \frac{(2\pi GM)^{2/3}}{3}\frac{\mathrm{d}M}{\mathrm{d}E}T^{-5/3}.$$ (5.153)

The 'standard' $t^{-5/3}$ light curve then follows if the energy distribution $\mathrm{d}M/\mathrm{d}E$ is uniform.[11] Subsequently, Lodato et al. (2009), Guillochon and Ramirez-Ruiz (2013) and Golightly, Nixon & Coughlin (2019b) showed that a more realistic stellar density profile (i.e. one that is not constant) implies that the power-law index of the fallback rate only reaches a value close to $-5/3$ at late (but still observable)

---

[11] Note that Rees (1988) did not show that this should be the case, but assumed it for simplicity. Later, numerical simulations by Evans and Kochanek (1989) suggested a uniform energy distribution was reasonable. Rees (1988) actually gave the index for this 'standard' light curve as $-5/2$, and this was corrected to $-5/3$ by Phinney (1989).

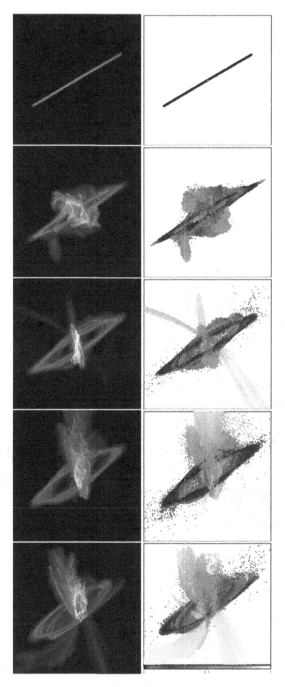

Figure 5.33 Density rendering (left-hand panels) and particle plots coloured by eccentricity magnitude (right-hand panels) of five snapshots of the evolution of a circumbinary gas disc with $e = 0.9$ and initially $\theta = 150°$. Credit: Aly et al. (2015).

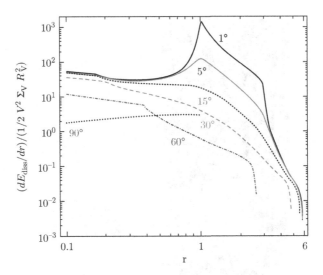

Figure 5.34 Differential energy available to be dissipated as an electromagnetic signal plotted as a function of dimensionless radius $r = R/R_V$ and expressed in units of the dimensional estimate $(1/2)V^2 \Sigma_R R_V^2$. Here θ is the angle of the GW kick to the disc plane, which is the plane of the pre-merger SMBH binary. Velocity kicks close to this plane can release orders of magnitude more energy than the dimensional estimate. Credit: Rossi et al. (2010).

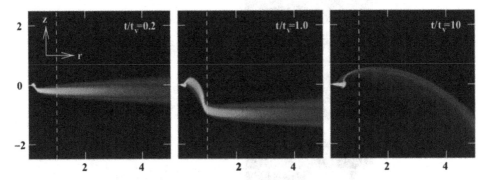

Figure 5.35 Visualization of the density in the (r, z) plane ($0 < r < 5$, $-2.5 < z < 2.5$) following a kick perpendicular ($\theta = 90°$) to a disc with a surface density profile $\propto r^{-3/2}$. By the end of the simulation the unbound material has been lost while the remaining bound disc has largely settled back into the equatorial ($z = 0$) plane. The dashed vertical line in each image marks the cylindrical radius $r = 1$. Credit: Rossi et al. (2010).

times, and after a significant amount of the bound debris has been accreted. If the star has a core of mass $M_c < M_2$ with much higher density (as in a red giant) that is not disrupted, Coughlin and Nixon (2019) show that the late-time fallback rate power-law index n_∞ is $\simeq -9/4$, and not $-5/3$, almost completely independent of the mass fraction M_c/M of the core. This happens because the debris returning

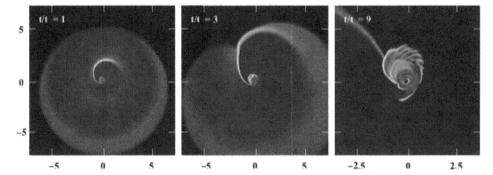

Figure 5.36 Rendering showing the evolution of the surface density of the disc following an in-plane kick. In this high-resolution simulation, the disc extends up to dimensionless radius $R/R_V = r = r_{ub} = [\cos\theta + (1 + \cos^2\theta)^{1/2}]$. During the peak phase of energy dissipation (left-hand panel), energy is dissipated at successively larger disc radii as an outward-moving wave propagates through the gas. The outer part of the disc is unbound and escapes ballistically. After the wave reaches the outer edge of the disc (centre panel), the rate of decay of the energy dissipation rate steepens markedly. At late times (right-hand panel, spatial scale and shading adjusted to show structure in the innermost regions), low angular momentum gas continues to accrete onto the bound remnant of the original disc, releasing energy at a low level and forming a highly non-axisymmetric accretion flow. Credit: Rossi et al. (2010).

at late times originates asymptotically close to the region where the core's gravity dominates over that of the black hole. It is always affected by the core's gravity for any non-zero core mass. This changes the relation between orbital energy and period significantly from (5.152), so the late-time light-curve index n_∞ differs from the 'standard' $-5/3$.

The study of TDEs has several aspects – a number concern the differences found when using full stellar models generated by the MESA (Modules for Experiments in Stellar Astrophysics) code for comparison with idealized prescriptions such as polytropes. The disrupted star may have significant spin, which can be prograde or retrograde with respect to the orbit. Golightly, Coughlin and Nixon (2019a) show that in the retrograde case the tidal force from the black hole has to spin down the star before disrupting it, causing delayed and sometimes only partial disruption events. But if the star's spin is prograde, this works with the tidal force, and the material falls back sooner and with a higher peak rate. Self-gravity can lead to the formation of clumps in the unbound debris from the disrupted star and could potentially be accelerated by slingshot to form a new class of low-mass hypervelocity objects (Coughlin et al., 2016).

5.9 Quasi-Periodic Eruptions

A phenomenon probably closely related to TDEs, but which may have wider implications for the feeding of low-mass SMBH, is the discovery of quasi-periodic

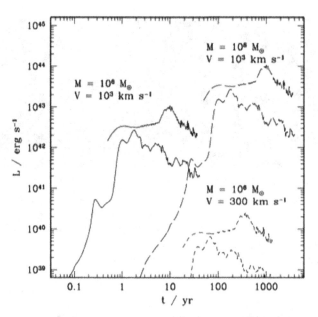

Figure 5.37 The predicted bolometric light curve of the kicked disc after scaling the numerical results to represent three different classes of systems. The solid curves show the light curves for $\theta = 15°$ (upper curve) and $\theta = 90°$ (lower curve) runs, following scaling to $M = 10^6 M_\odot$, $V = 10^3$ km s^{-1} and a disc mass ratio $q = 6 \times 10^{-3}$. Both runs are for a surface density profile with $p = 3/2$. The short-dashed curves are for identical parameters except for a lower kick velocity of $V = 300$ km s^{-1}. The long-dashed curves are for a system with $M = 10^8 M_\odot$, $V = 10^3$ km s^{-1} and a disc mass ratio $q = 6 \times 10^{-3}$. Credit: Rossi et al. (2010).

eruptions (QPEs). Miniutti et al. (2019) found large-amplitude (factors ~ 100) quasi-periodic X-ray eruptions (QPEs) from the low-mass black hole ($M_1 \sim 4 \times 10^5 M_\odot$) in the Seyfert 2 galaxy GSN 069 (see Figure 5.39). Each QPE lasts slightly more than 1 hr, with a typical recurrence time $\simeq 9$ hr. The emission has a very soft blackbody spectrum with peak temperature and luminosity $T \simeq 10^6$ K, $L \simeq 5 \times 10^{42}$ erg s^{-1}. The corresponding blackbody radius $R_{bb} = 9 \times 10^{10}$ cm is slightly larger than the gravitational radius $R_g = GM_1/c^2 = 6 \times 10^{10}$ cm of the black hole.

By now several similar systems are known; see Arcodia et al. (2021). The host galaxies here are not active, showing that QPEs do not require an AGN host. This in turn tends to rule out models in which an orbiting star periodically crosses an AGN accretion disc. The low mass of the black holes in all galaxies showing QPEs does suggest a connection with tidal capture, but the quasi-periodic behaviour requires that there is at most partial disruption of the captured star. In support of this, Shu et al. (2018) did detect a probable tidal capture in GSN 069, but we will see that

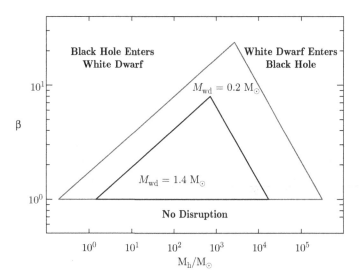

Figure 5.38 Domain where the black hole's tide can disrupt a white dwarf. The penetration factor $\beta = R_T/R_g$ is plotted as a function of the black hole mass M_h. A black hole of mass $\gg 10^5 M_\odot$ can swallow any white dwarf without first disrupting it. Credit: Rosswog, Ramirez-Ruiz and Hix (2009).

this particular event is unlikely to be connected with the QPE source, which must evidently have originated in an unobserved earlier event.

The quasi-periodic repetitions and very large amplitudes seen in GSN 069 naturally suggest a picture where mass overflows from a star as it fills its tidal lobe in an elliptical 9 hr orbit about the black hole. This picture imposes very tight constraints on the orbiting star. It must periodically fill its tidal lobe and transfer the right amount of mass to power the QPEs, presumably because of orbital decay by gravitational wave emission. These two constraints must then be compatible with a sensible stellar structure for the star.

To work out the tidal constraint we note that the orbital semi-major axis is

$$a = 1 \times 10^{13} m_{5.6}^{1/3} P_9^{2/3} \text{ cm} \qquad (5.154)$$

with $m_{5.6}$ the black hole mass M_1 in units of $4 \times 10^5 M_\odot$ and P_9 is the orbital period in units of 9 hr. The Roche lobe of the orbiting star, of mass M_2, is

$$R_{\text{lobe}} \simeq 0.46 (M_2/M_1)^{1/3} a(1 - e) \simeq 6.2 \times 10^{10} m_2^{1/3} (1 - e) \text{ cm}. \qquad (5.155)$$

Here $m_2 = M_2/M_\odot$ and e is the eccentricity (Sepinsky, Willems & Kalogera 2007; Dosopolou 2016a, 2016b). The star's radius $R_2 = r_2 R_\odot$ must equal R_{lobe} at pericentre, so it obeys the tidal constraint

$$r_2 = 0.89 m_2^{1/3} (1 - e). \qquad (5.156)$$

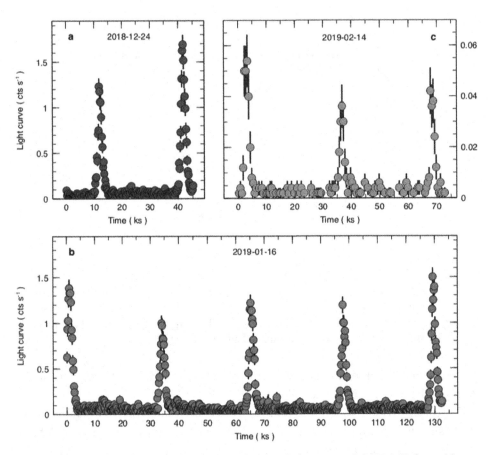

Figure 5.39 Background-subtracted 0.4–2 keV light curve of GSN 069 from (a) XMM3, (b) XMM4, and (c) Chandra, from December 2018 onwards. Credit: Miniutti et al. (2019).

The gas lost from the orbiting star each time it passes pericentre circularizes at

$$R_{\rm circ} \sim a(1-e) \simeq 10^{13}(1-e)\,{\rm cm} \qquad (5.157)$$

and forms an accretion disc with radius $R_d \sim R_{\rm circ}$, which must produce the QPEs. Averaging this luminosity over the 9 hr orbit tells us what the mass transfer rate must be. With black hole accretion efficiency $\eta \simeq 0.1$ this gives a mean mass transfer rate

$$-\dot{M}_2 \simeq 10^{-4}\,{\rm M_\odot\,yr^{-1}}. \qquad (5.158)$$

So the donor star must transfer mass to the black hole on a timescale

$$t_{\dot{M}} \sim -M_2/\dot{M}_2 \sim 10^4 m_2\,{\rm yr}. \qquad (5.159)$$

No known star has nuclear or thermal timescales as short as this, and dynamical mass transfer (i.e. a near-TDE event) can only persist for a few orbits. So to give the relatively long-lasting system observed, the binary must instead lose orbital angular momentum to gravitational radiation on the timescale t_M. Promisingly, this is is very efficient here because of the short orbital period and presumably high eccentricity. Then the system resembles a greatly speeded-up and eccentric version of short-period cataclysmic variable (CV) evolution (see Frank, King & Raine (2002), section 4.4). The much higher black hole accretor mass compared with the $\lesssim 1 M_\odot$ white dwarf accretors in CVs makes the mass transfer rate still larger.

The quadrupole gravitational radiation loss rate is given by

$$\frac{\dot{J}}{J}_{\text{GR}} = -\frac{32}{5}\frac{G^3}{c^5}\frac{M_1 M_2 M}{a^4}f(e), \tag{5.160}$$

where J is the orbital angular momentum, $M = M_1 + M_2 \simeq M_1$ the (effectively constant) total mass, and

$$f(e) = \frac{1 + \frac{73}{24}e^2 + \frac{37}{96}e^4}{(1-e^2)^{7/2}} \tag{5.161}$$

(Peters & Mathews 1963). The semi-major axis a and eccentricity decrease together on this timescale, related by

$$a = \frac{c_0 e^{12/19}}{1-e^2}\left(1 + \frac{121}{304}e^2\right)^{870/2299} \tag{5.162}$$

(Peters, 1964), where c_0 is a constant set by the initial value of a.[12] So for extreme eccentricities $e \sim 1$ (i.e. $1 - e << 1$) we have

$$a \propto \frac{1}{1-e^2} \sim \frac{1}{2(1-e)}, \tag{5.163}$$

so we assume

$$a = \frac{1-e_0}{1-e}a_0, \tag{5.164}$$

where $a_0 = 1 \times 10^{13}M_{5.6}^{1/3}P_9^{2/3}$ cm and $e_0 \simeq 1$ are the semi-major axis and eccentricity at the present epoch. In this evolution the pericentre separation

$$a(1-e) \simeq a_0(1-e_0) \tag{5.165}$$

remains almost constant when $e \sim 1$. This is as we expect since the GR emission is almost all produced in a point interaction at pericentre, so the orbit must always pass through this point as it evolves.

---

[12] Strictly, this relation is very slightly modified because of the mass transfer from the white dwarf to the black hole.

Using (5.163), the orbital angular momentum

$$J = M_1 M_2 \left(\frac{Ga}{M}\right)^{1/2} (1 - e^2)^{1/2} \simeq M_1 M_2 \left(\frac{Ga_0}{M}\right)^{1/2} (1 - e_0^2)^{1/2} \qquad (5.166)$$

obeys $J \propto M_1 M_2$, and we get

$$\frac{\dot{J}}{J} = \frac{\dot{M}_1}{M_1} + \frac{\dot{M}_2}{M_2} = \frac{\dot{M}_2}{M_2}\left(1 - \frac{M_2}{M_1}\right) \simeq \frac{\dot{M}_2}{M_2}, \qquad (5.167)$$

giving the current mass transfer rate as

$$-\dot{M}_2 \simeq 1 \times 10^{-7} m_{5.6}^{2/3} P_9^{-8/3} \frac{m_2^2}{(1 - e_0)^{5/2}} \, \mathrm{M}_\odot \, \mathrm{yr}^{-1}, \qquad (5.168)$$

where we have set $e = 1$ except in factors $(1 - e)$.

Equation (5.168) gives the evolutionary mean mass transfer rate. This is the average over the time t_{lobe} the tidal lobe takes to move through one density scaleheight of the star. This is typically about $10^{-4} R_2$ the inner Lagrange point (Ritter, 1988), so

$$t_{\mathrm{lobe}} \sim 10^{-4} \frac{R_2}{\dot{R}_2} \sim 10^{-4} \frac{M_2}{|\dot{M}_2|} \sim 0.1 \, \mathrm{yr.} \qquad (5.169)$$

In mass-transferring stellar-mass binaries t_{lobe} is far longer than the observing time-scale, so one must be careful in comparing the theoretical mean mass transfer rate (5.168) with direct observation – the instantaneous mass transfer rate can be significantly affected by short-term effects which cancel out over longer timescales. But here, because the mass transfer timescale is so short, the currently observed accretion rate is a good indicator of the long-term evolutionary mean: the tidal lobe moves of order a stellar scaleheight through the density stratification on the donor star on an observably short timescale.

This leaves us with two constraints ((5.156) and (5.168)), which specify M_2 and R_2. For a plausible interpretation, these values must be consistent with a physically reasonable mass–radius relation. Because the mass-transfer timescale $t_M \sim 10^4$ yr is so short, this already tells us that the relation of the radius to the mass must either be set by the adiabatic reaction of a non-degenerate star to adiabatic (rapid) mass loss or indicate a degenerate star (e.g. a white dwarf).

Using the observed mean mass transfer rate $\sim 10^{-4} \mathrm{M}_\odot \, \mathrm{yr}^{-1}$, (5.168) gives

$$1 - e_0 \simeq 0.14 m_2^{4/7}. \qquad (5.170)$$

Then (5.156) gives

$$R_2 = r_2 \mathrm{R}_\odot = 8.7 \times 10^9 m_2^{0.91} \, \mathrm{cm.} \qquad (5.171)$$

This radius is so small for any reasonable stellar mass m_2 that a low-mass white dwarf is the only possibility. The mass–radius relation for these stars is

$$R_2 \simeq 1 \times 10^9 (m_2/0.5)^{-1/3} \text{ cm.} \tag{5.172}$$

Then (5.171), (5.172) imply

$$M_2 = 0.32 M_\odot, \tag{5.173}$$

and from (5.170) we find that the current eccentricity

$$e_0 = 0.97 \tag{5.174}$$

is close to unity, as required for self-consistency (because we assumed $a(1 - e) \simeq$ constant).

As the evolution continues on the timescale

$$t_{evol} = \frac{M_2}{-\dot{M_2}} \sim 2000 \text{ yr} \tag{5.175}$$

the white dwarf expands further because it loses mass, and from (5.168) the mass transfer rate drops sharply. The system will transfer mass at ever-slowing rates almost indefinitely, but will become unobservably faint on a timescale of a few times t_{evol}. The pericentre separation $p = a(1 - e_0) \simeq 3 \times 10^{11}$ cm is about $7.5 R_g$, where the gravitational radius $R_g = GM_2/c^2$, so the system has an appreciable Einstein precession. The standard formula

$$\Delta\phi = \frac{6\pi GM}{c^2 a(1 - e)} \tag{5.176}$$

for the pericentre advance then gives

$$\frac{\Delta\phi}{2\pi} \simeq \frac{3 R_g}{p} \sim 0.4. \tag{5.177}$$

A possible origin for the capture of this very low-mass white dwarf is that it was the degenerate helium core of a low-mass ($\sim 1 M_\odot$) red giant left behind after partial tidal disruption (i.e. an event below the horizontal line in Figure 5.38). But another idea is attractive – Cufari, Coughlin and Nixon (2022) suggest that the white dwarf was originally a member of a close binary system disrupted by the SMBH through the Hills mechanism. It must have been the more massive star, but may have already transferred some of its mass to the central black hole. In line with either picture, subsequent UV observations (Sheng et al., 2021) find that the transferred matter is extremely nitrogen-rich and carbon-poor, as expected for a white dwarf.

Although a possible TDE event was observed from GSN 069 (Shu et al., 2018) there is no reason to suppose that it was the one producing the orbiting star in GSN 069, and indeed this seems unlikely.

The picture discussed here suggests that QPE sources may be only the most observable component of a significantly larger infall rate to the moderate-mass black hole in this galaxy. Similar or even far larger infall rates in other galaxies would not be observable at all for a central SMBH mass $M \gtrsim 10^7 M_\odot$, as then the infalling stars are swallowed whole by the black hole before they fill their tidal lobes, so that there is no accretion disc. At low redshift the Soltan relation shows that unobservable accretion like this cannot be the source of the bulk of the total SMBH mass. But it seems possible that it could be a significant contributor for smaller SMBH and at higher redshifts.

6

The Black Hole Scaling Relations

6.1 Introduction

Throughout this book we have adopted the now commonly accepted view that the centre of almost every galaxy contains a supermassive black hole. In Chapter 1 we noted that the hole mass M correlates with physical properties of the host galaxy. M is a fairly constant fraction of the stellar bulge mass M_b (see Figure 1.12), that is,

$$M \sim 10^{-3} M_b \qquad (6.1)$$

(Häring & Rix, 2004), and there is a tight relation of the form

$$M \simeq 3 \times 10^8 \mathrm{M}_\odot \sigma_{200}^\alpha \qquad (6.2)$$

between the SMBH mass and the velocity dispersion $\sigma = 200\sigma_{200}$ km s^{-1} of the host galaxy's central bulge. Here $\alpha \simeq 4.4 \pm 0.3$ (Ferrarese & Merritt 2000, Gebhardt et al. 2000; see Kormendy & Ho 2013 for a review).[1] These scaling relations between galaxies and their central black holes make it plain that SMBH are a vital element in understanding the formation of large-scale structures in the Universe.

At first sight the relations (6.1), (6.2) may seem surprising, as the small size (eqn (1.15)) of the radius of influence means that any SMBH has an almost completely negligible direct gravitational effect on its host galaxy. But one property of the hole

---

[1] In interpreting the M–σ relation (6.2) we should remember that measuring the SMBH mass M usually requires observers to resolve its sphere of influence, whose radius is

$$R_{\mathrm{inf}} \simeq \frac{GM}{\sigma^2} \simeq 8 \frac{M_8}{\sigma_{200}^2} \text{ pc}, \qquad (6.3)$$

with $M = 10^8 M_8 \mathrm{M}_\odot$. Then for a given σ, it is easiest to measure the largest hole masses. This implies a possible selection effect in the M–σ relation (6.2) – it may actually represent an upper limit to the SMBH mass M for a given σ (see Batcheldor, 2010) rather than a direct proportionality.

can potentially have a major effect on the host. The Soltan relation tells us that the hole grew mainly through luminous accretion of gas, so it has released accretion energy

$$E_{BH} \simeq \eta Mc^2 \sim 2 \times 10^{61} M_8 \text{ erg}, \tag{6.4}$$

where $\eta \simeq 0.1$ is the accretion efficiency. This is far larger than the binding energy

$$E_{bulge} \sim M_b \sigma^2 \sim 8 \times 10^{58} M_8 \sigma_{200}^2 \text{ erg} \tag{6.5}$$

of the host galaxy, whose bulge stellar mass is $M_b \sim 10^3 M$.

The large excess of E_{BH} over E_{bulge} suggests that the hole's energy release can influence the host galaxy bulge, even though any direct gravitational effect is negligible. Specifically, as accretion grows the hole mass, feedback of even a small part of the energy release can significantly affect the galaxy bulge gas. If this disrupts or drives off the gas, this may limit the hole's mass growth. This suggests we should compare E_{BH} with

$$E_{gas} = f_g E_{bulge}, \tag{6.6}$$

where $f_g < 1$ is the gas fraction rather than with E_{bulge} itself. For f_g close to the cosmological mean value $f_g \sim 0.16$ and a black hole close to the $M-\sigma$ relation we find

$$E_{BH} \sim 1600 E_{gas} \tag{6.7}$$

(here the right-hand side has an implicit factor $\sim \sigma_{200}^4/M_8$). Clearly the feedback process cannot be very efficient in removing gas, as otherwise the galaxy might never have managed to form at all or would have lost all its gas long before the black hole could have grown to the critical mass represented by (6.1), (6.2). Explaining the scaling relations with feedback is evidently a tightly constrained problem – we have to understand how such a small but precise fraction of E_{BH} is fed back to the galaxy. Ultimately we know that galaxies can become 'red and dead', that is, retain very little gas so that the formation of stars effectively halts, apart from those forming from gas recycled from older stars.

We discuss the physics of various forms of feedback, and their possible relevance for the scaling relations, in this and the next chapter. We will see that one particular form of feedback offers a promising explanation for the scaling relations.

Several suggested explanations of the scaling relations have appeared in the published literature. Table 6.1 lists those currently actively discussed.

Table 6.1 *Models for the SMBH–galaxy scaling relations*

Model	Comments
Feedback – Eddington winds (Ch. 6)	Predicts M–σ, massive outflows
Feedback – radiation, electron scattering (Sec. 7.2)	Forms initial cavity, warm absorber
Feedback – energy-driven only (Sec. 6.11)	M_σ too small by 10^{-3}
Feedback – jets (Sec. 7.3)	All bulge gas affected? M–σ?
Feedback – radiation, dust (Sec. 7.2)	Dust survival? M–σ?
Assembly (Sec. 6.10)	Disagrees with M–σ for dwarf galaxies

6.2 Black Hole Winds

The central question for the feedback picture is the physics of the connection with accretion. As a black hole accretes, it releases gas binding energy through dissipation in its accretion disc, most strongly near its inner edge at the ISCO. This initial release is in the form of electromagnetic radiation, and direct absorption or scattering of this is an obvious possible form of feedback. We will discuss this later (Section 7.2), but note that since we observe AGN precisely through this emitted radiation, much of it must escape for most of the lifetime of the SMBH.

A second form of feedback is also obvious – it is clear that the accretion luminosity of the SMBH can drive mechanical outflows. These come in two types. All types of accreting systems are at least intermittently able to produce narrowly collimated outflows – jets (see, e.g., Figure 1.10) as well as quasi-spherical winds. Jets can have feedback effects (see Section 7.3), but are probably not the origin of the scaling relations, most likely because in many cases much of the host galaxy gas is not strongly affected by them – we will discuss this further in Section 7.3. But we saw at the end of Section 4.10 that SMBH with super-Eddington mass supply are likely to produce quasi-spherical winds. Here collisions and feedback affecting all of the interstellar gas are inevitable.

Direct observation of outflows from AGN gives clear examples of this form of feedback through the detection of blueshifted X-ray absorption lines in the iron K band. These show the presence of highly ionized outflows with velocities $v \sim 0.1$–$0.25c$ (Chartas et al., 2002; Pounds et al., 2003; Reeves, O'Brien & Ward, 2003; see Figure 6.1). The characteristic P Cygni shape of this spectral line, with emission at electromagnetic frequency $\nu = E/h$ (where E is the X-ray energy at line centre), and a blueshifted absorption component at a frequency $\nu + \Delta\nu$, is very similar to

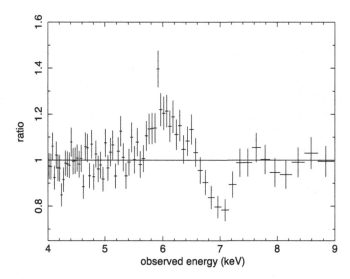

Figure 6.1 The P Cygni profile of the Fe XXV line from the Seyfert galaxy PG1211+143. This is characteristic of a wide-angle outflow. The similar equivalent widths of the emission and blueshifted absorption components imply that the outflow has a large covering factor. Credit: Pounds and Reeves (2009).

what is seen in observations of hot stars (hence the name P Cygni) and immediately tells us that the source is emitting a powerful quasi-spherical wind. This appears in absorption when viewed against the bright central AGN accretion disc, but in emission otherwise, giving the characteristic profile.

The wind speed is given by

$$\frac{v}{c} \simeq \frac{\Delta \nu}{\nu} = \frac{\Delta E}{E} \simeq 0.1, \tag{6.8}$$

and the fact that the emission and absorption components have comparable equivalent widths shows that the wind is roughly isotropic, occupying a large solid angle $\Omega = 4\pi b$, with $b \simeq 1$.

Using the rule of thumb that the terminal velocity of an outflow is similar to the escape speed from its launch radius R_l, we have

$$\left(\frac{2GM}{R_l}\right)^{1/2} \sim 0.1c, \tag{6.9}$$

so that $R_l \sim 200R_g$, that is, the wind comes from the close vicinity of the hole.

Observations (e.g., Tombesi, Cappi et al., 2010; Tombesi, Sambruna et al. 2010) now find many more of these 'ultrafast outflows' (UFOs), with velocities clustering near $v \sim 0.1c$, but ranging up to $v \sim 0.3c$. As they appear in blind searches with high frequency, it is now reasonable to argue that the majority of bright AGN emit

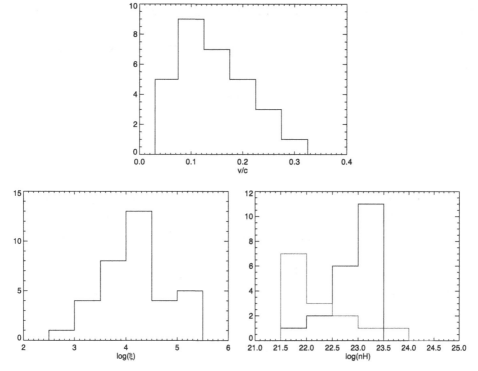

Figure 6.2 Range of outflow velocities v, ionization parameter ξ (erg cm s^{-1}), and column density N_H (cm^{-2}), from observations of type 1 AGN (Tombesi et al., 2011; Gofford et al., 2013). The histogram descending from left to right in the right-hand figure gives lower limits to the column density. Credit: King and Pounds (2015).

UFOs, at least from time to time (see Figure 6.2, and King & Pounds, 2015). This again agrees with the idea that the winds are quasi-spherical. The outflow rate is

$$\dot{M}_{\text{out}} \simeq 4\pi b N R^2 m_p v, \tag{6.10}$$

where N is the number density at spherical radius R. The X-ray spectrum itself gives a good estimate of the ionization parameter

$$\xi = \frac{L_i}{N R^2}, \tag{6.11}$$

where L_i is the ionizing luminosity, that is, of that part of the X-ray spectrum capable of producing the observed ionization states revealed by the spectrum. As L_i, ξ, and v are observed, we can find the outflow rate directly as

$$\dot{M}_{\text{out}} \simeq 4\pi b m_p \frac{L_i v}{\xi}, \tag{6.12}$$

and so work out the momentum and energy outflow rates $\dot{M}_{\text{out}}v, \dot{M}_{\text{out}}v^2/2$, where we can assume $b \sim 1$.

The observed values for the well-studied system PG1211+143 (see the discussion in King & Pounds, 2015) give mass loss, momentum, and energy rates

$$\dot{M}_{\text{out}} \sim 2.5 M_\odot \text{ yr}^{-1} \sim \dot{M}_{\text{Edd}},$$

$$\dot{M}_{\text{out}}v \sim 2 \times 10^{35} \text{dyne s}^{-1} \sim \frac{L_{\text{Edd}}}{c},$$

$$\frac{1}{2}\dot{M}_{\text{out}}v^2 \sim 4.5 \times 10^{44} \text{ erg s}^{-1} \sim 0.06 L_{\text{Edd}}.$$

Here $L_{\text{Edd}} = 5 \times 10^{45}$ erg s^{-1} is the Eddington luminosity for the known mass $M \simeq 4 \times 10^7 M_\odot$ of the SMBH in PG1211+143, and $\dot{M}_{\text{Edd}} \simeq 1.3 M_\odot$ yr^{-1} is the corresponding Eddington accretion rate for an accretion efficiency $\eta = 0.1$.

Since PG1211+143 is typical, these estimates show that AGN with UFOs have mass loss rates comparable to the Eddington accretion rate – the SMBH must eject a large fraction of the mass being supplied to it. This outflow has momentum rate $\sim L_{\text{Edd}}/c$, that is, about that of the radiation produced by accretion at the Eddington rate. The resulting kinetic power (from (6.7)) is significantly smaller than L_{Edd}. We will see in the next section that just this kind of outflow is expected for black holes being fed mass at rates slightly exceeding $\dot{M}_{\text{Edd}}$.

6.3 Eddington Winds

We saw in the last section that a large fraction of observed AGN show quasi-spherical ultrafast outflows from time to time. These have total scalar momenta $\sim L_{\text{Edd}}/c$, where L_{Edd} is the Eddington luminosity of the SMBH. A simple argument suggests that SMBH mass growth is likely to occur episodically, with accretion rates close to the value $\dot{M}_{\text{Edd}} = L_{\text{Edd}}/\eta c^2$ which would naturally produce outflows (see (4.119)).

The Soltan relation (Soltan, 1982) shows that SMBH grow most of their mass by luminous accretion, that is, during AGN episodes. But only a small fraction of all galaxies are observed to be active. Then when SMBH grow as AGN, they must evidently do so as rapidly as possible. The maximum possible rate of mass infall from a galaxy bulge with velocity dispersion σ is the dynamical value

$$\dot{M}_{\text{dyn}} \simeq \frac{f_g \sigma^3}{G}, \tag{6.13}$$

where f_g is the gas fraction. This follows since the maximum gas mass that could be in dynamical equilibrium is $M_g \sim \sigma^2 f_g R/G$. If this is destabilized and so falls freely inwards on a dynamical timescale $t_{dyn} \sim R/\sigma$, the accretion rate is M_g/t_{dyn}. This gives (6.13) or

$$\dot{M}_{dyn} \simeq 280\sigma_{200}^3 \ M_\odot \ yr^{-1}, \tag{6.14}$$

where we have again taken $f_g = 0.16$. From

$$\dot{M}_{Edd} = \frac{L_{Edd}}{\eta c^2} = \frac{4\pi GM}{\kappa \eta c} \tag{6.15}$$

with $\eta = 0.1$, and black hole masses M close to the observed relation (6.2), we find

$$\dot{M}_{Edd} \simeq 4.4 \ \sigma_{200}^4 M_\odot \ yr^{-1} \tag{6.16}$$

and so an Eddington accretion ratio

$$\dot{m} < \frac{\dot{M}_{dyn}}{\dot{M}_{Edd}} \simeq \frac{64}{\sigma_{200}} \simeq \frac{54}{M_8^{1/4}}. \tag{6.17}$$

So even dynamical infall cannot supply mass at very highly super-Eddington rates to supermassive black holes. In reality the rate (6.13) is usually a large overestimate, as it implicitly assumes that all the gas instantly loses its angular momentum. As we have seen in Chapter 4, accretion through a disc slows things dramatically. Although AGN mass supply rates can reach Eddington ratios $\dot{m} \sim 1$, significantly larger ones can evidently only occur if the SMBH mass is far below the M–σ value for its host galaxy bulge. So we expect that the mass supply rate $\dot{M}_{supp}$ in SMBH growth episodes must have relatively modest Eddington ratios $\dot{m} \sim 1$.

This conclusion agrees with the observation that, at least at low redshift, there appear to be no AGN analogues of the stellar-mass ultraluminous X-ray sources (ULXs). Here a compact accretor (a stellar-mass black hole, or in many cases a neutron star) is supplied at mass transfer rates with Eddington ratios $\dot{m} \gg 1$ (as high as $\dot{m} \sim 10^3$) as the companion star overflows its tidal (Roche) lobe on a thermal timescale. ULXs have apparent X-ray luminosities $>> L_{Edd}$, probably as the result of geometric collimation in the resulting very dense outflow.[2]

Then to understand how UFOs work we need to consider SMBH supplied at rates $\dot{m} \gtrsim 1$. Although the dynamics and associated radiative transfer of the accretion-outflow system of this type of disc accretion are complex in detail, we can infer its fundamental properties from some simple considerations. First, the outflow is unlikely to be strongly anisotropic except in one respect – there must be very little gas flowing parallel to the accretion disc axis very close to the black hole, as it would need to have effectively zero angular momentum. So we expect

---

[2] The caveat above-concerning low-redshift AGN arises because almost all of the AGN found at redshifts $z \gtrsim 6$ appear to have luminosities $L \simeq L_{Edd}$ for their estimated masses. At the time of writing it is unclear if this is a selection effect, or if these systems have mass supply rates $\gg \dot{M}_{Edd}$, and their detected (rest-frame UV) emission is from a photosphere in the dense outflowing wind – see (6.27), (6.28) below.

two narrow channels around the disc axis where there is essentially no gas, and where radiation can escape freely. In all other respects we can regard the outflow as roughly spherical.

We assume that the outflow reaches a terminal speed v at sufficiently large radial distance r and justify this later. Then mass conservation implies a density

$$\rho = \frac{\dot{M}_{\text{out}}}{4\pi v r^2}. \tag{6.18}$$

The electron-scattering optical depth of the outflow, viewed from infinity down to radius $r = R$, is

$$\tau = \int_R^\infty \kappa \rho \, \mathrm{d}r = \frac{\kappa \dot{M}_{\text{out}}}{4\pi v R}. \tag{6.19}$$

Now using (1.8) to eliminate κ, we find

$$\tau = \frac{1}{\eta} \frac{R_g}{R} \frac{c}{v} \dot{m}, \tag{6.20}$$

where $\dot{m} = \dot{M}_{\text{out}}/\dot{M}_{\text{Edd}}$ since $R_g = GM/c^2$. Then the surface $\tau \simeq 1$ defines the scattering photosphere

$$R_{\text{ph}} = \frac{1}{\eta} \frac{c}{v} \dot{m} R_g \tag{6.21}$$

outside which most photons escape without further scattering. They cannot accelerate the outflow after this point, so for consistency, the velocity here must be the escape velocity, that is,

$$R_{\text{ph}} \simeq \frac{c^2}{v^2} R_g, \tag{6.22}$$

so that

$$v \simeq \frac{\eta}{\dot{m}} c, \tag{6.23}$$

justifying the assumption of constant velocity made at the start of this calculation.

We can rewrite (6.23) in the form

$$\dot{M}_{\text{out}} v = \frac{L_{\text{Edd}}}{c}, \tag{6.24}$$

which says that the scalar outflow momentum rate is just that of an Eddington radiation field (of course, by symmetry the total vector momentum rates of both the outflow and the radiation field are zero). Thomson scattering is front–back symmetric, that is, the scattered photon has the same probability of being scattered forwards or backwards with respect to its original path. So on average a stream of photons encountering an electron loses *all* its momentum to the electron. Then for $\dot{m} \sim 1$ (so that $L \sim L_{\text{Edd}}$) this suggests that each photon of the disc radiation field scatters about once before escaping.[3] This is reasonable since most photons are emitted from the innermost part of the accretion disc, and even if initially directed away from the open channels along the disc axis, will find one of them and so escape after only a few scatterings. We would still expect a luminosity $\sim L_{\text{Edd}}$ to emerge from the scattering photosphere R_{ph}.

The total energy rate of the wind (its mechanical luminosity) is

$$L_{\text{mech}} = \frac{1}{2}\dot{M}_{\text{out}}v^2 = \frac{L_{\text{Edd}}}{2c}\frac{v}{c} = \frac{\eta}{2}L_{\text{Edd}} \simeq 0.05L_{\text{Edd}}, \tag{6.25}$$

considerably less than the radiative luminosity.

Although we have arrived at this picture by simple analytic arguments, detailed numerical simulations (e.g. Ohsuga & Mineshige, 2011) find very similar results. There are several important consequences of this picture.

1. If AGN (especially quasars) are sometimes fed at rates a little above $\dot{M}_{\text{Edd}}$ we expect them to have outflows with velocities

$$v = \frac{\eta}{\dot{m}}c \sim 0.1c. \tag{6.26}$$

We have already seen that X-ray observations find evidence of such outflows, revealed by P Cygni profiles, in the majority of AGN.

2. From (6.22) the large scattering photosphere

$$R_{\text{ph}} \simeq 1.5 \times 10^{15}M_8 \text{ cm} \tag{6.27}$$

implies that these systems should have a blackbody-like radiation component with effective temperature

$$T_{\text{phot}} = \left[\frac{L_{\text{Edd}}}{4\pi R_{\text{ph}}^2\sigma}\right]^{1/4} \simeq 3.6 \times 10^5 M_8^{-1/4}\dot{m}^{-1}(\eta/0.1)\,\text{K}. \tag{6.28}$$

This evidently appears in the rest-frame soft X-ray and EUV. If the interstellar photoelectric absorption allows, these are seen as ultrasoft AGN.

3. The wind moves with speed $\sim 0.1c$, so it can persist long after the AGN is observed to have become sub-Eddington, that is, for a time $\sim 10R/c$, where R is the radial extent of the wind (i.e. the shock radius, as we shall see). For $R \gtrsim 3$ pc this lag is at least a century, and far longer lags are possible, as we shall see. Then AGN showing other signs of super-Eddington phenomena (e.g. narrow-line Seyfert 2 galaxies) may nevertheless have sub-Eddington luminosities (NGC 4051 is an example: Denney et al., 2009).

---

[3] There appear to be rare cases where this single scattering limit does not hold. We consider these in Section 6.12.

From (6.26), (6.18) we estimate the ionization parameter

$$\xi = \frac{L_i}{NR^2} \tag{6.29}$$

of the wind. Here $L_i = l_i L_{Edd}$ is the ionizing luminosity, with $l_i < 1$ a dimensionless parameter specified by the quasar spectrum, and $N = \rho/\mu m_p$ is the number density. This gives

$$\xi = 3 \times 10^4 \eta_{0.1}^2 l_2 \dot{m}^{-2}, \tag{6.30}$$

where $l_2 = l_i/10^{-2}$ and $\eta_{0.1} = \eta/0.1$.

Equation (6.30) shows that the wind momentum and mass rates determine its ionization parameter. A given quasar spectrum has a predominant ionization state where the threshold photon energy defining L_i, and the corresponding ionization parameter ξ, together satisfy (6.30). It is clear that this requires high excitation: a low threshold photon energy (say in the infrared) would give a large value of l_2, but the high value of ξ then resulting from (6.30) would require the presence of very highly ionized species, incompatible with the assumed low excitation.

For any particular continuum spectrum there may be more than one solution of (6.30), and initial conditions would then specify which is realized for the given source. But for a typical quasar spectrum, an obvious self-consistent solution of (6.30) is $l_2 \simeq 1$, $\dot{m} \simeq 1$, and $\xi \simeq 3 \times 10^4$ – here the quasar would be currently radiating the Eddington luminosity. But as we noted after (6.8), we can also have cases where the quasar's luminosity has dropped after an Eddington episode, but the wind produced in that state is still flowing, with $\dot{m} \simeq 1$. Then the ionizing luminosity $10^{-2} l_2 L_{Edd}$ in (6.30) is smaller, giving a lower value of ξ. For example, a quasar of current luminosity $0.3 L_{Edd}$ would have $\xi \sim 10^4$. This implies a photon energy threshold producing helium- or hydrogenlike iron (i.e. $h\nu_{threshold} \sim 9\,keV$).

We conclude that

> *Eddington winds from AGN are likely to have velocities $\sim 0.1c$, seen as*
> *P Cygni profiles in helium- or hydrogenlike iron.*

A large number of systems with winds like this are known. The discussion here explains why in all cases the wind velocity is $v \sim 0.1c$, and further that the outflows are all found by identifying blueshifted resonance lines of Fe XXV, XXVI in absorption. Further, any observed wind with these properties automatically satisfies the momentum relation (6.24). This strongly suggests that it is launched by an AGN being fed mass at a slightly super-Eddington rate.

The escape of much of the accretion luminosity along fairly narrow channels parallel to the accretion disc axis means that systems viewed along these directions appear anomalously bright. This is the defining feature of the ultraluminous X-ray sources (ULXs), which are stellar-mass binaries with super-Eddington mass supply from a high-mass companion star on to a stellar mass black hole or neutron star. In this ejection picture, the central region of every disc with $\dot{M}_{\text{supp}} >> \dot{M}_{\text{Edd}}$ is identical in terms of the quantities $\dot{m} = \dot{M}_{\text{supp}}/\dot{M}_{\text{Edd}}$ and $r = R/R_{\text{sph}}$, while further out the disc structure is sensitive directly to the value of $\dot{M}_{\text{supp}}$. Simple geometric arguments (King, 2009a) then suggest that most of the central luminosity is emitted over a fractional solid angle $4\pi b$, where

$$b \propto \dot{m}^{-2} \tag{6.31}$$

because conditions far from the disc depend on $\dot{m}$, whereas all flows reduce to $\dot{m} \simeq 1$ near the black hole. The coefficient here is undetermined, but observations of ULXs allow one to infer a value. In a large class of these systems, the blackbody-like soft X-ray luminosity L_{soft} and blackbody temperature T_{bb} obey a universal relation

$$L_{\text{soft}} = CT_{\text{bb}}^{-4}, \tag{6.32}$$

where C is the same constant in all cases (both between different systems and also as individual systems vary). Combining with (6.32) for the observed value of C gives

$$b \simeq \frac{73}{\dot{m}^2} \tag{6.33}$$

(King, 2009a). The simple argument that $b \propto \dot{m}^{-2}$ evidently breaks down for modest Eddington factors $\dot{m} \lesssim 8$, and interpolations of the form

$$b = \frac{73}{73 + \dot{m}^2} \tag{6.34}$$

are often used for smaller $\dot{m}$.

This beaming effect is large in ULXs, where there are evolutionary phases of these accreting binaries where $\dot{m}$ can reach values ~ 10–10^3 or more and is responsible for their apparent super-Eddington luminosities $\gtrsim 10^{39}$–10^{41} erg s^{-1}, even though coherent X-ray pulsing shows that many of them involve accretion on to neutron stars, where $L_{\text{Edd}} \simeq 10^{38}$ erg s^{-1}.

This type of beaming is apparently less prevalent for AGN, probably because, as we have noted (see (6.17)), the required high mass supply rates $\dot{m} >> 1$ are unlikely for masses $M \gtrsim 10^6 M_\odot$. But it may be that some bright AGN have actually rather low-mass SMBH with beamed emission – note that for a given mass supply rate, $\dot{m}$, and so the tightness of any beaming, is larger for lower-mass black holes.

We can see from (6.26) that a larger Eddington factor $\dot{m}$ is likely to produce a slower wind. From comparison with ULXs we also expect the AGN radiation to be beamed away from a large fraction of the UFO, which should therefore be less ionized, and as a result more easily detectable than the small fraction receiving the beamed radiation. These properties – slower, less ionized winds – characterize BAL QSO outflows, perhaps suggesting that systems with larger $\dot{m} > 1$ appear as BAL QSOs. Zubovas and King (2013) tentatively confirm this idea.

4. Most importantly for the relation between SMBH and their host galaxies, a black hole fed with mass at a super-Eddington rate can feed back a significant momentum into its surroundings (see (6.24)). The associated feedback energy rate (sometimes called the mechanical luminosity) is

$$L_{\text{mech}} = \frac{1}{2}\dot{M}_{\text{out}}v^2 \simeq \frac{L_{\text{Edd}}}{c}\frac{v}{2} \simeq \frac{\eta}{2}L_{\text{Edd}} \simeq 0.05L_{\text{Edd}}. \tag{6.35}$$

This means that the bulge gas of the host galaxy feels only a fraction ~ 5 per cent of the accretion energy E_{BH}. But even this would still unbind it if it was efficiently communicated because the total SMBH accretion energy is so large compared with its binding energy. Applying the factor 0.05 from (6.35) to (6.7), we get

$$E_{\text{BH}}(\text{effective}) = 800E_{\text{gas}}. \tag{6.36}$$

Then if these outflows are the ultimate cause of the scaling relations there must be a further reduction in communicating their energy to the bulge gas.

6.4 The Wind Shock

Physically this points to a simple conclusion. The UFO typical speed $v \sim 0.1c$ is far higher than the sound speed in the interstellar gas of the host galaxy. Then at the spherical interface where the UFO gas collides with the interstellar gas there must be strong shocks driven into each of them (see Section 3.3). The result is the shock pattern shown in Figure 6.3.

The forward shock 'tells' the interstellar gas to start moving ahead of the 'piston' represented by the UFO, while the reverse shock slows the UFO gas so that it moves with the contact discontinuity where the wind and bulge gas meet (see Figure 6.3). The energy tending to unbind the interstellar gas is the kinetic energy of the UFO. This is conserved across the initial adiabatic reverse shock, which slows the wind but simultaneously heats and compresses it. As we have discussed, left unchecked, this energy would rapidly unbind the bulge gas and produce galaxy bulges very different from those we observe. But if instead the hot, adiabatically shocked UFO gas radiates most of this heat away, this could remove the tendency to unbind the interstellar gas.

This is evidently what must happen at this interface, so the main question is what cools the post-shock gas. It is easy to show that the usual two-body gas

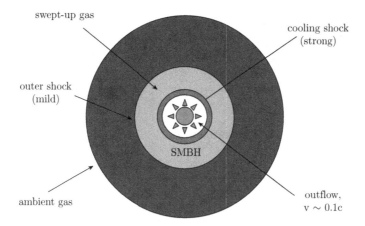

Figure 6.3 The impact of an Eddington wind on the interstellar gas of the host galaxy (1). A supermassive black hole (SMBH) accreting at just above the Eddington rate drives a near-spherical UFO wind (shown as the star) with radial velocity $\sim \eta c \sim 0.1c$. This collides with the interstellar gas of the host galaxy and is slowed in a strong reverse shock (narrow ring). The inverse Compton effect from the quasar's radiation field rapidly cools the shocked gas, removing its thermal energy and strongly compressing it over a very short radial extent. This gas may be observable in an inverse Compton continuum and lower-excitation emission lines associated with lower velocities (see Figure 6.4). The cooled gas exerts the pre-shock ram pressure on the galaxy's interstellar gas and sweeps it up into a thick shell ('snowplough', outside the narrow ring). This shell's motion drives a milder outer shock into the ambient interstellar medium (interface between the two outer rings). This forward shock ultimately stalls unless the SMBH mass M has reached the value M_σ satisfying the M–σ relation. For $M > M_\sigma$, the entire shock pattern expands radially to a radius R_{cool} (not shown), where the quasar radiation field is too dilute to cool the inner (wind) shock. The inner shock region now expands rapidly because of its high pressure, driving the interstellar gas out of the galaxy bulge and terminating the black hole mass growth at $M \simeq M_\sigma$. Credit: King (2010).

cooling processes (thermal bremsstrahlung and line emission) are far too weak to do this. Perhaps surprisingly, the most effective cooling of the shocked UFO gas is through the inverse Compton effect of the AGN radiation field (King, 2003), at least when the shock is not too far from the AGN. As first pointed out by Ciotti and Ostriker (1997), this is physically reasonable for any shocked gas near the SMBH. For a UFO shock, the adiabatic shock temperature (from 3.50, with $v \sim 0.1c$) is $T \sim 2 \times 10^{11}$ K, far higher than the effective temperature $\sim 1\,\mathrm{keV}/k \sim 10^7$ K of the AGN radiation field. The hot post-shock gas is bathed in a gas of much cooler photons from the quasar. The restriction that the shock must be fairly close to the AGN ensures that each hot electron in the shocked gas encounters enough of these photons quickly enough to cool it (see (6.41)).

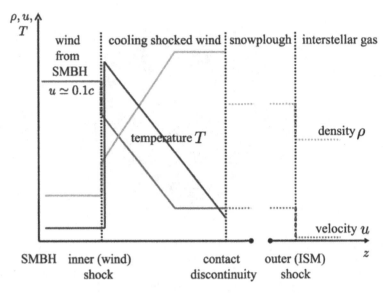

Figure 6.4 The impact of an Eddington wind on the interstellar gas of the host galaxy (2). Close-up view of the cooling shock shown as the narrow ring in Figure 6.3. A schematic view of the radial dependence of the gas density ρ, velocity u, and temperature T. The shock is locally plane, so all quantities are functions of the local thickness z. The black hole wind (UFO) is slowed at the inner shock. The gas temperature rises strongly, and the density and velocity, respectively, increase (decrease) by factors $\sim$ four. Immediately outside (larger z) this adiabatic shock, the inverse Compton effect of the cooler photons from the quasar radiation strongly reduces the temperature and slows and compresses the wind gas still further. This cooling region is very narrow compared with the shock radius (see Figure 6.3) and is observable through an inverse Compton continuum and X-ray emission lines whose excitation decreases with velocity (see Figure 6.6). The shocked wind sweeps up the interstellar gas of the host galaxy as a 'snowplough'. This region is much more extended than the cooling region (cf. Figure 6.3) and itself drives an outer shock into the ambient interstellar medium (ISM) of the host. The solid curves are in descending order at the left: velocity of the wind from the SMBH vicinity, wind gas density and wind gas temperature. The horizontal dotted curves at the right are, in descending order, the ISM gas density ρ and ISM gas velocity u. The vertical dotted lines denote the three discontinuities: inner shock, contact discontinuity, and outer shock. Credit: King (2010).

To see how this works we note that the Compton cooling time of an electron of energy E is in the AGN radiation field $L \simeq L_{\text{Edd}}$ is

$$t_C = \frac{3m_e c}{8\pi \sigma_T U_{\text{rad}}} \frac{m_e c^2}{E}, \qquad (6.37)$$

where m_e is the electron mass, σ_T the Thomson cross-section, and

$$U_{\text{rad}} = \frac{L_{\text{Edd}}}{4\pi R^2 c} \qquad (6.38)$$

is the AGN radiation energy density at radius R. This gives

$$t_C = \frac{2}{3} \frac{cR^2}{GM} \left(\frac{m_e}{m_p}\right)^2 \left(\frac{c}{v}\right)^2 \simeq 10^7 R_{\text{kpc}}^{-2} M_8^{-1} \text{ yr}, \tag{6.39}$$

where M_8 is the SMBH mass in units of $10^8 M_\odot$.

Cooling becomes important when this timescale is shorter than the flow time

$$t_{\text{flow}} \sim \frac{R}{v_{\text{out, m}}} \simeq 7 \times 10^6 R_{\text{kpc}} \sigma_{200} M_8^{-1/2} \left(\frac{f_g}{f_c}\right)^{-1/2} \text{ yr}, \tag{6.40}$$

where $v_{\text{out, m}}$ is the post-shock flow velocity we shall find in (6.64), when the condition is self-consistently fulfilled, and $f_c \sim 0.16$ is the cosmological baryon/(dark matter) density ratio. Using this we find that cooling is important for

$$R < R_C \sim 520 \sigma_{200} M_8^{1/2} v_{0.1}^2 \left(\frac{f_g}{f_c}\right)^{1/2} \text{ pc}, \tag{6.41}$$

where $v_{0.1}$ is the wind (UFO) velocity in units of $0.1c$.

We will see that the effect of an Eddington wind on the galaxy is vastly different depending on whether or not the Compton cooling of the shock discussed here is effective or not. In direct observations, the energy the inverse Compton process removes from the shocked gas should appear as a soft continuum with characteristic photon energy $\sim 1\,\text{keV}$. As we have noted Pounds and Vaughan (2011) report a possible detection of this spectral component in the Seyfert 1 galaxy NGC 4051. As required for consistency, the luminosity of this component is comparable to the expected mechanical luminosity of the wind in that system. But the collision of a UFO with the interstellar gas is a comparatively rare and short-lived event, so the lack of other detections is not unexpected.

Theoretically, the main question at issue here is that Compton cooling acts in the first instance on electrons, whereas almost all of the post-shock energy is initially in the ions. To drain the thermal energy from the shocked gas, there must be efficient coupling between electrons and ions so that the ions are themselves cooled. In most ionized gases, Coulomb collisions between electrons and ions force the ion and electron temperatures T_i, T_e to move effectively in step – this coupling is faster than the rate that Compton cooling takes energy from the electrons. But in extreme situations, as with the very strong shock impact of the UFO wind on the bulge gas, it is unclear that Coulomb collisions remain effective. Experience suggests that in such cases some other forms of coupling (often called microinstabilities) maintain the temperature equality $T_i \simeq T_e$. But this assumption has been questioned, which would radically change the effect of feedback (Faucher-Giguère & Quataert, 2012; but see Zubovas & Nayakshin, 2014). Fortunately there is direct observational evidence that the shocked wind gas does indeed cool, as originally suggested. If cooling occurs, the gas that has passed through the reverse wind shock is compressed to higher densities, which must make two-body radiation processes important. The free–free (thermal bremsstrahlung) and Compton cooling times are

$$t_{ff} \simeq 3 \times 10^{11} \frac{T^{1/2}}{N} \, s = 20 \frac{R_{16}^2}{M_7 \dot{m}} \, \text{yr} \tag{6.42}$$

and

$$t_C = 10^{-4} \frac{R_{16}^2}{M_8} \, \text{yr}, \tag{6.43}$$

respectively (see King, Zubovas & Power, 2011: here T, N are the post-shock temperature and number density, R_{16} is the shock radius in units of 10^{16} cm, M_7 is the black hole mass in units of $10^7 M_\odot$, and $\dot{m} \sim 1$ is the Eddington factor of the mass outflow rate).

As pressure is almost constant in an isothermal shock the density of any cooling gas must rise as $N \propto T^{-1}$. Further, the thermal bremsstrahlung (free–free) cooling time is

$$t_{ff} \propto \frac{T^{1/2}}{N} \propto T^{3/2}, \tag{6.44}$$

which decreases sharply while the Compton time does not change. Free–free emission (and all other two-body atomic cooling processes) become stronger than Compton once T has decreased sufficiently below the original shock temperature $T_s \sim 1.6 \times 10^{10}$ K. From (6.42), (6.43), this happens when

$$\left(\frac{T}{T_s}\right)^{3/2} < 5 \times 10^{-5} \tag{6.45}$$

or

$$T < 2 \times 10^7 \, \text{K}. \tag{6.46}$$

Then the temperature of ionized species at around a few keV is fixed by atomic cooling rather than Compton cooling. The strong recombination continua in NGC 4051 (Pounds & Vaughan, 2011; Pounds & King, 2013) are direct evidence for this additional cooling, giving an emission measure for the related flow component. The onset of strong two-body cooling confines the lower-ionization and lower-velocity gas to a relatively narrow region in the outer part of the post reverse-shock flow (see Figure 6.5). In a spherical flow with fractional solid angle b, a cooling shell thickness $\Delta R \sim 7 \times 10^{14}$ cm, and shell radius $R \sim 10^{17}$ cm, the emission measure given by the observed lines is well reproduced for $b \sim 0.5$. Significantly, the predicted strong correlation between ionization parameter and velocity is seen (see Figure 6.6). This is a clear signature of a cooling shock. Then NGC 4051 (uniquely) shows three signatures of a cooling shock: an inverse Compton continuum, an

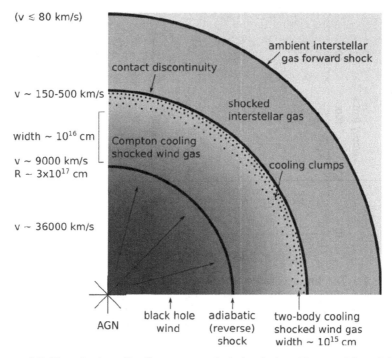

Figure 6.5 The physics of a Compton-cooled shock (see Figures 6.3 and 6.4) where an Eddington wind (UFO) impacts the interstellar gas of the host galaxy. Here the quasar's radiation field is assumed to cool the shocked gas by the inverse Compton effect. This strongly compresses and slows it before it moves a significant radial distance, making the shock region geometrically thin. The increase in density means that two-body radiative cooling becomes important in the most compressed gas, allowing the formation of lower-excitation ions with lower velocities. This offers an observational test for the presence of Compton cooling: it must produce cooler species with correlated slower velocities. Soft X-ray observations of NGC 4051 detect this two-body emission, as here the lighter metals dominate both absorption and emission (see Figure 6.6). Credit: Pounds and King (2013).

ionization–velocity correlation, and the appearance of two-body processes in the spectrum.

X-ray observations of these winds give another important quantity. The flux at low photon energies ($\lesssim 1$ keV) is progressively reduced by line-of-sight photoelectric absorption from K-shells of abundant elements (typically C, N, and O). Continuum fitting, and assuming cosmic abundances of these elements with respect to H, gives an equivalent hydrogen column density

$$N_H = \int N \, \mathrm{d}R, \tag{6.47}$$

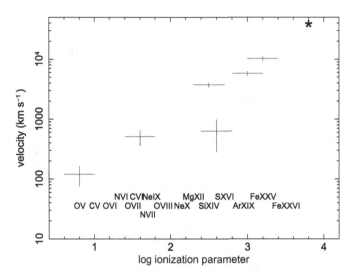

Figure 6.6 The linear correlation between velocity and excitation expected if Compton cooling is effective in reducing the ion temperature after the wind shock, represented by the star (see Pounds and King, 2013). This is direct evidence that this shock is effectively isothermal, severely reducing its velocity and the transfer of energy to the interstellar gas. Credit: Pounds and King (2013).

where N is the hydrogen number density. The integration here is over the line of sight through the whole outflow at a given time. Generally we have $N_H <$ 10^{24} cm^{-2}, as higher values would mean that the outflow was optically thick to electron scattering. This is unlikely on physical grounds (see the remark after (6.50)), and in any case might well make the AGN unobservable.

The value of the column density N_H is dominated by the higher densities ($N \sim R^{-2}$) at small radii, while the ionization parameter $\xi = L_i/NR^2$ is affected by the current AGN luminosity and the time variation of the outflow.

The column density N_H is often observed to vary rapidly and turns out to be a powerful diagnostic of the flow history and dynamics. Using (6.24) in the mass conservation equation

$$\dot{M}_{\mathrm{w}} = 4\pi \, br^2 v \rho(r), \tag{6.48}$$

where $\rho(r)$ is the mass density, the equivalent hydrogen column density of the wind is

$$N_H \simeq \int_{R_{\mathrm{in}}}^{\infty} \frac{\rho}{m_p}\mathrm{d}r = \int_{R_{\mathrm{in}}}^{\infty} \frac{\dot{M}_{\mathrm{w}}}{4\pi r^2 bv}\mathrm{d}r = \frac{L_{\mathrm{Edd}}}{4\pi \, bm_p R_{\mathrm{in}} cv^2}. \tag{6.49}$$

Here R_{in} is the inner radius of the flow, m_p is the proton mass, and the last step uses (6.24). Using the definition of L_{Edd} gives the wind's electron-scattering optical depth

$$\tau = N_H \sigma_T \simeq \frac{GM}{bv^2 R_{in}}, \tag{6.50}$$

where $\sigma_T \simeq \kappa m_p$ is the Thomson cross-section. For a continuous wind, that is, one with R_{in} equal to the typical launch radius $R_{launch} \simeq GM/bv^2 = (c^2/2bv^2)R_s \simeq 50R_s$, the scattering optical depth τ is self-consistently ~ 1 (see King & Pounds, 2003, eqn 4).

Observed values of N_H (Tombesi et al., 2011; Gofford et al., 2013) are generally in the range $N_{22} \sim 0.3$–30, where $N_{22} = N_H/10^{22}$ cm^{-2}, and always less than the value $N_H \simeq 1/\sigma_T \simeq 10^{24}$ cm^{-2} for a continuous wind. This is unsurprising, as systems like this would be obscured by electron scattering at all photon energies and perhaps difficult to see. Such systems might be common, but we probably cannot detect them – seeing a UFO system needs a smaller N_H, so from (6.50) the inner surface R_{in} of the wind must be larger than R_{launch}. This can only happen if all observed UFOs are time-variable so that we see them only some time after the UFO wind has switched off.

Then any UFO is a series of sporadically launched quasi-spherical shells and not a continuous outflow. From (6.49) the N_H value of each shell is given essentially by the gas density near its inner edge (see (6.49)), so we probably detect only the inner edge of the most recently launched shell. This has $R_{in} = vt_{off}$, where t_{off} is the time since the end of the launching phase of the most recent wind episode. Equation (6.50) gives

$$t_{off} = \frac{GM}{bv^3 N_H \sigma_T} \simeq \frac{3M_7}{bv_{0.1}^3 N_{22}} \text{ months}, \tag{6.51}$$

where $v_{0.1} = v/0.1c$.

Figure 6.2 shows that all observed UFOs have $N_{22} \sim 0.3 - 30$. Most of the SMBH masses here are $\sim 10^7 M_\odot$. So from (6.51) most UFO wind-launching events stopped weeks or months before the UFO was observed, and we observe a coasting shell. This is initially surprising, as we find UFOs through their characteristic P Cygni profiles, which would actually be stronger for UFOs with $N_{22} > 100$, but none have been detected. But since we can only observe UFOs 'post-launch', (6.51) implies that observing a UFO like this would correspond to catching it within days of launch. X-ray observations of AGN are generally too sparse for this to be common. The apparent upper limit to the observed N_H is then a consequence of low observational coverage, and means that most UFOs are short-lived. This presumably indicates inherently time-variable SMBH accretion on timescales of weeks to months (see (6.51)).

The lower limits to N_H in Figure 6.2 mean that for N_H smaller than some critical value, any blueshifted absorption lines must become too weak to detect.

Inherently the strongest are the resonance lines of H- and He-like iron, with absorption cross-sections $\sigma_{Fe} \simeq 10^{-18}$ cm^2. The number abundance of Fe relative to H is $Z_{Fe} = 4 \times 10^{-5}$. Then the condition for one of these lines to have detectable optical depth is $Z_{Fe} N_H \sigma_{Fe} > 1$ or $N_{22} > 2.5$. This is similar to the lowest observed values. By (6.51), UFOs whose launching ended more than a few months in the past decrease their N_H and become undetectable after a few years. We will see that a UFO event travels $\sim 10M_7$ pc or more before colliding with the host galaxy's interstellar gas. This needs a time $t_{coll}/v \sim 300 M_7 v_{0.1}^{-1}$ yr. As a further restriction on observability, UFOs may be unobservable simply because they are too strongly ionized, making the H- and He-like iron lines too weak to detect.

All this means that inferences from UFO detections do not necessarily give a good idea of the state of the AGN at its launch. Although we have argued that luminosities $\sim L_{Edd}$ are required, an AGN with only short super-Eddington episodes but a long-term average luminosity $< L_{Edd}$ would launch UFOs but almost never appear super-Eddington.

This discussion shows that UFO coverage is very sparse. We cannot see a continuous wind at all, and only see episodic wind shells shortly after launch, for a tiny fraction $t_{off}/t_{coll} \sim 10^{-3} v_{0.1}^{-2} N_{22}^{-1}$ of their time of flight ($\sim$ 300–3000 yr) to a collision with the host bulge gas. Far more AGN must produce UFOs than we observe, and the UFO sources we do detect must have far more episodes than we see. Interpreting observations to constrain feedback is far from straightforward. In particular, it is very likely that we cannot directly observe the most powerful feedback of all, from AGN accreting at rates $\gtrsim \dot{M}_{Edd}$.

We have studied the black hole wind in some detail, as we know that it must have a significant effect on the host galaxy interstellar medium (ISM). In this section we shall assume that the wind and the ISM are roughly spherically symmetric; we discuss the effects of deviations from this simple picture in Section 6.7.

The pattern of the UFO–ISM interaction (Figures 6.4, 6.7) is qualitatively similar to the collision of a stellar wind with the surrounding interstellar medium (see, e.g., Dyson & Williams, 1997). The UFO (top curve at left in Figure 6.4) slows abruptly at the inner (reverse) shock. Here the temperature approaches $\sim 10^{11}$ K, assuming that ions and electrons reach equipartition (see the discussion of the last section). The pressure of the shocked wind gas makes it push into the ISM, sweeping this up at the contact discontinuity between the UFO and the ISM. This swept-up ISM gas advances supersonically into the unperturbed ambient ISM, driving an outer (forward) shock into it (see Figure 6.7, top).

The fundamental process here is the reverse shock injecting energy into the host ISM. As we saw earlier, the effect of the shock is very different depending on whether or not cooling can remove most of the energy from the shocked gas on a timescale shorter than its flow time. If it can, the resulting narrow, strongly

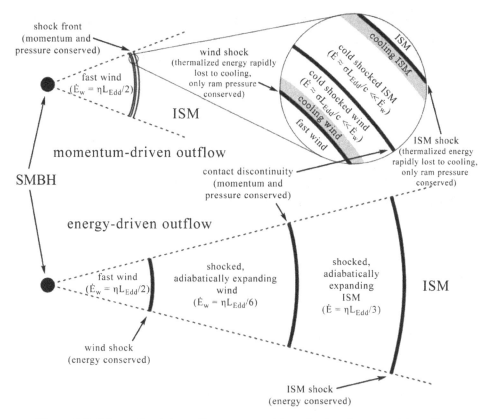

Figure 6.7 Schematic picture of momentum-driven (top) and energy-driven (bottom) outflows. In both cases a fast wind (velocity $\sim 0.1c$) impacts the interstellar gas of the host galaxy, producing an inner reverse shock slowing the wind, and an outer forward shock accelerating the swept-up gas. In the momentum-driven case (top), corresponding to the UFOs discussed in Sections 6.2–6.4, the shocks are very narrow and rapidly cool led to become effectively isothermal. Only the ram pressure is communicated to the outflow, leading to very low kinetic energy $\sim (\sigma/c)L_{Edd}$. In an energy-driven outflow (bottom), the shocked regions are much wider and do not cool. They expand adiabatically, communicating most of the kinetic energy of the wind to the outflow (in simple cases divided in a ratio of about 1:2 between the shocked wind and the swept-up gas). The outflow radial momentum flux is therefore greater than that of the wind. Momentum-driven outflow conditions hold for shocks confined to within ~ 1 kpc of the AGN and establish the M–σ relation (8.18) (King, 2003; 2005). Once the supermassive black hole mass attains the critical M–σ value, the shocks move further from the AGN and the outflow becomes energy-driven. This produces the observed large-scale molecular outflows which probably sweep the galaxy clear of gas. Credit: Zubovas and King (2012a).

cooling region is often idealized as a discontinuity itself, and called an *isothermal shock* (see Dyson & Williams, 1997). As momentum must be conserved across the shock, the shocked UFO gas transmits just its ram pressure (6.24) to the host ISM.

The resulting outflow of the ISM gas is called *momentum driven*. This amounts to transferring only a fraction $\sim \sigma/c \sim 10^{-3}$ of the mechanical luminosity $L_{\mathrm{mech}} \simeq 0.05 L_{\mathrm{Edd}}$ (see (6.35)) to the ISM. So in this case, only energy

$$E_{\mathrm{mom}} \sim \frac{\sigma}{c} E_{\mathrm{BH\ wind}} \sim \frac{\sigma}{c} \frac{\eta^2}{2} Mc^2 \sim 5 \times 10^{-5} Mc^2 \sim 0.1 E_{\mathrm{gas}} \qquad (6.52)$$

is injected into the bulge ISM, that is, about 10 per cent of the bulge gas binding energy $f_g M_b \sigma^2$ for black holes close to the M–σ relation (there is an implicit factor σ_{200}^5 / M_8 on the right-hand side).

We see that efficient shock cooling produces momentum-driven flows which do not threaten to blow away all the bulge gas and suppress accretion. Provided that the wind shocks cool in this way, the galaxy is a stable environment for black hole mass growth.

There is an obvious opposite limit, where the post-shock gas retains all the mechanical luminosity

$$E_{\mathrm{wind}} \simeq 0.05 L_{\mathrm{Edd}} \simeq 100 E_{\mathrm{gas}} \qquad (6.53)$$

(cf. (6.35)) thermalized in the shock. Instead of being confined to a narrow iso-thermal shock region, the shocked UFO gas expands adiabatically into the bulge ISM gas in a geometrically extended region (see the lower part of Figure 6.7), quite unlike the momentum-driven outflow ('isothermal') case.

This is 'energy-driven flow' and is much more violent than momentum-driven flow. Since it retains all the energy in the UFO wind, an energy-driven flow would sweep the galaxy bulge gas away for an SMBH mass far smaller than observed. The estimate (6.7) is for an SMBH mass near the M–σ relation; a hole with mass a factor of 100 below this would already unbind the bulge in doubling its mass. A black hole in an energy-driven environment is unlikely to reach the observed SMBH masses close to the M–σ value.

Efficient shock cooling is the only plausible way out of this existential diffi-culty, and this is only possible through the inverse Compton effect. We conclude that until the SMBH mass reaches the M–σ value, black hole wind shocks only occur within the critical cooling radius R_C given by (6.41). Quite generally then, momentum-driven flows are confined to a small region $R < R_C$, while energy-driven flows must be large-scale. The arrival of the SMBH at the M–σ mass must control the switch from relatively gentle momentum-driven outflow, allow-ing SMBH growth, to violent energy-driven outflow sweeping the galaxy bulge clear of gas, and cutting off further SMBH growth.

In this picture all but a small fraction of the mechanical luminosity (6.35) of the black hole wind for $M \leq M_\sigma$ is eventually radiated away as an inverse Compton continuum with characteristic photon energy ~ 1 keV. As we have noted,

Pounds and Vaughan (2011) report a possible detection of this spectral component in the Seyfert 1 galaxy NGC 4051, and the luminosity of this component is comparable to the expected mechanical luminosity (6.35) of the wind in that system.

Cooling shocks are called 'isothermal' (see Section 3.3) because the gas temperature rapidly returns to something like its pre-shock value. Momentum conservation requires that the gas is also strongly slowed and compressed as it cools. So the post-shock velocity of the X-ray emitting gas should correlate with its temperature (or roughly, ionization) while Compton cooling is dominant. Once this has compressed the gas sufficiently, two-body processes such as free–free and bound–free emission must begin to dominate since they go as the square of the density, and their cooling times also decrease with temperature (Pounds & King, 2013). The discussion after (6.46) shows that there is direct observational evidence for both of these effects in NGC 4051. So this object (uniquely) shows three signatures of a cooling shock: an inverse Compton continuum, an ionization–velocity correlation, and the appearance of two-body processes in the spectrum.

6.5 The M–σ Relation

The discussion of this chapter has focused on the possibility that feedback from SMBH accretion is the ultimate cause of the black hole scaling relations. We have already noted that the wind impact implies a pair of shocks each side of the contact discontinuity between the wind and the host ISM (see Figure 6.7). The shock pattern is very different in the cases where the shocked black hole wind cools or not (upper and lower panels 03, respectively; Figure 6.7).

We noted that only in the first case – a cooling shock giving momentum-driven flow – the bulge is likely to retain enough gas to grow SMBH masses to observed values. Accordingly, we assume that inverse Compton cooling from the AGN radiation field is effective and puts the flow in the momentum-driven regime. The gas region between the wind shock and the contact discontinuity, where it impacts and sweeps up the host ISM, is very narrow (upper panel of Figure 6.7). The outer shock accelerating the ISM is also strongly cooled so that the snowplough region of swept-up ISM is narrow as well. Then we can treat the whole region between the inner and outer shocks as a single narrow, outward-moving gas shell, whose mass grows as it sweeps up the host ISM.

We assume that the host galaxy bulge has an isothermal profile of velocity dispersion σ, with mass density

$$\rho(r) = \frac{\sigma^2}{2\pi G r^2},$$ (6.54)

mainly in the form of stars so that the mass within radius R is

$$M(R) = \frac{2\sigma^2 R}{G}. \tag{6.55}$$

For a roughly constant gas fraction f_g, the black hole wind sweeps up a gas shell of mass

$$M_g(R) = \frac{2f_g\sigma^2 R}{G} \tag{6.56}$$

at radius R. This shell is subject to gravity from the SMBH and stars within it, plus the ram pressure of the black hole wind, so has the equation of motion

$$\frac{d}{dt}[M_g(R)\dot{R}] + \frac{GM_g(R)[M + (1-f_g)M(R)]}{R^2} = \frac{L_{Edd}}{c}, \tag{6.57}$$

where M is the SMBH mass, and we note that the mass inside the shell is reduced from the value (6.55) by a factor $(1 - f_g)$. From (6.55), (6.56) and the definition of L_{Edd} this reduces to

$$\frac{d}{dt}(R\dot{R}) + \frac{GM}{R} = -2\sigma^2\left(1 - \frac{M}{M_\sigma}\right), \tag{6.58}$$

where

$$M_\sigma = \frac{f_g(1-f_g)\kappa}{\pi G^2}\sigma^4, \tag{6.59}$$

which for the usual small gas fractions f_g is

$$M_\sigma \simeq \frac{f_g\kappa}{\pi G^2}\sigma^4. \tag{6.60}$$

This gives

$$M_\sigma \simeq 3 \times 10^8 \sigma_{200}^4 M_\odot \tag{6.61}$$

for the cosmological gas fraction $f_g \simeq 0.16$, where $\sigma_{200} = \sigma/(200\,\mathrm{km\,s^{-1}})$, and even for extremely gas-rich galaxies can never exceed

$$M_\sigma(\mathrm{max}) = \frac{\kappa\sigma^4}{4\pi G^2} \simeq 4.7 \times 10^8 \sigma_{200}^4 M_\odot. \tag{6.62}$$

Accordingly, in this book we adopt the simple form (6.60) for M_σ throughout.

We get a first integral by multiplying through by $R\dot{R}$ and integrating once:

$$R^2\dot{R}^2 = -2GMR - 2\sigma^2\left[1 - \frac{M}{M_\sigma}\right]R^2 + \mathrm{constant}. \tag{6.63}$$

Then for large R we have

$$\dot{R}^2 \simeq -2\sigma^2\left[1 - \frac{M}{M_\sigma}\right]. \tag{6.64}$$

This has no solution for $M < M_\sigma$. Evidently if the SMBH mass is below the value M_σ the swept-up shell of interstellar gas cannot reach large radius because the Eddington thrust of the black hole wind is too small to lift the shell's growing weight against the the gravity of the bulge within it. The shell eventually becomes too massive and so tends to fall back and probably fragment. But if $M > M_\sigma$, (6.64) shows that the shell ultimately coasts outwards with constant radial speed $\dot{R} = (2\epsilon)^{1/2}\sigma$, where $M/M_\sigma = 1 + \epsilon$. If the accretion rate of the AGN remained unchanged, the shell would reach the cooling radius R_C (eqn (6.41)) in a time $\sim 10^6 \epsilon^{-1/2}$ yr. Then UFO gas behind it would no longer cool as it shocked, but instead initiate energy-driven flow.

It is clear that the nature of SMBH feedback on the bulge gas changes markedly once M reaches the critical value M_σ. With the gas fraction f_g fixed at the cosmological value $f_c = 0.16$, the expression

$$M_\sigma \simeq \frac{f_g \kappa}{\pi G^2} \sigma^4 \simeq 3.2 \times 10^8 \mathrm{M}_\odot \sigma_{200}^4 \tag{6.65}$$

is remarkably close to the observed relation (6.2), even though it contains no free parameter. We will see in Section 6.8 why observations of the full ensemble of galaxies with measured M and σ tend to give an exponent for σ slightly larger than the value 4 derived here. This agreement strongly suggests that SMBH growth stops at this point, and we will discuss this in detail. First we consider the derivation of M_σ more closely.

We took the simplest possible description of a galaxy spheroid as an isothermal sphere (eqn (6.55)). In reality any galaxy bulge is more complicated than this, so we consider the effects of removing some of the assumptions here. First, if the potential is still taken as spherically symmetric but the cumulative mass $M(R)$ is not simply linear in R, we always find a first integral of the equation of motion (6.57) simply by multiplying through by $M(R)\dot{R}$. This gives an analogous condition to (6.64) for a swept-up momentum-driven shell to reach large radii. McQuillin and McLaughlin (2012) do this explicitly for three widely used density distributions (Hernquist, 1990; Navarro, Frenk & White, 1996, 1997; Dehnen & McLaughlin, 2005). Clearly there is now no constant velocity dispersion σ, but its role is played by the peak circular speed in the potential in each case. This gives relations like (8.18), but now between SMBH mass and asymptotic circular speed in galaxy spheroids. McQuillin and McLaughlin (2012) show that these results are in practice scarcely distinguishable from (6.65).

If the bulge is not spherically symmetric it is still true that the black hole always communicates its presence only through the ram pressure of its wind. Then the bulge gas experiences strong radial forces in the solid angles exposed to this wind (this is not true of gas pressure, as we shall see in Section 6.7). If SMBH accretion is chaotic (see Section 5.3), the orientation of the accretion disc with respect to the host galaxy changes randomly with each new accretion episode, tending to isotropize the long-term effect of momentum feedback. We will see in Section 6.6 that there is a rapid and very marked increase in the spatial scale as the SMBH reaches the critical mass. These considerations probably explain why the simple spherically symmetric derivation of the critical M_σ mass appears to work in the general context.

The result (8.18) is so close to observations that it strongly suggests that SMBH grow asymptotically to this mass because feedback cuts off their gas supply at this point. This is intuitively reasonable – we can imagine that driving most of the bulge gas away severely limits future SMBH growth. Evidently the current accretion disc event driving the black hole wind can continue, but since its mass is limited by self-gravity to $(H/R)M \lesssim 10^{-3}$, this does not lead to significant mass growth. We shall see that to expel finally the bulge gas on the large scale observed, the hole must undergo several more episodes like this and grow its mass fractionally, still leaving (8.18) as a reasonable estimate.

This last point emphasizes that we still have to explain precisely how the gas is expelled. Observations show that black hole accretion occurs preferentially in gas-rich galaxies, which suggests that the black hole must largely clear the galaxy bulge of gas to terminate its growth. So the M–σ relation marks the point where outflows undergo a global transition from momentum driving to energy driving.

6.6 Clearing Out a Galaxy: The Energy-Driven Phase

We know from (6.7) that an energy-driven outflow injects more than enough energy to drive away all the interstellar gas from the galaxy bulge, halting further SMBH growth. Here we examine how this works in detail.

Once M even slightly exceeds M_σ the outflow geometry becomes very different (see Figure 6.7) because the shocks are now outside the cooling radius R_C. The shocked wind region is no longer efficiently cooled and narrow, but large and expanding because of its strong thermal pressure (lower panel of Figure 6.7). This expansion pushes the wind shock inwards to the cooling radius R_C (Zubovas & King, 2012a), where it remains. (If it tries to move within R_C, momentum driving instantly pushes it out again, since $M > M_\sigma$.) At the same time the shocked wind pushes strongly outwards into the bulge gas.

This expansion rapidly evens out the internal pressure of the shocked wind gas as it expands at its sound speed $\sim 0.03c$. This allows us to regard the gas pressure P as uniform over this region (but changing with time), a condition sometimes called *isobaric*. As before, the contact discontinuity at the outer edge of the shocked wind sweeps up the surrounding shocked ISM, but it now has the equation of motion

$$\frac{d}{dt}\left[M_g(R)\dot{R}\right] + \frac{GM_g(R)M(R)}{R^2} = 4\pi R^2 P. \tag{6.66}$$

Here the pressure P is much larger than the ram pressure ρv^2 term in (6.57). In the second term on the left-hand side we neglect the contribution $GM_g M/R^2$ of the black hole gravity, as $R \gg R_C > R_{\rm inf}$. The system of equations is closed by the energy equation. (In the momentum-driven case we did not require this explicitly because the assumed isothermal shock conditions replaced it with the defining condition that all the wind energy not associated with the ram pressure was rapidly lost to radiation.) In the energy-driven case the energy equation constrains the internal energy, and so the pressure P, by specifying the rate that energy is fed into the shocked gas, minus the rate of PdV working on the ambient gas and against gravity:

$$\frac{d}{dt}\left[\frac{4\pi R^3}{3}\cdot\frac{3}{2}P\right] = \frac{\eta}{2}L_{\rm Edd} - P\frac{d}{dt}\left[\frac{4\pi}{3}R^3\right] - 4f_g\frac{\sigma^4}{G}\dot{R}. \tag{6.67}$$

We take a specific heat ratio $\gamma = 5/3$ and use (6.25) for the energy input from the outflow and (6.55) to simplify the gravity term $GM(R)M(R)/R^2$. Now we eliminate P between (6.58) and (6.67) and replace the gravity terms as before, using the isothermal expression for $M(R)$. We parametrize the AGN luminosity as $lL_{\rm Edd}$ to allow for small deviations from the Eddington value. Then we have

$$\frac{\eta}{2}lL_{\rm Edd} = \dot{R}\frac{d}{dt}\left[M_g(R)\dot{R}\right] + 8f_g\frac{\sigma^4}{G}\dot{R} + \frac{d}{dt}\left\{\frac{R}{2}\frac{d}{dt}\left[M_g(R)\dot{R}\right] + 2f_g\frac{\sigma^4}{G}R\right\}, \tag{6.68}$$

which leads to

$$\frac{\eta}{2}lL_{\rm Edd} = \frac{2f_g\sigma^2}{G}\left\{\frac{1}{2}R^2\dddot{R} + 3R\dot{R}\ddot{R} + \frac{3}{2}\dot{R}^3\right\} + 10f_g\frac{\sigma^4}{G}\dot{R}. \tag{6.69}$$

This equation specifies the motion of the contact discontinuity between wind and interstellar gas (Figure 6.7) in the energy-driven case and replaces the equation (6.57) appearing in the momentum-driven case.[4]

We can assume $M = M_\sigma = f_g\kappa\sigma^4/\pi G^2$ in $L_{\rm Edd}$ since as soon as the SMBH mass reaches this value the ISM is rapidly removed, preventing any significant increase. Equation (6.69) has a constant-velocity solution $R = v_e t$, where

$$2\eta lc = 3\frac{v_e^3}{\sigma^2} + 10v_e. \tag{6.70}$$

---

[4] There are slight transcription errors – detectable by discrepant dimensions, and the lack of the suffix g from $M_g(R)$ – in earlier published versions (King, 2005; King & Nixon, 2015) of (6.67) (a missing factor $\dot{R}$ from the last term on the right-hand side) and (6.68) (an unwanted extra factor $\dot{R}$ inside the derivative in the term in {}), but the final expression (6.69) is correct in both papers. I am indebted to Zezhong Liang for pointing this out.

Assuming $v_e \ll \sigma$ here leads to a contradiction ($v_e \simeq 0.01c \gg \sigma$), so the v_e^3 term dominates the right-hand side, giving

$$v_e \simeq \left[\frac{2\eta l\sigma^2 c}{3}\right]^{1/3} \simeq 925 l^{1/3}\sigma_{200}^{2/3} \text{ km s}^{-1}. \tag{6.71}$$

This solution is an attractor: Figure 6.8 shows that all solutions quickly converge to it, whatever their initial conditions. This shows explicitly that when the shocked wind from the black hole does not cool, its pressure expansion accelerates the swept-up gas shell velocity from the momentum-driven value to this new higher velocity. If driving by the AGN is switched off entirely at some point, we expect the swept-up shell to continue moving out for a time before halting.

We can follow this by setting $l = 0$ in (6.69) so that now

$$\frac{1}{2}R^2\dddot{R} + 3R\dot{R}\ddot{R} + \frac{3}{2}\dot{R}^3 + 5\sigma^2\dot{R} = 0. \tag{6.72}$$

The independent variable t does not appear explicitly in this equation, so we can reduce its order by one with the substitution $\dot{R} = p$. This gives $\ddot{R} = pp'$ and $\dddot{R} = p^2 p'' + pp'^2$, where the primes stand for differentiation with respect to R. Equation (6.72) becomes

$$\frac{1}{2}R^2(pp')' + 3Rpp' + \frac{3}{2}p^2 + 5\sigma^2 = 0, \tag{6.73}$$

or

$$\frac{1}{4}R^2 y'' + \frac{3}{2}Ry' + \frac{3}{2}y + 5\sigma^2 = 0, \tag{6.74}$$

where $y = p^2$. Writing $y_1 = y + 10\sigma^2/3$ makes this equation algebraically homogeneous, that is,

$$R^2 y_1'' + 6Ry_1' + 6y_1 = 0, \tag{6.75}$$

which has the linearly independent solutions $y_1 \propto R^{-2}, R^{-3}$. Then in terms of the original variables we have

$$p^2 = \dot{R}^2 = \frac{A_2}{R^2} + \frac{A_3}{R^3} - \frac{10\sigma^2}{3}. \tag{6.76}$$

We choose the constants A_2, A_3 to fit the boundary conditions $\ddot{R} = 0, \dot{R} = v_e$ at $R = R_0$, where R_0 is the radius reached by the contact discontinuity between black hole wind and bulge gas when the central quasar turns off, which gives

$$\dot{R}^2 = 3\left(v_e^2 + \frac{10}{3}\sigma^2\right)\left(\frac{1}{x^2} - \frac{2}{3x^3}\right) - \frac{10}{3}\sigma^2, \tag{6.77}$$

where $x = R/R_0 \geq 1$. Figure 6.8 confirms that all numerical solutions of (6.69) ultimately follow this solution. Since v_e is only weakly dependent on the AGN

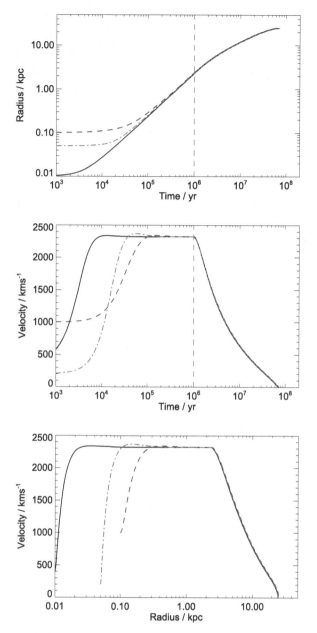

Figure 6.8 Evolution of an energy-driven shock pattern for the case $\sigma = 200\,\mathrm{km\,s^{-1}}, f_g = 10^{-2}$ computed numerically from the full equations (6.69) and (6.88). Top: radius of the forward shock vs time, middle: velocity vs time, bottom: velocity vs radius. The curves refer to different initial conditions: solid – $R_0 = 10\,\mathrm{pc}$, $v_0 = 400\,\mathrm{km\,s^{-1}}$; dashed – $R_0 = 100\,\mathrm{pc}$, $v_0 = 1000\,\mathrm{km\,s^{-1}}$; dot-dashed – $R_0 = 50\,\mathrm{pc}$, $v_0 = 200\,\mathrm{km\,s^{-1}}$. All these solutions converge to the attractor (6.71). The vertical dashed line marks the time $t = 10^6\,\mathrm{yr}$ when (for this case) the quasar driving is switched off. All solutions then follow the analytic solution (6.77). A case where the quasar remains on for a Salpeter time $\sim 4 \times 10^7\,\mathrm{yr}$ would sweep the galaxy clear of gas. Credit: King, Zubovas, and Power, 2011.

luminosity (as $v_e \sim l^{1/3}$), we can see that changes of the AGN luminosity have very little effect on the energy-driven outflow once it is launched. From Figure 6.8 we see that even if the AGN turns off completely, it takes about about 2×10^6 yr before the outflow velocity is halved, and the radius of the outflow continues to increase. This stability is a direct result of the shocked wind gas's huge reservoir of thermal energy.

Importantly, this makes it likely that many outflows will be observed at epochs when their mechanical luminosities are significantly lower than the maximal value (6.25), and we will see that the observed values do spread on the lower side of this estimate.

The outflow stalls at a radius $R_{\rm stall}$ where $\dot{R} = 0$. From (6.77) this requires

$$\frac{1}{x^2} - \frac{2}{3x^3} = \frac{10\sigma^2}{9(v_e^2 + 10\sigma^2/3)}. \tag{6.78}$$

Since $x >> 1$ and $v_e >> \sigma$, we find

$$x^2 \simeq \frac{9}{10}\left(\frac{v_e^2}{\sigma^2} + \frac{10}{9}\right) \simeq \frac{9v_e^2}{10\sigma^2}, \tag{6.79}$$

so

$$R_{\rm stall} \simeq 0.95 \frac{v_e}{\sigma} R_0. \tag{6.80}$$

Again with $x >> 1$, we can write (6.77) as

$$R\dot{R} = [C - DR^2]^{1/2}, \tag{6.81}$$

where

$$C = 3\left(v_e^2 + \frac{10}{3}\sigma^2\right)R_0^2 \text{ and } D = \frac{10}{3}\sigma^2. \tag{6.82}$$

Then (6.81) gives the time before stalling at

$$t_{\rm stall} = \int_{R_0}^{(C/D)^{1/2}} \frac{{\rm d}R^2}{2[C - DR^2]^{1/2}}, \tag{6.83}$$

so

$$t_{\rm stall} = \frac{1}{D}(C - DR_0^2)^{1/2} \simeq \frac{R_0 v_e}{2\sigma^2} \simeq \frac{R_{\rm stall}}{2\sigma}. \tag{6.84}$$

Since the SMBH was accreting for a time $t_{\rm acc}$ before reaching R_0, and the contact discontinuity was moving with speed $\simeq v_e$ for this time, we have

$$R_0 \simeq v_e t_{\rm acc}. \tag{6.85}$$

Then from (6.84) we get

$$R_{\rm stall} \simeq \frac{v_e^2}{\sigma} t_{\rm acc} \tag{6.86}$$

and so

$$t_{\text{stall}} = \frac{R_{\text{stall}}}{2\sigma} = \frac{v_e^2}{2\sigma^2} t_{\text{acc}}, \tag{6.87}$$

which shows explicitly that the outflow moves for a long time after the AGN turns off, before eventually stalling.

The quantity $R(t)$ specifies the motion of the contact discontinuity between the black hole wind gas and the interstellar gas of the host galaxy bulge. An outer shock runs ahead of this interface into the ISM with a higher velocity, and the mass outflow rate is fixed by how quickly this shock entrains new interstellar gas ahead of the contact discontinuity. In its own frame the shock causes a velocity jump by a factor $(\gamma + 1)/(\gamma - 1)$ (where $\gamma \sim 5/3$ is the specific heat ratio; see (3.44)). Transforming to the observer's frame, where the ISM ahead of the shock is at rest, this fixes the shock velocity as

$$v_{\text{out}} = \frac{\gamma + 1}{2} \dot{R} \simeq 1230 \sigma_{200}^{2/3} \left(\frac{lf_c}{f_g} \right)^{1/3} \text{ km s}^{-1}, \tag{6.88}$$

where we have used $\gamma = 5/3$ in the last form, and $f_c \simeq 0.16$ is the cosmological value of f_g. This gives a shock temperature of order 10^7–10^8 K for the forward (ISM) shock (far lower than the temperature $\sim 10^{10-11}$ K for the wind shock). The outer shock and the contact discontinuity must have been very close to each other at the point where energy-driven flow took over from momentum-driven flow (see Figure 6.7), with R close to R_C. After a time t the outer shock is at

$$R_{\text{out}}(t) = \frac{\gamma + 1}{2} R(t) = \frac{\gamma + 1}{2} v_e t, \tag{6.89}$$

which is $\gg R_C$. The resulting mass outflow rate is

$$\dot{M}_{\text{out}} = \frac{dM(R_{\text{out}})}{dt} = \frac{(\gamma + 1)f_g \sigma^2}{G} \dot{R}. \tag{6.90}$$

For comparison the mass rate of the black hole wind, taking $M = M_\sigma$, is

$$\dot{M}_{\text{w}} \equiv \dot{m} \dot{M}_{\text{Edd}} = \frac{4 f_c \dot{m} \sigma^4}{\eta c G}. \tag{6.91}$$

This is much smaller than the outflow rate $\dot{M}_{\text{out}}$ it pushes, so the flow is mass-loaded. We define the mass-loading factor as the ratio of the mass flow rate in the shocked ISM to that in the wind:

$$f_{\text{L}} \equiv \frac{\dot{M}_{\text{out}}}{\dot{M}_{\text{w}}} = \frac{\eta(\gamma + 1)f_g}{4\dot{m}} \frac{\dot{R}c}{f_c \sigma^2}. \tag{6.92}$$

This gives

$$\dot{M}_{\text{out}} = f_{\text{L}}\dot{M}_{\text{w}} = \frac{\eta(\gamma + 1)f_g}{4}\frac{\dot{R}c}{f_c\,\sigma^2}\dot{M}_{\text{Edd}}. \qquad (6.93)$$

If the AGN radiates at about its Eddington luminosity $\sim L_{\text{Edd}}$ we have $\dot{R} = v_e$ and (6.71) gives

$$f_{\text{L}} = \left(\frac{2\eta c}{3\sigma}\right)^{4/3}\left(\frac{f_g}{f_c}\right)^{2/3}\frac{l^{1/3}}{\dot{m}} \simeq 460\sigma_{200}^{-2/3}\frac{l^{1/3}}{\dot{m}}, \qquad (6.94)$$

and

$$\dot{M}_{\text{out}} \simeq 4060\,\sigma_{200}^{10/3}l^{1/3}\,\text{M}_\odot\,\text{yr}^{-1} \qquad (6.95)$$

for typical parameters $f_g = f_c$ and $\gamma = 5/3$. The effect of mass-loading means that this is even bigger than the dynamical infall rate $\dot{M}_{\text{dyn}}$ (eqn (6.13)). The total gas mass in the bulge is roughly $M_g \sim 10^3 f_g M_\sigma$, using the SMBH–bulge mass relation $M_b \sim 10^3 M$. We see that if the outflow persists for a time

$$t_{\text{clear}} \sim M_g/\dot{M}_{\text{out}} \sim 1 \times 10^7\sigma_{200}^{2/3}l^{-1/3}\,\text{yr}, \qquad (6.96)$$

it must sweep away a large fraction of the bulge gas. The actual outflow time needed for this depends on both the type and the environment of the galaxy, and because the estimate (6.96) is comparable to the Salpeter time, the final SMBH mass may increase by small but measureable amounts above the critical value M_σ. We will see later in this section that this gives three parallel but slightly offset M–σ relations, and means that power-law fits $M \propto \sigma^\alpha$ for large samples covering several galaxy types tend to find exponents α slightly larger than the canonical value $\alpha = 4$.

Equations (6.23), (6.95) imply

$$\frac{1}{2}\dot{M}_{\text{w}}v^2 \simeq \frac{1}{2}\dot{M}_{\text{out}}v_{\text{out}}^2, \qquad (6.97)$$

and most of the wind kinetic energy ultimately powers the mechanical energy of the outflow, as we would expect for energy driving. The continuity relations across the contact discontinuity show that if the quasar is still active, the shocked wind retains 1/3 of the total incident wind kinetic energy $\dot{M}_{\text{w}}v^2/2$, giving 2/3 to the swept-up gas represented by $\dot{M}_{\text{out}}$.

Equation (6.97) means that the swept-up gas must have a scalar momentum rate greater than the Eddington value L_{Edd}/c since we can rewrite it as

$$\frac{\dot{P}_{\text{w}}^2}{2\dot{M}_{\text{w}}} \simeq \frac{\dot{P}_{\text{out}}^2}{2\dot{M}_{\text{out}}}, \qquad (6.98)$$

where the $\dot{P}$'s are the momentum fluxes. With $\dot{P}_w = L_{Edd}/c$, we have

$$\dot{P}_{out} = \dot{P}_w \left(\frac{\dot{M}_{out}}{\dot{M}_w}\right)^{1/2} = \frac{L_{Edd}}{c} f_L^{1/2} \simeq 20 \frac{L_{Edd}}{c} \sigma_{200}^{-1/3} l^{1/6}. \tag{6.99}$$

The solutions (6.71) and (6.77) describe the motion of the contact discontinuity between the shocked wind gas and the interstellar gas it sweeps up in the snow-plough (see Figures 6.3 and 6.7). This interface is strongly Rayleigh–Taylor (RT) unstable[5] (see Figure 6.9) because the huge expansion of the shocked wind gas makes its mass density much lower than the swept-up interstellar gas outside it so that in gravitational terms we have a heavy fluid lying on a lighter one. The shocked bulge gas has temperature

$$T_{out} = \frac{3}{16} \frac{\mu m_H v_e^2}{k} \simeq 4.7 \times 10^7 \sigma_{200}^{4/3} \left(\frac{lf_c}{f_g}\right)^{1/3} \text{ K} \tag{6.100}$$

and number density

$$N_{out} = \frac{2 f_g \sigma^2}{\pi G \mu m_H R^2} \simeq 60 f \sigma_{200}^2 R_{kpc}^{-2} \text{ cm}^{-3}. \tag{6.101}$$

The RT instability drives strong overturning motions on every scale. These mix shocked, cooling bulge gas with the more dilute hot wind gas inside it, maintaining the isobaric uniform pressure condition within the whole outflow. This decreases with time as t^{-2} since $R \propto t$. The outflow has optical depth

$$\tau = \kappa N_{out} m_H R_{out} \simeq 0.2 f \kappa \sigma_{200}^2 R_{kpc}^{-1}. \tag{6.102}$$

This shocked interstellar gas must have a two-phase structure (see Figures 6.10, 6.11). We define the pressure-based ionization parameter

$$\Xi = \frac{L}{4\pi R^2 N_{out} kc T_{out}} = \frac{P_{rad}}{P_{gas}} \tag{6.103}$$

(Krolik, McKee & Tarter, 1981), where P_{rad}, P_{gas} are the radiation and gas pressures. Here

$$L = l L_{Edd} = \frac{4 f_c \sigma^4 c}{G} l \tag{6.104}$$

is the AGN luminosity. Using (6.100), (6.101) this gives

$$\Xi = 0.07 f^{-1/3} l. \tag{6.105}$$

Figures 4 and 5 in Krolik, McKee and Tarter (1981) show that for values of Ξ less than ~ 0.5 the outflowing gas can only be stable in a cold phase. The ionization parameter stays fixed as the outflow expands since both radiation and gas

[5] See AF for a discussion.

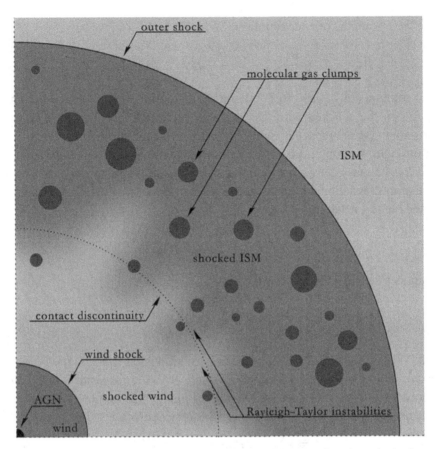

Figure 6.9 The multiphase nature of the energy-driven outflow launched when
the central SMBH reaches the critical M_σ mass. The AGN drives a powerful quasi-spherical wind with speed $v \sim 0.1c$ from its accretion disc. This
wind is strongly shocked just outside the inner Compton cooling radius (radius
$R_C \sim 0.5M_8/\sigma_{200}$ kpc, where M_8 is the black hole mass in units of $10^8 M_\odot$).
The expanding shocked wind sweeps up and drives a forward shock into the
host interstellar gas. The contact discontinuity between the expanding wind gas
and swept-up ISM is Rayleigh–Taylor unstable, so these two components mix
together in the outflow and help maintain a constant pressure. The outflow cools
radiatively, and most of it freezes out into clumps of cold molecular material. So
despite its high velocity $v \sim 1000$ km s^{-1}, most of the outflow is in molecular
form. The very high temperature of the wind shock near R_C implies that these outflows should be intrinsically luminous gamma-ray sources. Credit: Zubovas and
King (2014).

pressure decrease as R^{-2}. Then reaching an equilibrium requires the gas fraction f to decrease: as the gas cools and molecular clumps detach from the hot
flow, the latter's density drops and f eventually decreases enough (by a factor of
~ 400) that a two-phase equilibrium is established. This predicts that almost all
the outflow should be in molecular form, with only tenuous hot gas filling the

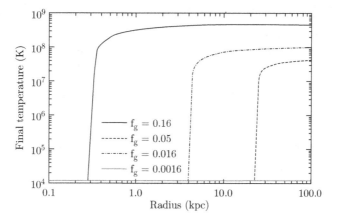

Figure 6.10 Temperature of shocked interstellar gas one dynamical time after the shock heats it, as a function of radius R. For low gas fractions and at large radii, radiative two-body cooling is inefficient, so the gas stays hot ($T > 10^7$ K). There is a sharp transition at a particular radius for every gas fraction; within this radius, gas is cool ($< 10^5$ K), and the resulting two-phase instability leads to the formation of molecules. Credit: Zubovas and King (2014).

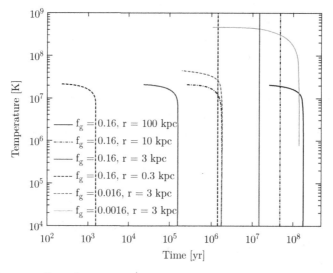

Figure 6.11 Cooling of pressure-confined gas in an outflow as a function of time for a range of gas fractions and outer shock radii. The gas is assumed to be stationary (an appropriate assumption since cooling is faster than dynamical time). The cooling process is slow at first, but then accelerates rapidly as the gas begins to contract and Compton heating becomes inefficient. Vertical lines show dynamical timescales at certain radii. Credit: Zubovas and King (2014).

rest of the outflow. Detailed simulations including molecular cooling (Richings & Faucher-Giguère 2018) confirm this.

The work of this section predicts that after reaching the M–σ mass, continuing SMBH accretion drives high-speed ($\sim 1000\,\mathrm{km\,s^{-1}}$; eqn (6.23)) outflows with prodigious mass rates ($\sim 4000 M_\odot\,\mathrm{yr^{-1}}$; eqn (6.95)). These are largely in molecular form, with corresponding momentum rates $\sim 20 L_{\mathrm{Edd}}/c$.

This is closely similar to what observations find. A significant number of galaxies have high-speed molecular outflows, which consistently have $\dot{M}_{\mathrm{out}} v_{\mathrm{out}}$ larger than L_{Edd}/c for their likely SMBH masses: Cicone et al. (2014) find that momentum rates $20L/c$ are common. This is an inevitable consequence of mass-loading ($f_L > 1$). These high momentum rates offer a way for galaxies to resist the accretion that cosmological simulations suggest still continues at large scales (Costa, Sijacki & Haehnelt, 2014). Specifically, observations show abundant evidence for molecular outflows with speeds and mass rates similar to (6.23) and (6.95). Feruglio et al. (2010), Rupke and Veilleux (2011), and Sturm et al. (2011) find large-scale (kpc) flows with $v_{\mathrm{out}} \sim 1000\,\mathrm{km\,s^{-1}}$ and $\dot{M}_{\mathrm{out}} \sim 1000\ M_\odot\,\mathrm{yr^{-1}}$ in the nearby quasar Mrk 231. Other galaxies show similar phenomena (see Tacconi et al., 2002; Veilleux, Rupke & Swaters, 2009; Riffel & Storchi-Bergmann, 2011a, 2011b; Sturm et al., 2011; see tables 1 and 2 in Zubovas & King, 2012a for a detailed comparison with the theoretical predictions). In each case it appears that AGN feedback is the driving agency. There is general agreement for Mrk 231, for example, that the mass outflow rate $\dot{M}_{\mathrm{out}}$ and the kinetic energy rate $\dot{E}_{\mathrm{out}} = \dot{M}_{\mathrm{out}} v_{\mathrm{out}}^2/2$ are too large to be driven by star formation, but comparable with values predicted for AGN feedback.

This all suggests that energy-driven outflows from SMBH that have just reached their M–σ masses should be able to sweep galaxy spheroids clear of gas. An observational test of this is the expected peak mechanical luminosity (cf. (6.25))

$$L_{\mathrm{mech}} \sim \frac{\eta}{2} L_{\mathrm{Edd}} \simeq 0.05L \qquad (6.106)$$

of the fastest outflows, where $L = l L_{\mathrm{Edd}}$ is the observed AGN luminosity. Figure 12 of Cicone et al. (2014) shows that observation does largely confirm the relation (6.106)). If the AGN are close to their Eddington luminosities (i.e $L \propto M \propto \sigma^4$ and $l \simeq 1$), the clearout rate $\propto \sigma^{10/3}$ (eqn (6.95)) should scale linearly with the driving luminosity. Figure 9 of Cicone et al. (2014) shows evidence for this correlation, with normalization close to that predicted. But we should also remember that the AGN luminosity L can vary rapidly, whereas the outflow velocity and mass rate are much more stable, and that outflows retain significant velocities quite long after the AGN driving stops (see the discussion after (6.77) and Figure 6.8). So we should expect the observed mechanical luminosities $L_{\mathrm{mech,\ obs}}$ to lie in the range

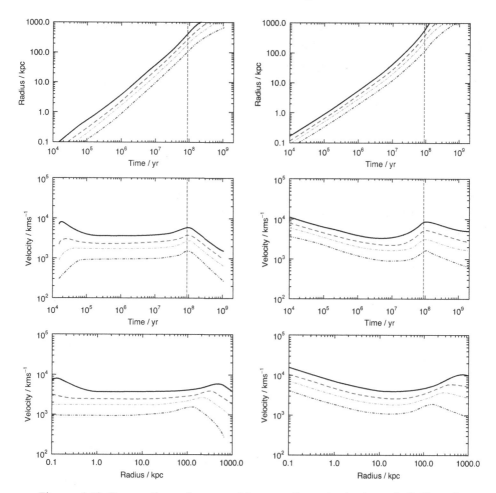

Figure 6.12 Propagation of energy-driven outflows in isothermal (left) and Navarro–Frenk–White (NFW) (right) haloes. The four outflows have $f_g = 3 \times 10^{-3}, 0.01, 0.03$, and 0.16 for the solid, dashed, double-dashed, and double dot-dashed curves, respectively. The SMBH is active for 90 Myr (dashed vertical line in the top and middle panels). As shown in this chapter, the outflow properties do not strongly depend on initial conditions (radius and velocity). Top panels: radius of the contact discontinuity as a function of time. Middle panels: outflow velocity as a function of time. Bottom panels: outflow velocity as a function of radius. Credit: Zubovas and King (2012b).

$$L \lesssim L_{\text{mech, obs}} \lesssim 0.05 L_{\text{Edd}}. \qquad (6.107)$$

Observations largely confirm this range (e.g. figure 12 of Cicone et al., 2014). Figure 6.19 shows a case where the agreement is striking.

It is important to note that in active galaxies where a UFO wind is observed, but a strong molecular outflow either is not observed or does not show a strong momentum enhancement over the UFO (i.e. $(\dot{M}v)_{\text{molecular}}$ is not $>> (\dot{M}v)_{\text{UFO}} \simeq L_{\text{Edd}}/c$), it is likely that the SMBH has not yet reached the M–σ mass. Examples of systems like this are given by Bischetti et al. (2019); Sirressi et al. (2019); and Reeves and Braito (2019). In principle, determining the SMBH masses and the relevant σ in such cases offers a direct test of the fundamental point that the M–σ relation signals the change between momentum and energy-driven feedback by a black hole wind.

As with the momentum driving that specifies the M–σ relation, we should consider the effects on energy-driven outflows of changing the spheroid potential from the simple isothermal one to more complex and possibly realistic ones. As a representative case, we consider an NFW halo with a concentration parameter (Navarro, Frenk & White, 1997) $c = 10$, scale radius $R_s = 25$ kpc (so that the virial radius $R_V = 250$ kpc), and virial mass $M_V = 3.3 \times 10^{12} M_{\odot}$. This gives $\sigma_{\text{peak}} = 200$ km s^{-1} at $R_{\text{peak}} = 2.15 R_s = 54$ kpc. We take the initial SMBH mass to be $M_0 = 3.68 \times 10^8 M_{\odot}$, equal to the critical SMBH mass found in previous papers (e.g. King, 2010). We compare this NFW halo with an isothermal halo with $\sigma = 200$ km s^{-1}. We do not truncate either halo beyond R_V, but allow them to continue to infinity. First we derive a generic equation of motion, valid for any spherically symmetric mass profile, provided that the gas density traces the background potential density, that is, $M_g(R) = f_g M(R)$ for any R, where f_g is the gas fraction. With this change, we repeat the derivation of (6.66), (6.67), which gives

$$\frac{\eta}{2 f_g} L_{\text{Edd}} = \frac{3}{2} \frac{GM^2 \dot{R}}{R^2} + \frac{GM\dot{M}}{R} + \frac{3}{2} GM\dot{R}\ddot{R} + \frac{3}{2}\dot{M}\dot{R}^2 \qquad (6.108)$$
$$+ \frac{1}{2} MR\dddot{R} + \dot{M}R\ddot{R} + \frac{1}{2}\ddot{M}R\dot{R}.$$

Here

$$\dot{M} = \frac{\partial M}{\partial R}\dot{R} \quad \text{and} \quad \ddot{M} = \frac{\partial M}{\partial R}\ddot{R} + \frac{\partial^2 M}{\partial R^2}\dot{R}^2.$$

This reduces to (6.67) for an isothermal potential.

Figure 6.12 shows the results of numerical integrations comparing the isothermal and NFW potentials. This makes it clear that the isothermal potential is probably the worst-case scenario for the escape of an outflow, in the sense of giving upper limits on the required values for the AGN activity timescale and the final SMBH mass. It is easy to see this – the weight of the gas is $\propto \sigma^4$. This is constant with radius R for the isothermal case, but variable for an NFW halo, and always smaller for a given R, since in NFW we have $\sigma \leq \sigma_{\text{peak}} = \sigma_{\text{isothermal}}$.

6.7 Effects of a Galaxy Disc: Triggered Starbursts

We have so far considered the galaxy spheroid largely in isolation. The SMBH scaling relations relate its mass to properties of the central bulge alone, although we will see in the next section that there are small but systematic differences in the scalings related to the wider galaxy history and morphology. But the energy-driven outflows we considered in the previous section are global: they expand to scales far larger than their initial sizes, and unless the galaxy is an elliptical they inevitably encounter its disc as they expand. So we should consider this interaction. In a gas-rich galaxy the gas in the innermost region of the disc at radius R_0 must be close to self-gravitating. Assuming a roughly isothermal potential implies a gas density $\sim \rho_d \sim 2\sigma^2/R_0$. This exceeds the bulge gas density by the factor $\sim 1/f_g \sim 10$. Equation (6.69) shows that higher gas densities give lower radial outflow velocities, as they meet greater resistance. Then when a spherical outflow encounters a high-density gas disc it flows around it, over its plane upper and lower faces. But the pressure in the outflow is at least initially higher than in the disc: the pressure at the contact discontinuity from (6.58) is

$$P_{CD} = \frac{1}{4\pi R^2}\left[\frac{d}{dt}[M(R)v_e] + \frac{GM^2(R)}{f_g R^2}\right] = \frac{f_g \sigma^2(v_e^2 + \sigma^2)}{\pi GR^2} \tag{6.109}$$
$$\simeq 4 \times 10^{-7}\sigma_{200}^{10/3} l^2 R_{\text{kpc}}^{-2} \text{ dyn cm}^{-2},$$

and the pressure at the forward shock into the ISM is

$$P_{fs} = \frac{4}{3}\rho(R)v_e^2 = \frac{2f_g\sigma^2 v_e^2}{3\pi GR^2} \simeq \frac{2}{3}P_{CD}. \tag{6.110}$$

These values contrast with the mid-plane pressure in a disc close to self-gravitating, which is

$$P_{\text{disc}} \sim \rho c_s^2 \sim \rho\sigma^2 \sim 2\frac{\sigma^4}{GR_d^2}. \tag{6.111}$$

Here we have assumed the sound speed $c_s \sim \sigma$ and taken the self-gravity condition as $G\rho \sim \Omega^2$, with $\Omega = \sqrt{2}\sigma/R_d$ the Kepler frequency at disc radius R_d. So when the outflow shock flows over the disc faces at $R = R_d$, its pressure is a factor $\sim (v_e/\sigma)^2 \sim 25$ higher than the discs, and a significant pressure excess remains until the outflow shock has travelled out to radii $R > R_d v_e/\sigma \sim 5R_d$. So we know the outflow must significantly compress the disc gas. This must set off a burst of star formation in the disc (see Thompson, Quataert & Murray 2005, appendix B), and here it reaches

$$\dot{\Sigma}_* \sim 2000\epsilon_{-3}\sigma_{200}^{10/3} l^{2/3} R_{\text{kpc}}^{-2} \text{ M}_\odot \text{ kpc}^{-2} \tag{6.112}$$

(Zubovas et al., 2013). Here $\epsilon_* = 10^{-3}\epsilon_{-3}$ is the efficiency of massive stars in converting mass into radiation, and we have substituted for v_e using (6.71). Zubovas et al. (2013) show that this leads to a starburst of total luminosity

$$L_* \simeq 5 \times 10^{47} L_{46}^{5/6} l^{-1/6} \text{ erg s}^{-1}, \tag{6.113}$$

where L_{46} is the AGN luminosity in units of 10^{46} erg s^{-1}. Such systems would appear as ULIRGs (ultraluminous infrared galaxies).

This suggests that in a galaxy with both a bulge and a disc, the clearout phase following the arrival of the SMBH mass at M_σ eventually leaves the galaxy bulge without gas, but depending on parameters, is accompanied by a starburst in the disc. Observations of dusty QSOs appear to show this, with the black hole mass already on the M–M_b relation (6.1), and so fully grown. In an elliptical, on the other hand, clearout must leave the galaxy 'red and dead'.

For different combinations of parameters a wide variety of star formation episodes are possible (Zubovas et al., 2013) (see Figures 6.13, 6.14). For illustration we describe two of these: in the first case the initial SMBH mass is set equal to the theoretical M_σ value and allowed to grow and radiate at the Eddington limit indefinitely. In the second case the outflow is effectively channelled away from the plane of the galaxy, only affecting the disc within 4 kpc of the centre.

Since a galaxy disc is a major obstacle to an outflow, it follows that it may be able to divert a quasi-spherical outflow into a bipolar shape. This is particularly true in cases where the SMBH mass grows only a little, in a minor accretion event. Zubovas, King and Nayakshin (2011) suggest that the gamma-ray-emitting bubbles disposed symmetrically about the plane of the Milky Way (Su, Slatyer, & Finkbeiner, 2010; see Figures 1.6 and 1.7) may be the remnants of a relatively recent and rather weak event like this. An outflow of this type would be a natural effect of a short but bright accretion event on to Sgr A*, if this happened concurrently with the well-known star formation event in the inner 0.5 pc of the Milky Way $\sim$ 6 Myr ago. The resulting near-spherical outflow is focused into a pair of symmetric lobes by the greater gas pressure along the Galactic plane, as we have considered in this section. The outflow shocks against the interstellar gas of the Galaxy bulge and could produce cosmic rays which power the observed gamma-ray emission. Away from the Galactic plane the lobes have scale $\sim$ 5 kpc and gamma-ray luminosity $L_\gamma \simeq 4 \times 10^{37}$ erg s^{-1}. Their total energy content is at least 10^{54-55} erg.

Outside the Galactic plane, the present-day Milky Way has a much lower gas fraction $f \sim 10^{-2} f_c$, where $f_c \simeq 0.16$ is the cosmological value. With the measured SMBH mass $4 \times 10^6 M_\odot$, and velocity dispersion $\sigma \simeq 100$ km s^{-1}, the cooling radius R_C is a few parsecs, far smaller than the current lobe size, so the lobes must represent an energy-driven outflow, with typical radial velocity

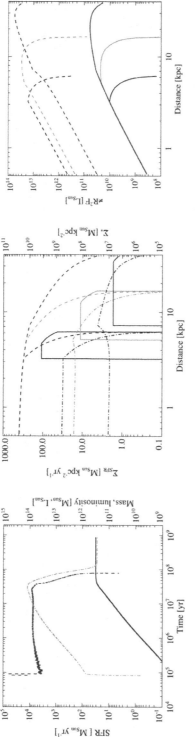

Figure 6.13 Properties of a starburst induced by AGN activity. The SMBH initially has a mass $3.68 \times 10^8 M_\odot = M_\sigma$ and grows and radiates at the Eddington rate. The outflow propagates in a static background NFW potential. Left-hand panel: time evolution of star formation rate (SFR) (dashed line, scale on the left), total mass of gas converted into stars (solid line) and luminosity of the newly formed stellar population (dot-dashed line; both scales on the right). The whole gas disc, $M_g = 3 \times 10^{11} M_\odot$, is converted in 50 Myr. Middle panel: radial plots at 3, 10, and 30 Myr of SFR density (solid, scale on the left), surface density of low-mass stars (dashed), and massive stars (dot-dashed; both scales on the right). Right-hand panel: radial plot of stellar luminosity $\pi R^2 F_*$ of the low-mass (solid) and massive (dashed line) stars; coding identical to the middle panel. Credit: Zubovas et al. (2013).

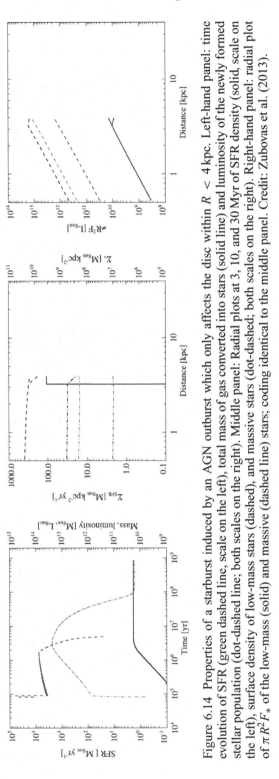

Figure 6.14 Properties of a starburst induced by an AGN outburst which only affects the disc within $R < 4 \, \mathrm{kpc}$. Left-hand panel: time evolution of SFR (green dashed line, scale on the left) and luminosity of the newly formed stellar population (dot-dashed line; both scales on the right). Middle panel: Radial plots at 3, 10, and 30 Myr of SFR density (solid, scale on the left), surface density of low-mass stars (dashed), and massive stars (dot-dashed; both scales on the right). Right-hand panel: radial plot of $\pi R^2 F_*$ of the low-mass (solid) and massive (dashed line) stars; coding identical to the middle panel. Credit: Zubovas et al. (2013).

$$v_e = 1640\sigma_{100}^{2/3} f_{-2}^{-1/3} \text{ km s}^{-1}, \tag{6.114}$$

where f_2 is the ratio f_g/f_c in units of 0.01. The outflow must currently be continuing, as assuming it has stalled requires it to have been launched by an accretion event at least 15 Myr ago, well before the most recent event at about 6 Myr in the past (Zubovas, King, & Nayakshin, 2011). The mass accreted in the quasar-like accretion event must have been $\sim 2 \times 10^3 M_\odot$ to power the lobes. Gamma-ray emission could come from cosmic rays created either by Sgr A* directly or through acceleration in the shocks with the external medium. The Galaxy disc remains unaffected, agreeing with the observational evidence that SMBHs do not correlate with galaxy disc properties. It is easy to show that this outflow would have been strongly resisted by the mass $\sim 5 \times 10^7 M_\odot$ of the Galaxy's central molecular zone, accounting for the narrow waist of the outflow, and its appearance as a pair of lobes placed symmetrically each side of the Galactic plane.

Alternative interpretations as a pair of jets require them to be directed symmetrically about this plane. As we have mentioned several times, observations clearly show that jets are quite randomly oriented with respect to the host galaxy, so this is implausible.

6.8 The Three M–σ Relations

In this chapter we have seen that the arrival of the black hole mass at the M–σ relation causes its feedback to change from momentum driving to energy driving. The energy-driven phase clearing the gas from a galaxy bulge is much more violent. But it is obvious that to eject the gas, the black hole must inject a non-negligible amount of accretion energy during this phase, which requires some black hole mass growth. Evidently, we need to estimate the mass increment ΔM needed for this, as a requirement $\Delta M \gg M_\sigma$ would mean we had failed to explain the M–σ relation.

Two things influence the mass increment ΔM. For an SMBH in a spiral galaxy with a relatively small bulge, ΔM is presumably significantly smaller than, for example, in an elliptical, where the much larger bulge mass means that energy driving by the central SMBH wind must continue for longer in order to expel the remaining gas. Second, galaxies in clusters, particularly those near the centre, may continue to accrete gas during the energy-driven phase and so need a larger ΔM.

For a typical bulge radius $R_b \sim 1$–2 kpc in a spiral galaxy, the time between reaching M–σ and finally clearing the bulge is

$$t_{\text{out}} \sim \frac{R_C}{\sigma} + \frac{R_b - R_C}{v_e} \sim \frac{500 \text{ pc}}{\sigma} + \frac{1 \text{ kpc}}{10^3 \text{ km s}^{-1}} \sim 3 \times 10^6 \text{ yr} \tag{6.115}$$

As this is smaller than the Salpeter time, we would expect the SMBH mass in spiral galaxies to be fairly close to the value (8.18).

But in an elliptical it is likely that the outflow has to drive the gas beyond the virial radius R_V. We make the standard assumption that R_V is the point where the mean density is 200 times the critical density for the Universe. Modelling the galaxy as an isothermal sphere then gives

$$R_V = \frac{\sigma}{5\sqrt{2}H}, \tag{6.116}$$

where

$$H(z) = 100h(z)\,\text{km s}^{-1}\,\text{Mpc}^{-1} \tag{6.117}$$

is the Hubble parameter at redshift z, and $h(z)$ is the dimensionless Hubble parameter. Then driving the gas beyond the virial radius requires that the stalling radius $R_{\text{stall}} > R_V$. From (6.86) this requires

$$t_{\text{acc}} > \frac{1}{7H}\frac{\sigma^2}{v_e^2}, \tag{6.118}$$

which gives

$$t_{\text{acc}} \gtrsim 10^8 \left(\frac{\sigma_{200}f_g}{\eta_{0.1}f_c}\right)^{2/3} \text{yr}, \tag{6.119}$$

where $\eta_{0.1} = \eta/0.1$.

This is about two Salpeter times, so in an elliptical the SMBH must grow more, to a final mass

$$M_{\text{final}} \sim e^2 M_\sigma \sim 7.5 M_\sigma. \tag{6.120}$$

The second factor affecting M, M_b is the galaxy environment. Many ellipticals are in clusters and can gain gas as they orbit through the intracluster gas. A sign of this is their extra luminosity: Brightest cluster galaxies (BCGs) are near the centres of cluster potentials, and several are known to contain unusually massive SMBH (McConnell et al., 2011), suggesting that this is a result of accreting gas from the cluster.

The extra black hole mass growth required to remove the bulge gas and so terminate SMBH growth, together with the extra mass a galaxy in a cluster may gain from its surroundings, implies three parallel but slightly offset M–σ relations $M \propto \sigma^\alpha$, with $\alpha \simeq 4$ for spirals, field, and cluster ellipticals (see Figure 6.15). In principle there is also a relation for cluster spirals, but these are rare. The figure shows that the spread in offsets means that an observed sample drawn from galaxies of all three types tends to give a slope slightly steeper than the individual ones for each type. This may be the reason for the slight discrepancy between

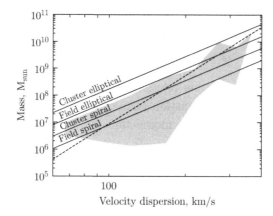

Figure 6.15 The four (in reality three, as cluster spirals are rare) M–σ relations (solid lines). The dashed line shows that their combined effect that fits to an observational sample drawn from all three types of galaxy tends to steepen the fitted slope of the M–σ relation. The solid lines have slopes $M \propto \sigma^4$ and the dashed lines have $M \propto \sigma^6$. The grey area is the approximate locus of data points in Figure 3 of McConnell et al. (2011); see also McConnell and Ma (2013). Credit: Zubovas and King (2012b).

the observed overall slope $\alpha = 4.4 \pm 0.3$ and the theoretical value of 4. All three types of galaxies obey similar M–M_b relations (6.1) within errors since a higher M_b causes growth of the SMBH above M_σ.

6.9 The Black Hole–Bulge Mass Relation

So far we have concentrated almost exclusively on understanding the causes of the M–σ relation and have said very little about the second scaling relation, that is, the linear relation $M \simeq 10^{-3} M_b$. We arrived at a picture where the SMBH mass cannot grow significantly beyond the critical value M_σ.

But the nature of the M–M_b relation must be very different. If feedback ensures that the velocity dispersion σ sets the black hole mass M, we cannot argue at the same time that M_b independently sets M. Although reaching M_σ means that gas is cleared out of the bulge and star formation is halted, this does not explain why the stellar mass at this point is $\sim 10^3 M_\sigma$.

So the connection between the black hole mass M and the stellar bulge mass M_b cannot be directly causal. Instead, whatever fixes M_b must also make $M_b \propto \sigma^4$. In fact this has long been known, at least for the largest spheroids – elliptical galaxies – in the form of the Faber–Jackson relation

$$L_* \sim 2 \times 10^{10} \mathrm{L}_\odot \sigma_{200}^4 \tag{6.121}$$

Faber & Jackson (1976), where L_* is the total stellar luminosity. Then for mass-to-light ratios ~ 5 we immediately get the stellar mass as

$$M_* \sim 1 \times 10^{11} M_\odot \sigma_{200}^4 \sim 10^3 M_\sigma. \tag{6.122}$$

It appears plausible that this relation is set by stellar momentum feedback ultimately suppressing star formation (see, e.g., Murray, Quataert, and Thompson, 2005). Star formation injects momentum into the gas trying to form stars at the rate

$$\dot{p}_* \sim \epsilon_* c \dot{M}_*, \tag{6.123}$$

where M_* is the total stellar mass, and $\epsilon_* \sim 10^{-3}$ is the total luminous energy released by a main-sequence star in units of its rest-mass energy. Then the total momentum produced by star formation is

$$p_* \sim \epsilon_* c M_*. \tag{6.124}$$

If this gives the star-forming gas a momentum $> \eta\sigma$ per unit mass, with $\eta \sim 1$, this inhibits star formation by preventing local collapse. We expect a bulge gas mass $M_0 \sim M_{g,\mathrm{vir}}$ before star formation begins, where $M_{g,\mathrm{vir}}$ is the virial value. Then star formation is self-limiting at the value $M_b < M_0 = M_{\max}$ such that $p_* \gtrsim \eta M_0 \sigma$, that is,

$$\epsilon_* c M_{\max} \simeq \eta M_0 \sigma \tag{6.125}$$

or

$$M_b \lesssim \eta M_0 \frac{\sigma}{\epsilon_* c} \lesssim 0.6 \eta M_0 \sigma_{200}. \tag{6.126}$$

With the usual assumption that matter is virialized within a radius such that the mean density is 200 times the critical value, we have a virial radius

$$R_V = \frac{\sigma}{5\sqrt{2}H}, \tag{6.127}$$

where $H(z) = 100 h(z)\,\mathrm{km\ s^{-1}\ Mpc^{-1}}$ is the Hubble constant at redshift z, so a virial mass

$$M_V = \frac{2\sigma^2 R_V}{G} = \frac{\sqrt{2}}{5H} \frac{\sigma^3}{G}, \tag{6.128}$$

which from (6.126) implies a bulge gas mass

$$M_b \lesssim 0.6 \eta f_g \sigma_{200} \frac{\sqrt{2}}{5H} \frac{\sigma^3}{G} = 2.4 \times 10^{11} \eta \frac{f_g}{0.16} \frac{\sigma_{200}^4}{h(z)} M_\odot. \tag{6.129}$$

Combining this with (6.61), we find

$$\frac{M_\sigma}{M_b} \gtrsim \frac{1.25 \times 10^{-3} h(z)}{\eta}. \tag{6.130}$$

We see that M/M_b is bigger at higher redshift, that is, galaxies completing their evolution at high z have SMBH with relatively larger masses compared to their bulges. At low redshift we have

$$M_b \sim \frac{0.14 f_g t_H \sigma^4}{\epsilon_* c G}. \tag{6.131}$$

t_H is the Hubble time, and

$$M_\sigma \sim \frac{1.8 \kappa \epsilon_* c}{\pi G t_H} M_b \sim 10^{-3} M_b \tag{6.132}$$

(Power et al., 2011), which is similar to observational estimates (cf. 6.10).

This derivation illustrates an important point: in the feedback picture, there is no physics in the M–M_b relation. This arises purely because both M and M_b are separately proportional to σ^4. They each result from momentum-driven feedback, but from black hole accretion and star formation, respectively.

This raises the possibility that in nucleated galaxies (i.e. those whose central regions are dominated by nuclear star clusters, with no detectable sign of the presence of a supermassive black hole) there should be an offset M–σ relation between the mass of the cluster and the velocity dispersion. It is straightforward to show that momentum feedback from a star cluster of a given mass is about 20 times less efficient than for Eddington-limited SMBH accretion (McLaughlin, King & Nayakshin, 2006). This should then lead to an offset $M_C - \sigma$ relation between total cluster mass M_C and velocity dispersion, that is,

$$M_c \simeq 20 M_\sigma \simeq 6 \times 10^9 \sigma_{200}^4 \mathrm{M}_\odot. \tag{6.133}$$

Typically these galaxies are small, with $\sigma < 120 \, \mathrm{km \, s^{-1}}$. Observations aimed at identifying a relation like (6.133) leave the possibility unclear. Ferrarese et al. (2006) do observe a similar relation, but other observations find relations with different powers of σ (2.73 $\pm$ 0.29, Leigh, Böker & Knigge 2012; and $\sim$2.1 $\pm$ 0.31, Scott & Graham, 2013).

6.10 The Assembly Picture of the Scaling Relations

So far we have considered the SMBH scaling relations only from the point of view of feedback. But a very different idea (Peng, 2007; Jahnke & Macciò, 2011) asserts that the scaling relations do not result from black hole feedback, but are essentially statistical. If the SMBH and their host galaxy spheroids are both products of mergers of large numbers of much smaller galaxies whose black hole and bulge masses are uncorrelated, the central limit theorem implies a linear relation $M \propto M_b$. The dispersion of this relation must tighten at larger M, M_b because these black holes and galaxies are on average the results of larger numbers of mergers. (In practice, the fit to the observed M–M_b relation predicted by Jahnke & Macciò (2011) is improved beyond what this argument alone would give, by including the effects of star formation, black hole accretion, and converting some of the stars formed in the disc component of each halo to bulge mass.)

This assembly picture does not directly give an M–σ relation, but for medium-to high-mass galaxies we do get a relation $M \propto \sigma^4$ by using the Faber–Jackson relation ((6.135), (6.122)) since it gives $M_b \propto \sigma^4$. Evidently the normalizations of the two scaling relations that result are given by the distributions of uncorrelated black hole and bulge mass pairs in the original seed galaxies. So provided that the assembly of random seed pairs produces an M–M_b relation of the right normalization (i.e. $M \simeq 10^{-3} M_b$, the observed M–σ normalization (6.61) is reproduced for these larger galaxies too).[6]

But the predicted outcomes of the assembly and feedback pictures differ sharply for low-mass galaxies ($M_b \lesssim$ few $\times\, 10^9 M_\odot$). The observed sample of central black holes in these galaxies is now large enough to explore this, and the known M–σ relation now extends down to low black hole masses $M \lesssim 10^6 M_\odot$ (e.g. Baldassare et al., 2020, figure 3). This is expected in the feedback picture, as there all galaxies limit the growth of their central black holes through the physics described in Section 6.5, giving $M \propto \sigma^4$. In addition, we should observe energy-driven winds expelling the gas that would otherwise grow the black hole mass beyond M_σ, and we shall see that this is also observed.

But in the assembly picture galaxies of sufficiently low mass do not experience enough mergers to produce a tight relation between M and M_b (see Figure 6.16). In general the predicted black hole masses lie above the extrapolation of the M–M_b relation and are noticeably bigger than observed values, with $M \gtrsim 10^7 M_\odot$ at stellar masses $M_* \lesssim 10^9 M_\odot$. These more massive black holes should have larger spheres of influence, where stars move faster, making it unlikely that observational selection effects make them hard to find. Evidently this difficulty appears because the upper limits on the initial seed black holes mean that the central black hole becomes too large after only only a few mergers. One could try to overcome this by adopting a lower value for the upper limit of the seed masses than the $10^4 M_\odot$ assumed by Jahnke and Macciò (2011). Clearly, the predicted mass scatter at low redshift must scale as $\sim 1/\sqrt{N}$, so compatibility with observed masses requires seed black hole masses $\lesssim 100 M_\odot$.

But as well as these very tight constraints, the assembly picture has a problem in reproducing the M–σ relation at low galaxy masses. As we detailed in Section 6.9, for medium to high galaxy masses the M–M_b relation then follows

---

[6] We will see how mergers fit with the feedback picture of these scalings in Chapter 8.

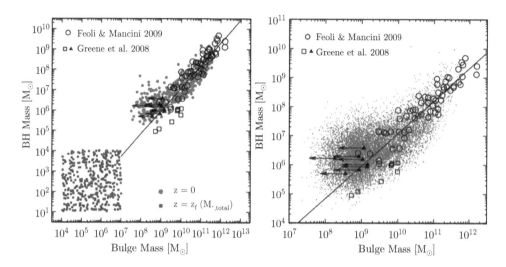

Figure 6.16 The assembly picture of the SMBH mass–bulge mass scaling relation. Black holes and bulges (round points) at low redshift $z = 0$ are formed from repeated mergers at high redshift of 'seed' black holes and associated bulge pairs, drawn randomly from the distribution of square points (lower left of the left-hand figure). The right-hand figure shows the resulting M–M_b relation in close-up, compared with observations. Credit: Jahnke and Macciò (2011).

by using the Faber–Jackson relation $L_b \propto \sigma^4$. But at low galaxy masses, the relation between stellar properties and the velocity dispersion changes. Galaxies with $\sigma \lesssim 100\,\mathrm{km\,s^{-1}}$ are instead observed to have

$$M_b \propto L_b \propto \sigma^2 \qquad (6.134)$$

(e.g. Kourkchi et al., 2012). Assuming continuity between the two relations (6.122), (6.134) at $\sigma \simeq 100\,\mathrm{km\,s^{-1}}$, this flatter σ^2 relation predicts an M_b four times larger than that from the σ^4 relation at $\sigma = 50\,\mathrm{km\,s^{-1}}$. The usual $M \sim 10^{-3} M_b$ gives

$$M_b \simeq 2 \times 10^9 \sigma_{50}^2 \mathrm{M_\odot}, \qquad (6.135)$$

where σ_{50} is the galaxy velocity dispersion σ in units of $50\,\mathrm{km\,s^{-1}}$.

This flatter relation $M_b \sim \sigma^2$ implies that dwarf galaxies all have roughly similar radii R_b: approximating them as isothermal spheres, that is,

$$R_b \simeq \frac{GM_b}{2\sigma^2}, \qquad (6.136)$$

gives

$$R_b \sim 1\,\mathrm{kpc}, \qquad (6.137)$$

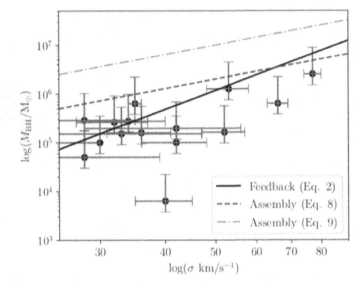

Figure 6.17 The M–σ relation for dwarf galaxies (Baldassare et al., 2020). The $M \propto \sigma^2$ relations (6.138), (8.18) predicted by the assembly picture are the dashed and dash-dot lines, while the solid curve is the original $M \propto \sigma^4$ relation (1.20) predicted by feedback. Credit: King and Nealon (2021).

effectively independent of σ or M_b. The reason for this is unclear, but low-mass galaxies do seem to appear with constant radii in cosmological simulations, possibly because the Jeans length is $\sim 1\,\mathrm{kpc}$ for gas temperatures $\sim 10^4\,\mathrm{K}$ typical of the warm ISM.

Assembly always gives a linear M–M_b relation, which then predicts an M–σ relation flattening to

$$M_\sigma \simeq 2 \times 10^6 \sigma_{50}^2 \, \mathrm{M}_\odot \tag{6.138}$$

for $\sigma \lesssim 100\,\mathrm{km\,s^{-1}}$, in contradiction to the observed $M_\sigma \propto \sigma^4$.

Kormendy and Ho (2012) find a larger normalization $M \simeq 5 \times 10^{-3} M_b$ for the M–M_b relation than Häring and Rix (2004), so this relation would become

$$M_\sigma \simeq 1 \times 10^7 \sigma_{50}^2 \, \mathrm{M}_\odot \tag{6.139}$$

in this case. We see from (6.55) that this normalization implies significantly larger radii for dwarfs compared with the sizes seen in figure 2 of Manzano-King, Canalizo and Sales (2019).

Figure 6.17 compares the two relations (6.138), (8.18) with the data from Baldassare et al. (2020). This figure also shows the original $M \propto \sigma^4$ relation (6.61) for comparison.

A second type of observation offers another test of the origin of the scaling relations. Spatially resolved kinematic measurements (Manzano-King, Canalizo & Sales, 2019) of dwarf galaxies in the stellar mass range $M_b \sim 6 \times 10^8 - 9 \times 10^9 \mathrm{M}_\odot$ find ionized gas outflows out to distances up to 1.5 kpc, with velocities above the escape value for their dark matter halos. There are line-ratio indications of AGN activity in most of the galaxies with outflows, and in most cases the outflow appears to be driven by the AGN rather than a starburst.

Mild outflows are allowable in the assembly picture, but have no particular significance for it. But as we have seen, powerful outflows with very specific properties are inevitable in the feedback picture. Once M reaches the value (1.20), all of the mechanical energy of the UFO wind is communicated to the host's bulge ISM in a forward shock, driving this gas away in an energy-driven outflow. In an isothermal potential this has speed (6.88), that is,

$$v_{\mathrm{out}} \simeq 1230\sigma_{200}^{2/3} \left(\frac{lf_c}{f_g}\right)^{1/3} \mathrm{km\,s}^{-1}, \tag{6.140}$$

where $l \sim 1$ is the ratio of the driving SMBH accretion luminosity to the Eddington value, and $f_c \simeq 0.16$ is the cosmological mean value of f_g. (The dark matter halo at larger radii has no effect on the baryonic physics fixing v_{out} and M_σ.) As before, the corresponding mass outflow rate is

$$\dot{M}_{\mathrm{out}} = 3700\sigma_{200}^{8/3} l^{1/3} \, \mathrm{M}_\odot \, \mathrm{yr}^{-1}, \tag{6.141}$$

where f_g has been taken equal to $f_c = 0.16$ (in King & Pounds, 2015, the corresponding equation (eqn (57)) gives the exponent of σ incorrectly as 10/3 rather than 8/3). As the energy-driven outflow described by (6.23, 6.141) begins to escape the baryonic part of the galaxy it must accelerate above the speed given by (6.23).

To apply this formalism directly to dwarf galaxies would need velocity dispersions σ, but these are not measured in the sample of Manzano-King, Canalizo & Sales (2019) However, figure 2 of that paper shows that (6.137) gives a good representation of the visible size of these galaxies. Then using (6.135) eliminates σ from (6.23) in favour of M_b, giving

$$v_{\mathrm{out}} \simeq 387 M_9^{1/3} x^{1/3} \, \mathrm{km\,s}^{-1}, \tag{6.142}$$

where $M_9 = M_b / 10^9 \mathrm{M}_\odot$, and

$$x = \frac{lf_c}{R_{\mathrm{kpc}} f_g}, \tag{6.143}$$

with $R_{\mathrm{kpc}} = R_b / \mathrm{kpc} \simeq 1$ the radius of the visible galaxy, and we expect $x \sim 1$. Table 1 (column 7) of Manzano-King, Canalizo & Sales (2019) shows that observed outflow velocities are significantly larger than given by (6.142). This is just as we expect if the outflows have already largely escaped the visible galaxies. This is very plausible given their large spatial scales, of order the half-light radii.

We also expect that the mass outflow rate in these galaxies should be rather higher than the values $\dot{M}_{\mathrm{out}} \simeq 100 M_{\odot}\,\mathrm{yr}^{-1}$ implied by (6.141) with $\sigma \sim 50\,\mathrm{km\ s}^{-1}$. If feedback is continuous, a galaxy will lose most of its gas in a total time

$$t_{\mathrm{deplete}} \sim \frac{f_g M_b}{\dot{M}_{\mathrm{out}}} \sim 10^6 \frac{M_9}{\sigma_{50}^{8/3}}\,\mathrm{yr}, \tag{6.144}$$

where $M_9 = M_b/10^9 M_{\odot}$, and we have used (6.55), (6.137). Again replacing σ_{50} with M_9 from (6.135), we find

$$t_{\mathrm{deplete}} \sim 2.5 \times 10^6 M_9^{-1/3}\,\mathrm{yr}. \tag{6.145}$$

This gives depletion times of a few million years for all dwarf galaxies in the energy-driven stage of AGN feedback that we expect when the SMBH mass approaches M_{σ}.

The Faber–Jackson-like relation (6.134) for dwarfs means that their total stellar masses vary as σ^2 rather than σ^4. Then, since the $M_{\sigma} \propto \sigma^4$ relation appears to extend down to dwarf galaxies, there is no longer a linear relation between M and M_b. From (6.61), (6.135) we instead get

$$M \simeq 4 \times 10^4 M_9^2 M_{\odot} R_{\mathrm{kpc}} \tag{6.146}$$

for SMBH masses close to M_{σ}, where factor $R_{\mathrm{kpc}} = R_b/(1\,\mathrm{kpc}) \sim 1$.

Figure (6.18) gives a comparison with observations and also with the M–M_b relation found by Schutte, Reines and Greene (2019). It appears that SMBH are less massive relative to their hosts at low galaxy masses. This in turn hints that in dwarf galaxies, the stellar feedback fixing M_b may be less able to remove gas before it makes stars. In line with this, Garratt-Smithson et al. (2019) find that gradual (rather than instant) stellar feedback delays the unbinding of most of the gas in dwarf galaxies by forming 'chimneys' in the dense shell surrounding the hot feedback region. These then vent the hot gas from the galaxy before its feedback effect can remove much of the star-forming gas.

It seems that the central black holes in dwarf galaxies play a similar active role in their evolution as do SMBH in more massive galaxies, even though they may be relatively less massive compared with their hosts. We conclude that the extension of the tight M–σ relation for massive galaxies down to dwarf galaxies with low-mass ($M \sim 10^5$–$10^6 M_{\odot}$) black holes is natural in the feedback picture of the scaling relations, but hard to reconcile with the assembly picture.

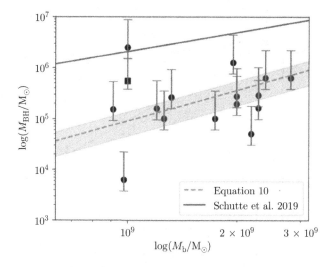

Figure 6.18 The quadratic M–M_b relation ((6.146), with best-fit value $R_b = 2.3 \pm 1.1$ kpc) for low-mass galaxies plotted against the data from table 1 of Baldassare et al. (2020) and references therein, together with the point from Davis et al. (2020). The linear M–M_b relation found by Schutte, Reines and Greene (2019) for the full SMBH sample for all galaxies is plotted for comparison. Credit: King and Nealon (2021).

The apparent inability of the assembly picture to predict the observed SMBH–galaxy relations at all masses arises because it implicitly assumes that the relation between M and M_b is fundamental and so must predict the M–σ relation. But the wide discrepancy between SMBH and bulge binding energies already noted in Section 6.1 suggests instead that the basic physical process is feedback, and the fundamental relation is between M and σ. As we have argued, the M–M_b relation between black hole and bulge mass is acausal, arising from the quite independent connection between M_b and σ set by stellar, rather than black hole, feedback. The discovery of powerful AGN-driven winds rapidly removing gas from dwarf galaxies gives direct support for the feedback picture.

6.11 Momentum or Energy?

In this chapter we have argued that the scaling relations between SMBH and their host galaxies result from the effects of feedback by quasi-spherical winds driven by black hole accretion. Observations strongly suggest that powerful high-speed molecular outflows sweep galaxy spheroids largely clear of gas. These outflows have significantly greater scalar momenta $\dot{M}v$ than the radiative value L/c of the central AGN. We have argued that this happens because, once the SMBH mass reaches the M–σ value, the shock interaction where the black hole wind impacts the

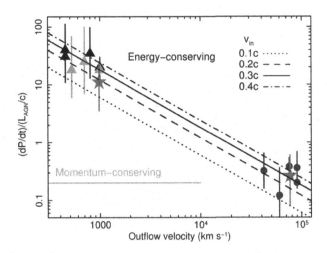

Figure 6.19 The momentum fluxes of the inner X-ray UFO wind and the molecular outflow observed in IRAS F11119+3257 (starred points) plotted versus their velocities, normalized to the momentum flux L/c of the AGN radiation. For comparison, the plot also shows the energetics of mildly relativistic UFO winds (round points) detected in a sample of quasars and of the molecular outflows detected in ULIRGs (ultraluminous infrared galaxies: triangles) with similar AGN luminosities. With the caveat that the fast X-ray wind is observed now, while the large-scale molecular outflow is probably an integrated effect of such winds over a much longer period of time, there is a very good quantitative agreement between observations and theory (cf. (6.24), (6.25), (6.26), (6.88)). Credit: Tombesi et al. (2015).

host galaxy ISM changes from momentum driving, acting on small spatial scales close to the SMBH, to energy driving, instead acting globally on the whole galaxy bulge and causing a high-energy clearout of its gas (see Figure 6.19). The M–σ relation marks the point where this transition occurs. Because the switch to energy driving removes most of the gas available for significant further mass growth, SMBH masses M cannot easily grow beyond $M_\sigma \propto \sigma^4$. Quite separately, momentum-driven outflow feedback limits the bulge stellar mass M_b to a value also $\propto \sigma^4$, accounting for the proportionality $M \sim 10^{-3} M_b$, with a coefficient increasing with the Hubble parameter.

We will argue in Section 6.14 that observations support this picture of a sharp transition from small-scale to large-scale feedback in several ways. But because it is basic to the whole picture of SMBH feedback presented here, we first consider this transition in more detail.

First, it is clear that the switch from momentum driving to energy driving depends on the specific nature of the gas cooling. The expression $M_\sigma = f_g \kappa \sigma^4 / \pi G^2$ for the critical mass is derived under the assumption that gas cooling is efficient, that is, it has a cooling time shorter than the derived gas flow timescale, enforcing momentum driving. This holds close to the SMBH since the AGN radiation field supplies enough soft photons for the inverse Compton effect to cool the post-shock electrons efficiently within the characteristic radius R_C. The switch occurs because as M approaches M_σ the effective potential changes, and even momentum-driven flow can push the shocks beyond R_C. Here the shocked wind gas can no longer cool within a flow timescale, and a much more vigorous energy-driven flow develops, which can sweep up and expel all the interstellar gas from the galaxy bulge.

This makes it clear that cooling of the ambient or swept-up interstellar gas in a galaxy is irrelevant to the question of energy or momentum driving despite suggestions to the contrary occasionally appearing in the literature: only the cooling of the shocked black hole wind gas is relevant. The switch from momentum driving to energy driving is fundamental, and we emphasize this by showing that any picture in which the driving remains purely in either one of the two states leads to contradictions.

We consider first the hypothetical case where a black hole wind acts on its surroundings by pure momentum driving alone, at all radii and all times, without any change to energy driving. For generality we consider a wind of speed v_w whose mechanical luminosity $\dot{M}_w v_w^2/2$ is a fixed fraction λ of L_{Edd} – we do not explicitly assume that the wind has the Eddington momentum, as in practice appears to hold for UFOs. Then the momentum feedback first becomes dynamically important when the black hole mass M_{crit} implies a wind thrust $\dot{M}_w v_w = 2\lambda L_{\mathrm{Edd}}/v$ comparable with the weight

$$W = \frac{GM(R)M_g(R)}{R^2} = \frac{4f_g \sigma^4}{G} \tag{6.147}$$

of the overlying gas in an isothermal potential (cf, (6.57)). With $L_{\mathrm{Edd}} = 4GM_{\mathrm{crit}}c/\kappa$ we get

$$M_{\mathrm{crit}} \sim \frac{v_w}{2\lambda c} M_\sigma. \tag{6.148}$$

Since $\lambda = \dot{M}_w v_w^2/2L_{\mathrm{Edd}}$ and $L_{\mathrm{Edd}} = \eta \dot{M}_{\mathrm{Edd}} c^2$, we get

$$M_{\mathrm{crit}} \sim \frac{v_w}{2\lambda c} M_\sigma = \frac{\eta c}{v_w} \frac{\dot{M}_{\mathrm{Edd}}}{\dot{M}_w} M_\sigma, \tag{6.149}$$

which is essentially the argument of Fabian (1999). For general parameters of the wind, the critical mass differs from M_σ: we get $M_{\mathrm{crit}} = M_\sigma$ only if $v_w = \eta c\dot{M}_{\mathrm{Edd}}/\dot{M}_w$, which is equation (6.26). This requires that the wind momentum is Eddington, that is, $\dot{M}_w v_w = \eta \dot{M}_{\mathrm{Edd}} c = L_{\mathrm{Edd}}/c$. So pure momentum driving gives the critical mass as M_σ if and only if the driving wind has the properties observed for UFOs, that is, the Eddington momentum.

But we should also ask what momentum driving predicts once the critical mass is reached. Assuming that the switch to energy driving fails, that is, pure momentum driving continues, the first integral (6.64) of the equation of motion now gives

$$\dot{R}^2 \simeq 2\sigma^2 \left[\frac{M}{M_{\text{crit}}} - 1 \right], \qquad (6.150)$$

where now the term in square brackets is positive. This equation gives the instantaneous connection between the black hole mass ($>M_\sigma$) and the outflow speed over all of the bulge except near $R = 0$ and so tells us how much the black hole mass has to grow in order to drive the swept-up ISM to a given radius R. If we assume that the central AGN always radiates the Eddington luminosity we can write $\dot{M} = M/t_S$, where $t_S \simeq 4 \times 10^7$ yr is the Salpeter time, and write $\dot{R} = \dot{M} dR/dM$ to integrate (6.150) as

$$R = 2\sqrt{2}\sigma t_S \arctan u \simeq 23\sigma_{200} \arctan u \text{ kpc}, \qquad (6.151)$$

where $u = (M_f/M_{\text{crit}} - 1)^{1/2}$ and M_f is the value of M when the gas shell has reached R. We see that to expel the gas and ultimately leave the galaxy bulge red and dead (i.e. $R > 20$ kpc or more) requires $u \gtrsim \pi/4$ or

$$M_f \gtrsim 2M_{\text{crit}} \sim \frac{2\eta c}{v_w} \frac{\dot{M}_{\text{Edd}}}{\dot{M}_w} M_\sigma. \qquad (6.152)$$

So pure momentum driving is unable to drive off the bulge gas without a significant increase of the black hole mass above M_{crit}. Several authors have reached this conclusion (see Silk & Nusser, 2010; McQuillin & McLaughlin, 2012). Often these papers also take $M_{\text{crit}} = M_\sigma$ and so conclude that pure momentum driving alone does not offer a convincing route to reproducing the observed M–σ relation. One might try to escape this conclusion by arguing from (6.152) that momentum driving would produce the observed result $M_f \simeq M_\sigma$ if $v_w \simeq 2\eta c/\dot{m}$. But if galaxy bulges accrete at the rates suggested by cosmological simulations it seems unlikely that any hypothetical momentum-driven outflows would have enough thrust to prevent infall and so could not turn off star formation definitively (see Costa, Sijacki & Haehnelt, 2014). We conclude that pure momentum driving, even given the lack of a likely shock cooling process, does not give a realistic picture of the interaction between SMBH and their hosts.

The direct opposite case from that considered in the last paragraph is pure energy driving by winds, where one assumes that radiative cooling is negligible everywhere. Early treatments often implicitly made this assumption (e.g. Silk & Rees, 1998). Then the equation of motion is (6.69), and we know that these outflows settle to an attractor solution with constant speed for any SMBH mass, however small this is, and so does not define a critical SMBH mass as required for a scaling relation. With $R = v_e t$ and using the definition of $L_{\rm Edd}$ (see (6.15)), the speed v_e is

$$v_e^3 = \frac{\pi G^2 c \eta M}{3 f_g \kappa \sigma^2}. \tag{6.153}$$

This expresses the fact that in principle the adiabatically expanding shocked gas pushes the interstellar gas away as a hot atmosphere for *any* SMBH mass – one can easily find the corresponding mass outflow rate by setting $\dot{M}_{\rm out} v_e^2 / 2 \sim L_w$ since we know that the outflow mechanical luminosity is a significant fraction of the driving wind mechanical luminosity $L_w = \eta l L_{\rm Edd}/2$.

To get an SMBH scaling relation – that is, a critical SMBH mass – for a purely energy-driven outflow, one must impose a further condition. This is often taken as the requirement $v_e \sim \sigma$, defining some kind of escape velocity. But as the outflow is driven by pressure, the ballistic escape velocity is not relevant, so it is not clear that this is a realistic contraint. Even if the AGN switches off completely, (6.77) and Figure 6.8 show that the residual gas pressure still drives outflow for a long time.

If we nevertheless impose this condition we find a critical mass

$$M_{\rm energy} = \frac{3 f_g \kappa}{\pi G^2 \eta c} \sigma^5 = \frac{3\sigma}{\eta c} M_\sigma = 0.02 M_\sigma = 6 \times 10^6 {\rm M}_\odot \sigma_{200}^5, \tag{6.154}$$

which is a factor $3\sigma/\eta c \sim 1/50$ too small in comparison with observations.

The paper by Silk and Rees (1998) considered the growth of a protogalaxy (i.e. gas with $f_g \sim 1$) around a supermassive seed black hole that formed earlier, but their argument applies to the coevolution of the SMBH and a host galaxy also if we take $f_g \sim 0.1$. The paper assumes that the wind sweeps the surrounding gas into a shell which has speed v_s. It implicitly neglects pressure work, and the fact that energy is shared between the shocked wind and the swept-up outflow. This gives

$$L_w = 2\pi r^2 f_g \rho(r) v_s^3, \tag{6.155}$$

as each new shell of mass $4\pi r^2 \rho(r) v_s$ now simply gains kinetic energy $v_s^2/2$ as it is swept up. Using the isothermal relation density (6.54), the assumption $v_s \sim \sigma$ gives

$$L_w \simeq \frac{f_g}{G} \sigma^2 v_s^3, \tag{6.156}$$

and if we make the additional assumptions that the gas velocity must asymptote to a value $\sim \sigma$, and that the wind energy L_w is of order $L_{\rm Edd}$, we get

$$M_{\rm SR} \simeq \frac{f_g \kappa}{4\pi G^2 f_w c} \sigma^5 \simeq 5 \times 10^4 \left(\frac{f_g}{0.16}\right) {\rm M}_\odot \sigma_{200}^5, \tag{6.157}$$

with $f_w = L_w/L_{\rm Edd}$. The neglect of pressure work and the energy-sharing between shocked wind and swept-up gas overestimates the wind driving efficiency, so this mass is even smaller than (6.154). It is clear that wind energy driving of this type does not correctly reproduce the M–σ relation, as it gives a critical mass too low by factors $\sim 4c/\sigma \sim 600$.

6.12 Feedback from Hyper–Eddington Mass Supply

We noted earlier (eqn (6.17)) that in general the mass supply to a galaxy's central black hole is not highly super-Eddington. But it seems possible that in some cases (e.g. for low black hole mass) that this may not always hold. Assuming again that most of this mass supply is blown away by the radiation pressure of the small part that accretes (and produces luminosity $\sim L_{\rm Edd}$), it is clear that we cannot assume the single-scattering limit and the consequent equality of radiation momentum $L_{\rm Edd}/c$ and the outflow momentum of a UFO wind, that is,

$$\dot{M}_w v \simeq \frac{L_{\rm Edd}}{c}, \tag{6.158}$$

which we used earlier to derive the usual M–σ relation. Instead, because most photons must scatter frequently we expect matter and radiation to be well-mixed, and replaced (6.158) by

$$\frac{1}{2}\dot{M}_w v^2 \simeq L_{\rm Edd}. \tag{6.159}$$

King and Muldrew (2016) consider this case. From (6.159) we get typical outflow velocities

$$v' = \left(\frac{2l' L_{\rm Edd}}{\dot{m}_w}\right)^{1/2} c, \tag{6.160}$$

where $\dot{m}_w s \simeq \dot{M}/\dot{M}_{\rm Edd}$ is the Eddington factor, and the AGN luminosity is $l' L_{\rm Edd}$. For $10 < \dot{m}_w < 100$ this gives $v' \sim 0.1$–$0.2c$, not very different from the momentum-driven UFOs. But the thrust exerted on the galaxy ISM is much larger:

$$\dot{M}_w v' = \frac{2L'_w}{v'} = \frac{2l'c}{v'}\frac{L_{\rm Edd}}{c} \sim 20l'\frac{L_{\rm Edd}}{c}. \tag{6.161}$$

In this case the argument that gave (6.60) now gives a significantly smaller critical SMBH mass

$$M \simeq \frac{1}{20l'}\frac{f_g \kappa}{\pi G^2}\sigma^4. \tag{6.162}$$

It appears that the accreting SMBH in a few galaxies, notably 1 Zwicky 1 (Ding et al., 2022), is like this: here the black hole mass is a factor ~ 20 below that expected from the M–M_b relation. The mass supply rate in this case appears to be strongly super-Eddington, but the origin of this is unclear – the black hole mass is not unusually low ($\sim 10^7 M_\odot$), for example.

6.13 Cosmological Simulations and M–σ

Cosmological simulations of large-scale structure formation often derive empirical M–σ relations. Because of the demands on numerical resolution this cannot involve the detailed small-scale physics used in this chapter. The SMBH's influence on the host galaxy is instead usually modelled by distributing energy over the gas of the numerical 'galaxy' at a certain rate, iterating the injected luminosity to fit the observed M–σ relation. This procedure generally yields values for the required mechanical luminosity of order $0.05L_{\mathrm{Edd}}$, remarkably close to the value expected from (6.25) for a black hole UFO wind with the Eddington momentum, that is, $\dot{M}_{\mathrm{out}} v = L_{\mathrm{Edd}}/c$.

But this success throws up a question. As we saw, absorbing the full mechanical luminosity $0.05L_{\mathrm{Edd}}$ of a UFO wind (i.e. not allowing shock cooling before the M–σ mass is reached) would give the energy-driven (7.10) or Silk–Rees masses (6.157), which are too small compared with observations by factors ~50 and 600, respectively.

This suggests that the simulations in reality only numerically couple the injected luminosity to the gas at rates comparable to the inefficient value that occurs in momentum driving. The physics determining this involves length scales very far below the resolution of any conceivable cosmological simulation. This in turn suggests that the weak coupling must be implicit in the assumed 'sub-grid' physics that is unavoidable in all such simulations (see, e.g., Costa, Sijacki & Haehnelt, 2014, appendix B).

6.14 Small versus Large-Scale Feedback

UFO winds are very commonly observed in AGN, even though we saw that many UFO events must be missed (see the discussion after (6.51)). UFOs give an obvious way for the central SMBH to make the host galaxy aware of its presence. This in turn suggests explanations for both the SMBH–galaxy scaling relations, and the expulsion of gas from galaxy spheroids that ends vigorous star formation. Evidently, this AGN feedback must sometimes work on small scales and sometimes at large scales. The work of this chapter suggests a natural association between momentum driving and small scales and between energy driving and large scales.

Wind momentum driving offers natural explanations of a list of small-scale phenomena:

1. Super-solar elemental abundances in AGN spectra. Because momentum-driven shells become too massive and fall back for any SMBH mass less than M_σ, momentum driving must compress the same gas many times before the black hole mass reaches M_σ. Repeated generations of massive stars must form from the same swept-up gas. These enrich the gas close to the SMBH with nuclear-processed material before momentum driving changes to energy driving as the M–σ mass is reached.

2. Dark matter cusp removal. The same repeated sweeping-up episodes, each with gas mass comparable to the SMBH mass, has a strong tendency to weaken dark matter cusps. With its larger baryonic mass, this is a more powerful version of the mechanism invoked by Pontzen and Governato (2012) (see also Garrison-Kimmel et al., 2013), who considered supernovae near the SMBH.

3. Quiescence of AGN hosts. Most of these galaxies do not show very rapid star formation in the central regions of their galaxy discs or much evidence for high-speed (~ 1000 km s^{-1} and massive ($\sim$ few 100M$_\odot$ yr^{-1}) outflows on large scales. This is compatible with wind driving by momentum but not by energy.
  Energy driving may be implicated in the following large-scale phenomena:

4. Metals in the circumgalactic medium. Metals are made in galaxies and so must be expelled to make the circumgalactic medium (CGM). This suggests that the timescale to reach the M–σ mass must be long enough to allow stellar evolution time to enrich a significant fraction of the galaxy bulge gas before energy driving expels it.

5. Mechanical luminosities of galaxy-scale molecular outflows. As we saw, these are observed to be as much as 5 per cent of their central AGN luminosities L and would naturally be somewhat lower in cases where the AGN driving is decreasing. This is very reasonable for energy driving with momenta close to the predicted $20L/c$.

6. Suppression of cosmological infall. Energy-driven outflows at large radii probably limit galaxy masses by preventing accretion continuing indefinitely (Costa, Sijacki & Haehnelt, 2014).

This list evidently favours a combination of momentum and energy driving, with a switch between the two regimes, as naturally arises in the picture developed here.

Despite these promising indications, a full picture of how black holes and galaxies influence each other still requires an understanding of how SMBH feeding works. In particular, we need to know what physical mechanism can produce a supply of gas with so little angular momentum that much of it can accrete onto the central supermassive black hole within a few Salpeter times – see Chapter 5 and particularly Section 5.2. The hole's gravity is far too weak to influence the galaxy on the mass scale needed. Only feedback can do this, perhaps supporting the suggestion that SMBH feedback may ultimately cause SMBH feeding (see Section 5.2).

7

SMBH Feedback in General

7.1 Introduction

The previous chapter gave an extensive discussion of momentum- and energy-driven feedback from quasi-spherical black hole wind, with the aim of explaining the SMBH–galaxy scaling relations. But an SMBH can physically affect its surrounding in a number of other ways that are important in various contexts, and we discuss these here.

7.2 Radiation Driving

We have already noted that in principle, by far the strongest feedback from a black hole is its accretion luminosity L. But we also noted that since we observe AGN through the radiation they emit, the effects of radiation driving may often be limited because the 'target' gas is optically thin. In contrast, feedback by quasi-spherical winds is effective even when its mechanical luminosity is much less than L, because collisions with surrounding matter are inevitable.

But active galactic nuclei correspond to phases when the SMBH is growing its mass through accretion of gas from a very small-scale disc around it, and there are good reasons to expect that at least at certain epochs a significant mass of gas often surrounds the hole in these episodes. First, a majority of AGN show significant signs of obscuration (see the discussion in Elvis, 2000). Second, to grow a black hole's mass in a reasonable time needs a large gas mass close to it, and since a disc's mass is severely limited by self-gravity, this gas reservoir must cover a large solid angle. Finally, simple estimates of the column densities of matter in a galaxy potential suggest that most AGN are likely to undergo major growth events (e.g. from mergers) in fairly dense gas environments.

Radiation feedback is evidently important when the SMBH's surroundings have high scattering optical depth (i.e. are strongly obscuring). We would, for example, expect this kind of situation after a galaxy merger, at the point when the two SMBH have themselves merged (or are at least close enough to the dynamical centre of the merged galaxy that we can regard them as a single accreting object). Then photons scatter many times, and so radiation pressure must become significant. In scattering slightly inelastically, the luminosity L does work against the gravitational force on the surrounding gas in the central spheroid of the galaxy. If we assume that the gas is distributed isothermally, that is, with density

$$\rho(r) = \frac{f_g \sigma^2}{2\pi G r^2}, \tag{7.1}$$

as is likely in most cases, then the gas mass within radius R obeys the usual relations

$$M_g(R) = \frac{2f_g \sigma^2 R}{G}, \tag{7.2}$$

and the total mass (including stars, and any dark matter) is

$$M(R) = \frac{2\sigma^2 R}{G}, \tag{7.3}$$

with σ the velocity dispersion and f_g the gas fraction relative to all matter (e.g. dark matter, and stars), which we assume has the cosmic value 0.16.

The trapped radiation sweeps the gas up progressively into a shell of inner radius R and mass $M_g(R)$. The relevant opacity source is Thomson scattering, and for a geometrically thin shell the electron-scattering optical depth at radius R is

$$\tau_{sh}(R) = \frac{\kappa M_g(R)}{4\pi R^2} = \frac{\kappa f_g \sigma^2}{2\pi G R}, \tag{7.4}$$

where $\kappa \simeq 0.34 \, \text{cm}^2 \, \text{g}^{-1}$ is the electron-scattering opacity. The value $\tau_{sh}(R)$ is an upper limit to the optical depth of a thicker shell, as then more of the gas is at larger radii, with smaller density. The undisturbed gas outside the shell R has optical depth no greater than

$$\tau(R) = \int_R^\infty \kappa \rho(r) \mathrm{d}r = \frac{\kappa f_g \sigma^2}{2\pi G R} = \tau_{sh}(R), \tag{7.5}$$

most of which is concentrated near its inner radius R. So the radiation encounters total optical depth

$$\tau_{tot}(R) \le \tau(R) + \tau_{sh}(R) \simeq \frac{\kappa f_g \sigma^2}{\pi G R} \tag{7.6}$$

whatever the thickness of the shell. Gas distributed like this is very optically thick near the black hole if R is small (see (7.8)). So the accretion luminosity L of the AGN is initially essentially trapped and isotropized by repeated scattering, which increases the interior radiation pressure P. This growing pressure pushes against the weight

$$W(R) = \frac{GM(R)M_g(R)}{R^2} = \frac{4f_g\sigma^4}{G} \tag{7.7}$$

of the swept-up gas shell at any radius R. But although the radiation pressure grows, its effective coupling to matter drops continuously because the shell's optical depth falls off as $1/R$ as it expands. Radiation begins to leak out of the cavity, until for $\tau_{tot}(R) \sim 1$ it is unable to drive the shell further. So radiation-pressure sweeping up of gas must stop at a 'transparency radius'

$$R_{tr} \sim \frac{\kappa f_g\sigma^2}{\pi G} \simeq 50\left(\frac{f_g}{0.16}\right)\sigma^2_{200} \text{ pc.} \tag{7.8}$$

Here (up to a logarithmic factor) the optical depth τ_{tot} is of order 1 and radiation begins to escape, acting as a safety valve for the trapped radiation pressure.

It is straightforward to follow this process, which is mathematically identical to the case of an energy-driven wind sweeping up the ISM which we dealt with in Section 6.6. A geometrically thin shell of swept-up gas has the equation of motion

$$\frac{d}{dt}[M_g(R)\dot{R}] = 4\pi R^2 P - W. \tag{7.9}$$

The radiation pressure does work on the surroundings, giving the energy equation

$$\frac{d}{dt}[VU] = L - 4\pi R^2\dot{R}P - W\dot{R}, \tag{7.10}$$

where $V = 4\pi R^3/3$ is the volume interior to the shell. This is filled with radiation of energy density $U = 3P$, and supplied with more energy at the rate L. We follow the derivation of Section 6.6, this time for a general relation $P = (\gamma - 1)U$ (the index $\gamma = 4/3, 5/3$ for the present case of radiation and the earlier case of a monatomic gas). We eliminate the pressure P from (7.10) by using (7.9). The result is

$$L = \frac{2f_g\sigma^2}{3G(\gamma - 1)}[R^2\dddot{R} + (3\gamma + 1)R\dot{R}\ddot{R} + (3\gamma - 2)\dot{R}^3] \tag{7.11}$$
$$+ \frac{6\gamma - 5}{3\gamma - 3} \cdot \frac{4f_g\sigma^4}{G}\dot{R}.$$

This reduces to the equation (6.69) given in Section 6.6 for energy driving by a wind if we set $\gamma = 5/3$ (the mechanical luminosity L of the wind is $\eta/2$ times the driving near-Eddington radiative luminosity $\sim L_{Edd}$ in this case).

In the current trapped-radiation case we have $\gamma = 4/3$, which gives

$$L = \frac{2f_g\sigma^2}{G}[R^2\dddot{R} + 5R\dot{R}\ddot{R} + 2\dot{R}^3] + \frac{12f_g\sigma^4}{G}\dot{R}. \tag{7.12}$$

As in the energy-driven wind case there is a constant-velocity solution $R = v_e t$, with

$$L = \frac{4f_g\sigma^2 v_e^3}{G} + \frac{12f_g\sigma^4}{G}v_e. \tag{7.13}$$

This equation defines a unique solution v_e, which is an attractor. We write L as

$$L = \frac{dE}{dR}v_e, \tag{7.14}$$

where E is the total radiation energy inside R. Then (6.71) becomes

$$\frac{dE}{dR} = \left(3 + \frac{v_e^2}{\sigma^2}\right)W. \tag{7.15}$$

For modest accretion luminosities L (i.e. well below the Eddington value for the final black hole mass M_σ), we must have $v_e \ll \sigma$. Then (7.15) shows that the total accretion energy needed to push the gas to the transparency radius $R_{\rm tr}$ is

$$E_{\rm tr} \simeq 3WR_{\rm tr} = \frac{12\kappa f_g^2\sigma^6}{\pi G^2} = 12f_g M_\sigma\sigma^2, \tag{7.16}$$

so the SMBH must accrete a fractional mass

$$\frac{\Delta M}{M_\sigma} \gtrsim \frac{E_{\rm tr}}{M_\sigma \eta c^2} = \frac{12f_g}{\eta}\left(\frac{\sigma}{c}\right)^2 \simeq 8.6 \times 10^{-6}\frac{\sigma_{200}^2}{\eta_{0.1}}, \tag{7.17}$$

where $\eta_{0.1}$ is the accretion efficiency in units of 0.1, to push the surrounding gas beyond the transparency radius. (Note that unlike in the energy-driven wind case ((6.77) and Figure 6.8) there is no further expansion beyond $R_{\rm tr}$ driven by residual pressure – here the remaining radiation pressure simply does not couple to the surrounding gas once the transparency radius is reached.)

The required mass ΔM is always far smaller than the SMBH mass itself, so we expect transparency to be achieved early in the life of the central SMBH, and easily maintained after this – for example after a merger. This gives a limit on the lifetime of the so-called Compton-thick property of AGN of

$$\Delta t \simeq \frac{\Delta M}{\dot{m}M_\sigma}t_{\rm Sal} \simeq \frac{400}{\dot{m}}\sigma_{200}^2 \text{ yr}, \tag{7.18}$$

where $\dot{m}$ is the ratio of the SMBH accretion rate to its Eddington value. In other words, Compton-thick conditions are relatively short-lived.

From the usual isothermal relations, the total mass inside $R_{\rm tr}$ is

$$M(R_{\rm tr}) = \frac{2\sigma^2}{G}R_{\rm tr} = 2\frac{f_g\kappa\sigma^2}{\pi G^2} = 2M_\sigma, \tag{7.19}$$

so the total mass inside $R_{\rm tr}$ is twice the final mass of the black hole. The gas mass inside $R_{\rm tr}$ is $2f_g M_\sigma \sim 0.32M_\sigma$, so comparable to the final hole mass. Radiation

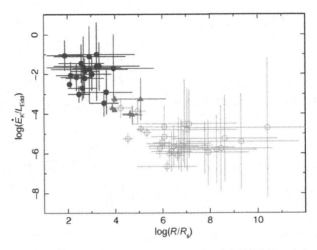

Figure 7.1 Comparison between the radial distance $\log(R/Rs)$ and the estimated outflow kinetic energy rate for an overlapping sample of WAs (hollow circles) and UFOs (solid circles) in nearby, bright AGN. For a black hole mass of $10^8 M_\odot$, the distance scale may be converted to parsecs by noting that $10^6 R_s$ is ~ 10 pc. Credit: Tombesi et al. (2013).

pressure sweeps this up into a shell at $R_{\rm tr}$. Here it is transparent to the SMBH accretion luminosity, but its large mass makes it a real obstruction to winds blown by the hole, such as the UFOs we discussed in Chapter 6, which must shock against it. The shocked UFO winds rapidly cool, decelerate and recombine, mixing with swept-up interstellar gas. So we expect the colliding wind gas to end with modest ionization, and much slower velocities than it had before the collision. These properties are very similar to those inferred for the so-called warm absorber (WA) components in AGN spectra, so it seems conceivable that WAs result from these wind impacts (King & Pounds, 2014).

Figure 7.1 compares the derived kinetic energy rates with radial distance constraints for a sample of both UFOs and WAs. A minimum radial distance comes from the assumption that the velocity is comparable to the local escape value, and a maximum from the relation

$$R = \frac{L_i}{N_H \xi}, \tag{7.20}$$

where L_i is the ionizing luminosity, and $N_H = NR, \xi = L_i/NR^2$ are the equivalent hydrogen column density and the ionization parameter given by spectral fitting. The plot is made independent of black hole mass by giving the energy rate in units of $L_{\rm Edd}$, and the radius in Schwarzschild radii R_s. As we would expect, UFOs cluster at $\sim 10^2 R_s$. Importantly, we see that WAs cluster between 10^6 and 10^7 Schwarzschild radii R_s. We can compare this with (7.8), which predicts $R_{\rm tr}/R_s \simeq 5 \times 10^6$, taking

$\sigma = 200\,\mathrm{km\,s^{-1}}$ for an SMBH mass of $10^8 M_\odot$. Figure 7.1 also illustrates the much higher kinetic power carried by the UFOs as a group, as we expect if high-speed winds carry AGN feedback. Tombesi et al. (2013) also find measured mass outflow rates are essentially constant with radius (their figure 2). Then the factor $\sim 10^4$ between the UFO and WA energy rates in Figure 7.1 suggests that these two components are the start and end points of the same mass-conserving outflows, with mean velocity differences of order ~ 100 caused by shocks at R_{tr}. As we have seen, the UFO velocities $\sim 0.1c$ are drastically reduced in strong shocks which are efficiently Compton cooled by the AGN radiation field, giving up almost all their kinetic energy. In some cases the UFOs may hit previous shocked ejecta or infalling gas well within R_{tr}. This latter seems likely for NGC 405 (see Pounds & King, 2013 and references therein for a discussion).

These arguments show that the trapped radiation pressure from an SMBH fed with a large mass of gas, for example in a merger event, probably affects the surrounding interstellar medium strongly. Much of this gas is swept into a dense shell with characteristic radius R_{tr}. Here photons can escape and stabilize the growing radiation pressure inside. The shell's mass is of order the final M–σ mass the hole will eventually reach, and its distance from the SMBH is comparable to that of the region characteristic of warm absorber behaviour. This agrees with the reasoning that the shell at R_{tr} is so massive that it halts outflows from the central SMBH that reach this radius, essentially transforming UFOs into warm absorbers as it does this.

The shell at R_{tr} is very likely to be Rayleigh–Taylor unstable, giving it a complex topology, so that photons can only escape it after multiple scattering. A sign of this may be the lack of any cold absorption accompanying the WA components. It appears very likely that a significant fraction of the AGN luminosity must be absorbed and re-radiated by gas of significant optical thickness near R_{tr}, giving a characteristic blackbody component with temperature

$$T_{\mathrm{tr}} = \left(\frac{l L_{\mathrm{Edd}}}{4\pi f R_{\mathrm{tr}}^2 \sigma_{\mathrm{SB}}} \right)^{1/4} \sim 100 \left(\frac{l M}{f M_\sigma} \right)^{1/4} \mathrm{K}, \tag{7.21}$$

where l is the re-radiated fraction of the Eddington luminosity L_{Edd}, $4\pi f$ is the solid angle of the obscuring shell, and we have written σ_{SB} to distinguish the Stefan–Boltzmann constant from the velocity dispersion. This low temperature may make this component hard to detect, and a full shell ($f = 1$) might completely hide an AGN.

We have so far considered only electron scattering as the coupling between radiation and matter. But particularly at larger distances from the nucleus, much of the cold diffuse matter in the galaxy bulge may be in the form of dust. The absorption coefficient of dust depends strongly on wavelength and is far higher than electron scattering in the ultraviolet, but decreases sharply in the infrared (e.g. Draine & Lee, 1984). Depending on parameters, one can have cases where a shell of gas is optically thick in the IR; thin in IR but thick in the UV; or optically thin in the UV.

The intermediate case (thin in IR but thick in the UV) often holds. Then the energy of an ultraviolet photon that is absorbed by a dust grain is re-emitted almost isotropically as a large number of infrared photons, which escape freely. The net effect is that dusty gas feels only the initial momentum of the incident UV photon, while most of the incident energy escapes. Then a spherical shell around an AGN experiences a radial force $\simeq L/c$, where L is the ultraviolet luminosity, as long as these opacity properties hold. This is dynamically similar to wind-powered flows in the momentum-driven limit, and this type of radiation-powered flow is often also called momentum-driven outflow, even though the physical mechanism is very different from that considered in Chapter 6 in connection with UFO winds shocking against the host ISM. In other opacity regimes it is possible that the IR photons are trapped. Then all of the incident energy ends up driving the shell outwards, so the dynamics are now energy driven.

The mathematical similarity between wind-powered and radiation-powered momentum driving gives an empirical estimate of an M–σ relation for the latter under the assumptions that observed AGN define the relation, and that their observed luminosities correspond to $L/L_{\mathrm{Edd}} \sim 0.1$–$1$. Murray, Quataert, and Thompson (2005) find $M_{\mathrm{crit}} = (L_{\mathrm{Edd}}/L)M_\sigma \sim 1$–$10 M_\sigma$ in this way, and there are optical depth effects that may bring this range closer to the observed value (Debuhr, Quataert & Ma, 2011). If trapping of IR photons is able to retain much of the incident radiant energy, spherical steady-state dust feels a radiation force $\tau L/c$, where τ is the radial optical depth of the dust. This form of radiation driving of optically thick dust can in principle produce outflows whose scalar momenta are boosted above that of the driving luminosity L/c by a factor $\sim\tau$ because photons may be re-absorbed several times. Ishibashi, Fabian and Arakawa (2021) review recent progress in this area.

For large optical depth the radiation field presumably approaches a blackbody form (see (7.21) in the discussion of the electron-scattering case), and one needs then to check that this does not contradict the opacity properties assumed for the dust.

7.3 Jets

Observations of all types of accreting systems frequently reveal a different type of outflow from the wide-angle ones considered in Chapter 6 and in Section 7.2.

Narrow collimated jets of gas are driven out at a velocity of order the local escape value at the accretor boundary, and are often bipolar (i.e. on both sides of the accretor – compare Figure 1.10). Jets from black hole accretors often have relativistic speeds with significant Lorentz factors.

It is clear that jets must have feedback effects, but evidently rather different from those of the wide-angle accretion disc winds we discussed in the previous section. In a feedback event the jet material must shock, and we can already deduce from the wide-angle case that if these shocks cool efficiently – the analogue of momentum-driven flow in the wide-angle case – the jet is likely to cut through the obstructing gas and remain fairly narrow. This is obviously unpromising for driving large-scale removal of gas from the vicinity of an AGN as discussed in the previous two chapters, but probably required to explain the double-lobe radio sources (see Figure 1.10), where the main dissipation occurs far outside the host galaxy of the driving SMBH. Conversely, if the jet shocks remain adiabatic, the shocked jet gas is likely to expand sideways from the original axis, and produce similar effects to the energy-driven phase of a wide-angle outflow, in an angle-averaged sense. It may be difficult to arrange that Compton cooling by the AGN – the main shock coolant in the wide-angle case – remains effective for a jet collision with ambient gas. Then the explanation of the M–σ relation discussed in Chapter 6 and in Section 7.2, which results from the switch from momentum driving to energy driving, cannot work for jet feedback.

Nevertheless, it is also clear that jets must produce significant dynamical effects, and the first observations suggesting the existence of jets in any context were early maps of double-lobe radio galaxies (e.g. Mitton & Ryle, 1969) showing regions of strong radio emission placed symmetrically on each side of the central galaxy (see Figure 7.2). Rees (1971) suggested that an unknown object in the galaxy nucleus fed energy to the radio lobes through collimated jets. (For the jets to retain enough energy to power the lobes, their encounters with the host galaxy ISM must presumably involve shocks that cool, as discussed in the previous paragraph.) The jets themselves were not detected in these early observations, except in the radio lobes, where they presumably interacted with ambient gas. Later observations (see Figure 7.3) show that this idea is correct by revealing the jets. We now know the unknown object was the central SMBH of a galaxy between the two lobes. But the central galaxies of these early radio sources were themselves not noticeably active (see Figure 7.2). Under the assumption of a steady state, this suggested that the jet power did not come from the instantaneous release of accretion energy, which would have made the central galaxy active, but instead from accretion energy that was stored in the form of black hole spin. This is similar to radio pulsars, where the emission is supplied by the rotation of the neutron star, and so for black holes

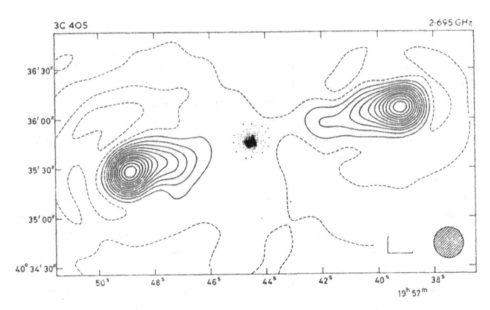

Figure 7.2 An early radio map of the Cygnus A radio galaxy, showing the two radio lobes on each side of the visible galaxy. Credit: Mitton and Ryle (1969).

would require an electromagnetic version of the Penrose process.[1] (We know from (2.31) that the spin can store at most 29 per cent of the total accreted energy, so in this picture jets are automatically significantly less energetic than emission from the accretion disc, for example.)

Radio pulsars extract the neutron star's rotational energy as the magnetic field of the neutron star interacts with a very low-density force-free charge-separated plasma,[2] and it is well established that they are able to produce highly energetic electrons which, for example, power the synchrotron emission from the Crab nebula. This suggested a picture in which a low-density plasma surrounding a spinning black hole might extract its spin energy electromagnetically and drive jets, perhaps in the form of high-energy particles, which would power the radio lobes. The 'no-hair' theorems tell us that unlike a pulsar, a spinning black hole cannot have an intrinsic magnetic field that is offset from the rotation axis and so forced to co-rotate with the spin, which would potentially give a way of converting rotational into electromagnetic energy. Instead, the accretion disc may generate a large-scale magnetic field, and perhaps still offer the chance of extracting black hole spin energy.

---

[1] We now know of many systems where jets are observed and the central galaxy is also simultaneously active. So 'inactive' systems like the early-discovered radio sources may simply reflect AGN variability, as the lifetimes of the radio lobes are likely to be longer than the AGN phases.

[2] This plasma actually consists of matter pulled from the neutron star surface, as the electric force $\propto \mathbf{E} \cdot \mathbf{B}$ is stronger than the neutron star surface gravity (Goldreich & Julian, 1969).

Figure 7.3 A recent combined visual and radio image of the elliptical galaxy Hercules A. Data from the Hubble Space Telescope's Wide Field Camera 3, and the Karl G. Jansky Very Large Array (VLA) radio telescope in New Mexico. The image clearly shows the jets powering the radio lobes. Credit: NASA, ESA, S. Baum and C. O'Dea (RIT), R. Perley and W. Cotton (NRAO/AUI/NSF), and the Hubble Heritage Team (STScI/AURA).

The first treatment of the extraction of black hole spin energy (Blandford & Znajek, 1977) assumed that the spinning black hole interacted with an ambient magnetic field connected to nearby matter, and adopted the force-free electrodynamic condition (3.126),

$$\rho_e \mathbf{E} + \frac{[c]}{c}\mathbf{j} \times \mathbf{B} = 0,$$

as holds in the low-density charge-separated magnetospheres of pulsars, in line with the lack of clear AGN activity in the central galaxies of double-lobe radio sources. One can solve the Einstein–Maxwell equations (King, Lasota & Kundt, 1975) for the magnetic multipoles in a Kerr spacetime and evaluate the quantity $\mathbf{E} \cdot \mathbf{B}$ along the magnetic fieldlines that thread the event horizon. This shows that in a charge-separated environment the black hole attracts electric charges of opposite signs near its poles and equator respectively (the absolute signs depend on whether the spin is parallel or antiparallel to $\mathbf{B}$).

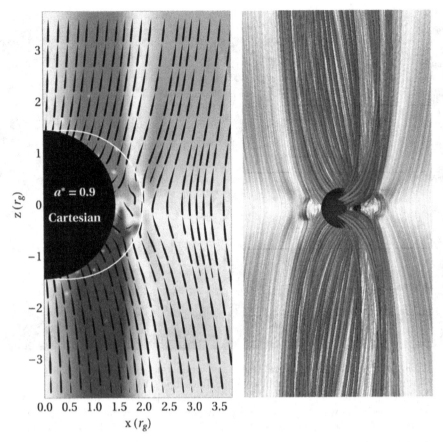

Figure 7.4 Numerical simulation of a Wald magnetosphere: a 3D magnetic field-line impression of a case with spin parameter $a = 0.9$. These results agree within numerical accuracy with the analytic results of King, Lasota, and Kundt (1975), and indicate the fact that the equator and pole of the hole accrete charges of opposite sign. Darker contours indicate a stronger twist of the magnetic field. Credit: Mahlmann et al. (2021a).

If we now assume that the black hole remains in a steady state, and so has net charge independent of time, the respective inflows of the opposite charge signs must balance, effectively implying the presence of a steady current across the black hole ergosphere. If this current is subject to dissipation, the energy can only have come from the spin energy of the black hole.

But if one drops the assumption of a steady state and allows for the possibility that the electric charge on the black hole can vary in time, an elegant argument by Wald (1974) (see Figure 7.4) shows that the hole quickly accretes a net charge $Q = 2BaM$ (in geometrized units), which then nullifies the apparent currents.

The charge Q gives an electromagnetic contribution to the metric coefficients of order $\sim 10^{-12}(M/10^8 M_\odot)$ compared with those of the black hole mass and spin. In other words, the charge has no discernible effect on the metric, but has a major effect on how charged particles move if they can do so independently, because electromagnetism is far stronger than gravity. A calculation of the effect of a charged current loop in an accretion disc aligned with the black hole spin (Petterson, 1975) explicitly supports Wald's argument. For a given current I and charge q, the resulting dipole magnetic field is effectively uniform and aligned with the black hole spin near the hole. As before, the $\mathbf{E} \cdot \mathbf{B}$ forces cause the hole to acquire a charge given by I and q such that charge accretion stops. To ensure charge neutrality as seen from infinity, one can now require that the charge on the hole exactly cancels the charge q of the loop. This relates I to q and shows that the black hole charge must have Wald's value $Q = 2BaM$, while the current loop has net charge $-Q$. These results mean that extracting spin energy from a black hole is in practice very difficult in a charge-separated environment.

We should now ask if this difficulty remains in denser environments – here particle collisions are frequent, so MHD conditions should apply instead of the force-free electrodynamic condition (3.126). Then we can estimate how a spinning black hole might cause a magnetic fieldline to do work on its surroundings as the hole spin tries to drag the fieldline through surrounding plasma. Of course, neither the event horizon nor the ergosphere of a black hole is a material surface, and indeed the event horizon is not a spacelike surface but null (lightlike), so fieldline-dragging in the conventional sense is impossible. Solving Maxwell's equations in the Kerr spacetime geometry and applying the boundary condition that there is no outgoing radiation at the horizon shows that a magnetic field on the scale $R_h \simeq R_g$ of the horizon decays on the light-crossing time R_h/c. We can incorporate this effect into MHD calculations by equating this timescale to the diffusion time $t_{\text{mag}} \sim R_h^2/\eta$ (from (3.124)), so assigning an effective magnetic diffusivity

$$\eta_h \sim R_h c \tag{7.22}$$

to the black hole. This implies that the hole is very far from being a perfect conductor: magnetic fields near the horizon decay in the shortest possible time. If a poloidal field B_{ph} extending within the ergosphere is dragged by the black hole rotation it induces a toroidal field $B_{\phi h}$ (see (3.123)). Setting $\partial \mathbf{B}/\partial t = 0$, taking spatial derivatives over a region of size $\sim R_h$, and using $v \sim R_h \Omega_h$ in (3.123), we see that the maximum toroidal field a black hole rotating with spin Ω_h can induce is

$$B_{\phi h}(\text{max}) \sim \frac{R_h}{\eta}(R_h \Omega_h)B_p \sim \frac{R_h \Omega_h}{c}B_{ph}, \tag{7.23}$$

where we have used (7.22). From the MHD stress tensor (3.95), the rate of working by the hole spin on the external medium, and so available for powering a jet, is

$$L_{\text{work}} \sim \left[\frac{4\pi}{\mu_0}\right] \left(\frac{B_{ph}B_{\phi h}}{4\pi}\right) \pi R_h^3 \Omega_h. \tag{7.24}$$

Then with (7.23) we get the maximum jet power driven by black hole fieldline-dragging as

$$L_h(\text{max}) \sim \left[\frac{4\pi}{\mu_0}\right] \frac{R_h^4 \Omega_h^2 B_{ph}^2}{4c}. \tag{7.25}$$

Evidently, disc gas orbiting near the event horizon can induce fields and do work against them in a similar way, giving a jet power

$$L_d \sim \left[\frac{4\pi}{\mu_0}\right] \left(\frac{B_{pd}B_{\phi d}}{4\pi}\right) \pi R_d^3 \Omega_d, \tag{7.26}$$

where $B_{pd}, B_{\phi d}$ are the poloidal and toroidal components of the disc surface field, and Ω_d is the angular velocity of the disc material at a disc radius R_d close to the inner edge, that is, $R_d \sim$ few $\times R_h$. Because the disc has much higher conductivity than the hole we have $B_{\phi d} \gtrsim B_{pd}$, so

$$L_d(\text{max}) \sim \left[\frac{4\pi}{\mu_0}\right] \frac{R_d^3 \Omega_d B_{pd}^2}{4}. \tag{7.27}$$

We will see an explicit example of a physical process in the innermost disc producing jet power of just this order (Equation (7.30)). From (7.25) and (7.27) we get

$$\frac{L_h(\text{max})}{L_d(\text{max})} \sim \left(\frac{B_{ph}}{B_{pd}}\right)^2 \left(\frac{R_h}{R_d}\right)^2 \frac{R_h \Omega_h}{R_d \Omega_d} \frac{R_h \Omega_h}{c} \sim \left(\frac{B_{ph}}{B_{pd}}\right)^2 \left(\frac{R_h}{R_d}\right)^{3/2} a^2, \tag{7.28}$$

where we have used $\Omega_h \sim ac/R_h$ and $R_d \Omega_d/c \sim (R_h/R_d)^{1/2}$ at the last step.

As $a < 1$ and R_d is order a few times R_h, we see that extracting significant energy from fieldline dragging by the black hole spin compared with that by the disc is a weak effect. This is not a surprising result, as we recall that in extracting the rotational energy of a radio pulsar uses the very strong field of the neutron star.

A possible remedy is to drag in stronger fields to the centre of the disc, where they can interact with the hole spin. But arranging this is problematic. We have already seen in Section 4.12 that finite diffusivity makes it difficult for a disc to drag a poloidal field inwards unless it is thick ($H \sim R$). Other possibilities are equally unlikely – Ghosh and Abramowicz (1997) study this question in detail, and point out that as any poloidal field threading the hole repels any like-directed poloidal field advected inwards by the disc, and cancels opposing fields, the field threading the hole cannot be much stronger than that threading the inner disc. Further, the currents generating the field that thread the hole must be in the disc rather than in the hole, so the field through the hole is just a continuation of the field in the disc.

Accordingly, the physics of electromagnetic extraction of black hole spin energy remains unclear. Numerical MHD simulations nevertheless frequently report jets. These are often ascribed to the Blandford–Znajek effect. But, as we have seen, this uses force-free electrodynamics, that is, an explicitly charge-separated plasma, and not MHD. In MHD there is no relation like (3.126); instead at every point there is a rest frame in which ρ_e and $\mathbf{j} \times \mathbf{B}$ are separately zero. Given the difficulties inherent in numerical MHD simulations (see Section 4.13), particularly in conserving energy, investigations with PIC codes may give insight. Most if not all MHD simulations of discs use polar coordinates, so the presence of an outflow along the symmetry axis is not altogether surprising (compare the discussion in Section 4.13), and it would be valuable to see if this appears in simulations performed in Cartesian coordinates.

Quite apart from the question of the energy source for jets, we have still not given a reason that any of the suggested driving mechanisms should produce narrowly collimated jets, rather than wide-angle outflows. Radiation pressure is unlikely to be the driving mechanism, as jets are preferentially seen in systems in significantly sub-Eddington states, and in any case is unpromising for explaining the observed tight collimation of the jets. The most popular idea for explaining collimation uses the ideal-MHD property of frozen-in magnetic fields. An idealized picture (see Figure 7.5) considers a poloidal fieldline (i.e. one with $B_\phi = 0$) attached at one end to a perfectly conducting rotator, such as an accretion disc, and at the other end to a near-stationary mass of perfectly conducting ambient plasma situated above the rotator, close to its axis. Since the field is frozen-in at both ends, the relative rotation winds up the initially poloidal field until it becomes strongly toroidal (oriented in the ϕ-direction). It then has a large magnetic pressure $\sim [4\pi/\mu_0]B_\phi^2/8\pi$ which accelerates the gas along the rotation axis against the accretor's gravity. This drives a shock wave ahead of and around the gas, which moves into the surrounding magnetic field. Far along the jet axis the field is effectively a set of purely toroidal loops which confine the accelerated gas close to the rotation axis through their fieldline tension (the so-called hoop stress). Eventually the system reaches a steady state in which the rotation twists the field into new toroidal loops, replacing those moving out with the jet. At some distance r from the jet base, the azimuthal component of the field (which goes as $\sim 1/r$) becomes stronger than the poloidal component ($\sim 1/r^2$), so the jet begins to buckle. This mechanism converts the rotational energy of the conductor into magnetic field pressure and then into outflowing gas kinetic energy, simultaneously confining the outflow close to the rotation axis.

As we saw in (3.88), one never needs to compute where the currents flow in MHD, as the curl of the magnetic field gives them directly and self-consistently.

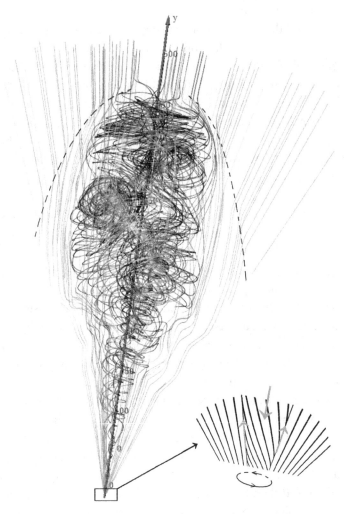

Figure 7.5 Jet formation by magnetic fields. A set of initially purely poloidal (i.e. $B_\phi = 0$) fieldlines is frozen-in at its inner end to a perfectly conducting rotator with angular frequency Ω, and at distance to the ambient plasma. The rotation creates toroidal magnetic loops, and the increasing magnetic pressure accelerates the ambient plasma along the rotation axis, eventually forming a jet. The jet is effectively a collection of toroidal field loops sliding outwards along the poloidal fieldlines and accelerating plasma along the jet axis through their pressure, while collimating it through their tension (hoop stresses). The volume current in the interior of the jet is compensated by the surface current along the interface with the surrounding magnetic field (inset: the arrows show the poloidal component of the current). The power of the jet is determined by two parameters: the rotational frequency of the central object, Ω, and the radial magnetic flux threading the object, Φ. Note that if the central accretor is a black hole it is unable to drag the fieldlines because of its extremely low conductivity (see (7.22)), so the required dragging must instead invoke plasma rotating near to it. Credit: R. Moll 2009, Max Planck Institute for Astrophysics.

But this point often causes confusion in discussions of jets. In the picture depicted in Figure 7.5, the current flows through the body of the outflow, across the head of the jet and then back along the boundary between the jet and its environment, rejoining at the base of the outflow, the head of the jet and along its boundary with the environment. Common pitfalls include supposing that the jet outflow is somehow the current itself, that the current flows in the direction of the rotation of the jet base, or that the currents are the source of the toroidal field component. As we have seen, it is the rotation of the magnetic field that produces the toroidal component, which then causes the currents to flow.

Although the picture sketched here is attractive, we should note that we still need to explain the origin of its basic ingredient, a magnetic fieldline attached to a central rotator. If the rotator is assumed to be the black hole, this returns us to the problems discussed earlier in this section connected with the hole's extremely low conductivity (see (7.22)).

We should also remember that jets are not unique to black hole systems, although this was where observers first found them. There is very clear observational evidence that *all* types of accreting system can produce jets – see Figure 1.11. It is unlikely that the 'central rotator' is the accreting object (neutron star or protostar), as their spin rates are far too low in many cases. It is much more likely that it is some region of the inner accretion disc. Occam's razor[3] weakens the case for a jet-driving mechanism unique to black holes. Accordingly, we should look for forms of the central rotator involving only the accretion disc.

A possible case of this occurs if accretion is misaligned with respect to the spin of the accretor. As we have argued, this is probably generic for SMBH accretion, as there is no preferred plane, but even in stellar-mass binaries some degree of misalignment is plausible. Then the disc annuli attempt to precess differentially about the symmetry axis of the accretor – by the Lense–Thirring (LT) effect for SMBH accretion, with precession frequency (2.23), that is,

$$\omega = \omega_{\mathrm{LT}} \simeq \frac{2ac}{R_g r^3}. \tag{7.29}$$

If disc viscosity results from the magnetorotational instability, as we argued in Section 4.12, it is natural to imagine that each annulus may carry a local magnetic field largely orthogonal to its plane. (This is the opposite limit of the case of a black hole with a misaligned magnetic field from that treated in Section 2.4. There the sources of the field were assumed to have far *higher* inertia than the hole, whereas here they have far lower inertia.)

---

[3] Put simply, 'don't invent two theories for the same thing'.

Then the precessional motion around the black hole spin axis, produces electromagnetic dipole power with time-averaged angular dependence

$$\frac{dL_{dip}}{d\Omega} = \frac{\omega^4 \mu^2 \sin^2 \beta}{8\pi c^3}(1 + \cos^2 \theta),$$ (7.30)

where $\mu = BR^3$ is the magnetic moment, θ the polar angle measured from the black hole spin axis, and β the angle between the disc normal and this axis. The total time-averaged power is[4]

$$L_{dip} = \frac{2\omega_{LT}^4 \mu^2 \sin^2 \beta}{3c^3}$$ (7.31)

emitted coherently at the precession frequency ω_{LT}. This gives

$$L_{dip} = \frac{64}{9\alpha}\left(\frac{R}{H}\right)\left(\frac{v_A}{c_s}\right)^2 \frac{\dot{M}c^2}{r^{17/2}} a^4 \sin^2 \beta,$$ (7.32)

using the Equation (4.90) for a thin accretion disc with accretion rate $\dot{M}$, viscosity parameter α, local scaleheight H, sound speed c_s, and Alfvén velocity v_A. The very steep $r^{-17/2}$ radial dependence means that the dipole power from a disc is essentially all from the innermost radius where the inclination is non-zero. This gives

$$\frac{L_{dip}}{L_{acc}} = \frac{64}{9\eta\alpha}\left(\frac{R}{H}\right)\left(\frac{v_A}{c_s}\right)^2 \frac{a^4 \sin^2 \beta}{r^{15/2}},$$ (7.33)

where $L_{acc}(R) \simeq \eta\dot{M}c^2 r^{-1}$ is the local gravitational binding energy release. This shows that even very small disc inclinations β can give significant dipole power for near-equipartition magnetic fields ($v_A \sim c_s$).

The precession does no work on the black hole spin, so all of the energy loss (7.31) must come from the gravitational binding energy released as the annulus contracts. The associated angular momentum is carried off by the circularly polarized emission along the black hole spin axis, which produces a torque of the form (2.26). This requires that $d\omega/dt$ is orthogonal to μ, so only the misaligned component of ω is reduced. The dipole emission causes the annulus to lose gravitational binding energy and align or counteralign with the black hole spin on the dipole power timescale. Energy is emitted as the annulus contracts, so that L_{dip} must be less than some fraction of L_{acc}. Equation (7.33) shows that $L_{dip} \sim L_{acc}$ for $r \sim 1$, almost independent of parameters (in particular, $\sin \beta$ can be extremely small). So there is a strong tendency to produce significant dipole emission near the ISCO ($r \sim 1$).

---

[4] Note that if we set $R_g \sim R_d, c/R_g \sim \Omega_d$, and $B \sim B_{pd}$, this equation recovers the estimate (7.27).

Significantly, the emission frequency is always extremely low, since from (7.34) we have

$$\omega < \frac{2c}{R_g} = 4 \times 10^{-3} M_8^{-1} \, \mathrm{s}^{-1}, \tag{7.34}$$

where $M_8 = M/10^8 \mathrm{M}_\odot$. The plasma frequency (see (3.68)) is

$$\omega_p = 2\pi \nu_p \simeq 4.5 \times 10^4 N^{1/2} \, \mathrm{s}^{-1}, \tag{7.35}$$

where N is the electron number density in cm^{-3}, implying

$$\frac{\omega}{\omega_p} \simeq \frac{10^{-7}}{N^{1/2} M_8}, \tag{7.36}$$

which is $\ll 1$ for any realistic number density N. The extremely low-frequency dipole emission is refracted towards the regions of lowest density, that is, away from the disc plane, and absorbed. Near the ISCO this means that energy is channelled towards the black hole spin axis. This suggests a promising way of driving jets launched close to the ISCO.

Astrophysical dipole emission is known to be important in radio pulsars, and these are indeed observed to produce jets directed along the spin axis of the neutron star (Markwardt & Ögelman, 1995) about which the magnetic field rotates. The dipole emission then taps the neutron star spin energy, and is believed to be the main cause of radio pulsar spin-down. For a black hole accreting from a misaligned disc, the inclined dipole *precesses*, so the dipole power must instead come from accretion energy.

Two more aspects of this picture of dipole power from a misaligned disc are encouraging. First, dipole emission is initially coherent, and may drive non-thermal emission and so high brightness temperatures in the jets. (This is well known for radio pulsars, if not yet fully understood.)

Second, the dipole power output peaks very sharply near the inner edge of the accretion disc, and dipole emission explicitly produces instantaneous (i.e. not time-averaged) power with amplitude modulated at Lense–Thirring frequencies. This does not require a particular geometry, so an observer at any direction in space can detect this modulation. This provides a natural origin for the quasiperiodic oscillations (QPOs) seen in X-rays from accreting systems. It has long been noted that observed QPO frequencies correspond to Lense–Thirring precessions near the inner edge of the accretion disc in both X-ray binaries and AGN, although a plausible physical mechanism for producing power at these frequencies has been lacking. Note that although the emitted power is $\propto \omega_{\mathrm{LT}}^4$, this does not imply a relation between observed QPO strength and frequency. The primary coherent radiation is always strongly absorbed and re-emitted at a much wider range of far higher electromagnetic frequencies.

The discussion here suggests that significant dipole energy loss from misaligned accretion disc annuli may drive jets. These must appear at small disc radii as they require a significant fraction of the accretion energy, with the strongest jets appearing near the ISCO. This type of jet cannot appear at any point where the disc aligns completely with the black hole spin, that is, $\beta = 0$, or if the hole does not spin ($a = 0$). These are the two types of stationary black hole state that Hawking's (1972) result allows. Then for this kind of jet to appear requires the dipole emission timescale t_{dip} to be shorter than the local timescale t_2 for viscous alignment, and the jets are stronger if the alignment timescale is not significantly shorter than the timescale for matter to reach the ISCO. King and Nixon (2018) show that

$$\frac{t_{\text{dip}}}{t_2} = \frac{3}{4a\alpha} \left(\frac{H}{R}\right)^2 \left(\frac{c_s}{v_A}\right)^2, \tag{7.37}$$

where the equality of sound speed c_s and Alfvén speed v_A corresponds to an equipartition magnetic field. Then the appearance of strong jets requires

$$\frac{H}{R} \lesssim (a\alpha)^{1/2} \frac{v_A}{c_s} \lesssim 0.1 - 0.37, \tag{7.38}$$

where the ratio is evaluated for $\alpha \sim 0.1$, $v_A \lesssim c_s$, and $0.1 < a < 1$.

This shows that strong jets are unlikely in discs with large aspect ratios H/R. In particular, significant radiation pressure requires $H/R \sim 1$, so jets are unlikely at luminosities close to Eddington. Observations of X-ray binary state changes (e.g. Belloni, 2010) show that jets appear in low states and disappear in high states. The low-frequency QPOs usually interpreted as LT precession often appear in the 'low-hard' state, where H/R cannot be very large, as a near-spherical flow cannot show significant precession (Nixon & Salvesen, 2014).

7.4 Can Jets Precess?

Observations of AGN jets often provoke suggestions that they may precess. The is a natural idea, given that the extreme Galactic binary system SS433 has powerful jets, seen in both the Hα emission line and the radio, and there is abundant evidence (see Figures 7.6, 7.7, and 7.8) that these jets precess with a period of 162 days. This example, together with tentative observational evidence (see Figure 7.9), has led to suggestions that the jets from radio galaxies and AGN also precess.

But the Hα-emitting jets seen in SS433 are very unusual. First, the speed of matter in the jets is $0.26c$, instead of a value much closer to c expected for a stellar-mass black hole or even a neutron star accretor. Second, unlike radio jets seen from AGN and radio galaxies, they are baryon-loaded, explaining the Hα emission. And finally, in addition to precessing on a 162-day period, the jets show

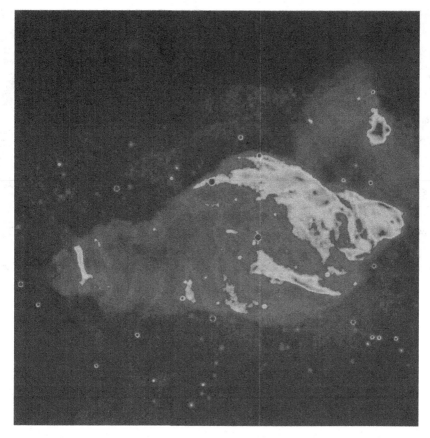

Figure 7.6 Representation of the W50 nebula surrounding the compact source of near-relativistic jets SS 433 at the centre. The 'ears' of this nebula are inflated by precessing jets, revealed by the very large periodic red- and blueshifts of the observed Hα emission line. The spherical central region is caused by a wide-angle outflow, seen in observations as the 'stationary' Hα line (actually showing the 13-day binary period). Credit: Dubner et al. (1998).

nodding motions at close to one-half of the 13-day binary period. Begelman, Volonteri, and Rees (2006) show that these features strongly suggest that the relativistic jets produced by the accretor are deflected by a precessing outer part of the accretion disc. This situation persists because the (presumed black hole) accretor's spin is misaligned with respect to the orbital motion of the companion star, but it is fortuitous that the angle of misalignment allows the jets to be deflected by the outer disc. This kind of arrangement then appears fairly accidental, so is unlikely to be common. In line with this, although it is inferred that most stellar-mass X-ray binaries produce jets, at least at certain epochs, SS433 does appear to be fairly

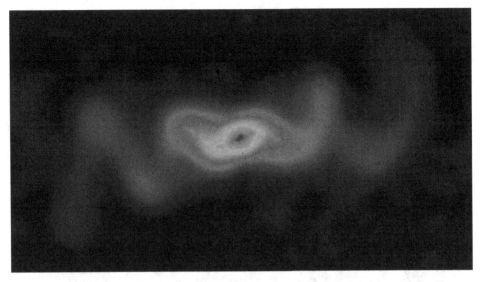

Figure 7.7 The corkscrew morphology of the SS433 radio jets. Credit: Blundell and Bowler (2004).

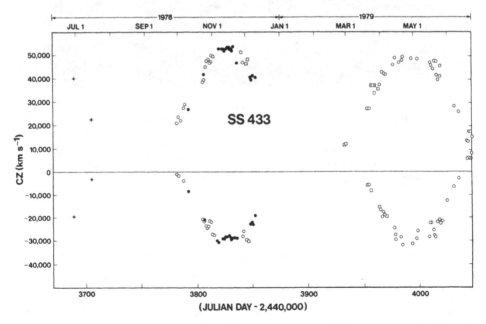

Figure 7.8 The enormous Doppler shifts of the jets in SS433 as seen in the Hα lines, showing the 162-day precession period. Note that the velocities of the jets are equal (so they lie in the plane of the sky) at a velocity that is too large to be the system's space velocity. This is caused by the *transverse* Doppler effect. Credit: Margon et al. (1979).

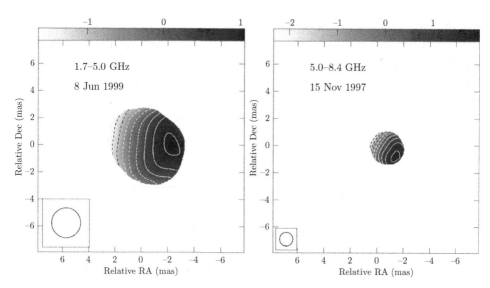

Figure 7.9 Spectral-index radio images of the active galaxy M81 at two different epochs. The change is sometimes ascribed to precession of the radio jet. Credit: Martí-Vidal et al. (2011).

unique – more than four decades after its discovery, no similar system has emerged, despite strenuous efforts.

Of course, for AGN it is impossible to infer jet precession in real time, so suggested precessions refer to cases where observed jets appear to have changed their axes over time. In reality we will see that it is difficult to make jets precess, except by the very rare process apparently at work in SS433, which is unlikely in AGN. The reasons are simply: (a) the angular momentum of any single realistic accretion event is always smaller than the angular momentum of the hole; and (b) a processing disc–hole system *aligns* on roughly the precession timescale, so the amplitude of the precession is significantly reduced after only one arc.

We consider first the evolution of a misaligned disc around a spinning black hole in the regime where warps propagate diffusively – that is, the dimensionless viscosity coefficient and disc aspect ratio obey $\alpha > H/R$ – see Section 5.3. We discuss the wavelike case ($\alpha < H/R$) at the end of this section.

The Lense–Thirring (LT) effect of a spinning black hole makes tilted disc orbits precess around its angular momentum vector (see (7.34)) at rates decreasing strongly with radius. Viscosity communicates the differential precession through the disc, and this acts to co- or counteralign the disc with the plane of the hole (see Section 5.3). The inner disc settles quickly into the equatorial plane of the hole while the outer parts remain misaligned, and the two parts are joined by a warped region – this is the Bardeen–Petterson effect (Bardeen & Petterson, 1975). A warp like this can remain stationary if an external torque (e.g. from a misaligned binary companion, as in SS433) maintains the tilt at the outer edge of the disc. Without an external torque the warp propagates outwards until the whole disc lies in the equatorial plane, with the entire hole–disc system aligned along its original total angular momentum vector.

In Section 5.6 we saw that for larger warps given by more extreme disc tilts the disc can break into distinct planes. Although this may alter the detailed evolution of the spin and disc angular momenta it cannot affect the final aligned state, as all the torques are internal to the hole–disc system.

To consider precessions we define $\mathbf{J}_d$, $\mathbf{J}_h$ and $\mathbf{J}_t = \mathbf{J}_d + \mathbf{J}_h$ as the disc, hole, and total angular momentum vectors, respectively, with magnitudes J_d, J_h, and J_t. Here $\mathbf{J}_d$ is defined loosely as that part of the disc angular momentum that can interact with the hole on timescales of interest.

During alignment, $\mathbf{J}_h$ precesses around $\mathbf{J}_t$ with an initial amplitude θ_i

$$\cos\theta_i = \frac{\mathbf{J}_h \cdot \mathbf{J}_t}{J_h J_t}. \tag{7.39}$$

The angle θ_i is small (so $\mathbf{J}_t$ and $\mathbf{J}_h$ point in similar directions) if either (a) the disc already points in a similar direction to the hole, or (b) $J_d \ll J_h$, making $\mathbf{J}_h \simeq \mathbf{J}_t$.

If $J_d \ll J_h$ alignment obviously cannot move the hole spin vector very far. The inner disc must have become anchored to the equatorial plane of the hole very rapidly, so alignment cannot move the inner disc very far either. We conclude that if $J_d \ll J_h$, the Lense–Thirring effect cannot drive a precessing jet.

So we might at best hope to find precessions in cases where $J_d \gtrsim J_h$. But this does not generate repeated jet precession either. The *initial* precession can be large if, for example, $J_d >> J_h$, but the alignment and precession timescales for the disc are similar (Lodato & Pringle, 2006; see Figure 7.10), so after only one precession time the hole is significantly aligned with the disc. Then in a tilted disc with $\alpha > H/R$ (i.e. with warps propagating in the diffusive regime) the Lense–Thirring effect alone cannot drive repeated jet precession.

We still have to consider three cases not treated so far. When disc warps propagate as waves, that is, for $\alpha < H/R$, there is a strong tendency for the disc to achieve a near-steady state as the wave propagates out to the disc edge and is reflected back inwards, which is not promising for driving precessing jets (see Nixon & King, 2013). In cases where the warp is sufficiently extreme that the disc breaks, the individual broken disc regions precess independently at their local LT frequencies. It is unclear whether such broken components could drive jets, but if so, this could give rather short-lived jet precession. Finally, we have not so far

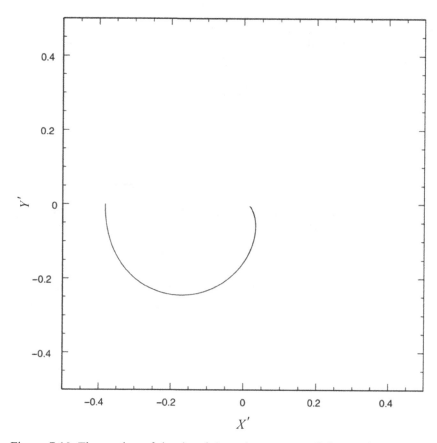

Figure 7.10 The motion of the tip of the unit vector parallel to $\mathbf{J}_h$ in the plane (x', y') orthogonal to the total system angular momentum $\mathbf{J}_h + \mathbf{J}_d$. The hole spin aligns after about half a turn in this plane, because the precession timescale is of the same order as the alignment time. Here $\mathbf{J}_h = \mathbf{J}_d$, the ratio of 'vertical' to 'horizontal' viscosity is $\nu_2/\nu_1 = 10$, and disc gas is fed in at $2\times$ the initial warp radius. Credit: Lodato and Pringle (2006).

mentioned the possibility of disc warping by radiation torques (see Pringle, 1997, and Section 5.10 of Frank, King & Raine, 2002). This again is a largely unexplored area – the main question is whether discs warped by these torques would drive jets at all.

7.5 Do Jets Move?

We showed in the previous section that the dynamics of the disc–hole system in the diffusive warp propagation regime $\alpha > H/R$ prevent sustained LT precessions. We can now ask if any accretion event can change the direction of the hole spin at all.

First, it is clear that the angular momentum J_d of any single realistic accretion disc event is always smaller than the hole angular momentum J_h. We have $J_d \sim M_d(GMR_d)^{1/2}$, $J_h = GM^2 a/c$, which implies

$$\frac{J_d}{J_h} = \frac{1}{a}\frac{M_d}{M}\left(\frac{R_d}{R_g}\right)^{1/2}.$$

(7.40)

Using (4.100) and (2.8) we get

$$\frac{J_d}{J_h} < \frac{10^{-3}}{a}\left(\frac{3\times 10^{16}}{1.5\times 10^{13}M_8}\right)^{1/2} \lesssim \frac{4.5\times 10^{-2}}{aM_8^{1/2}}.$$

(7.41)

This shows that repeated disc events can cause the SMBH spin axis to move, significantly if $aM_8 \lesssim 0.1$, that is, the hole mass is $\lesssim 10^8 M_\odot$, and/or the hole spin is low. We saw in Section 5.4 that the effect of repeated randomly oriented accretion events is to keep the SMBH spin low. Then (7.41) shows that the hole spin direction is likely to be strongly influenced by the orientation of the most recent event. This in turn suggests that successive jet ejections can move across the sky in a stepwise fashion, but only on the timescale of the events themselves.

8

Black Hole Demographics

8.1 Introduction

The earlier chapters of this book have discussed in some detail the physics of supermassive black holes and accretion on to them, and the resulting feedback. We have been able to do this without fully describing the connection to the host galaxy in detail, instead simply using boundary conditions which may give a plausible idea of this relationship. But it is clear that this procedure has limitations, and we must recognize that the nature of the connection between SMBH and their hosts is still unclear. In particular we still do not fully understand how gas gets close enough to the hole to accrete via a disc on a reasonable timescale. Without an answer to this question we cannot attempt a coherent first-principles discussion of how SMBH and their hosts evolve together.

There are a number of candidates for the mechanism feeding gas to the vicinity of the black hole. We discussed foraging in Section 5.2, and there are other proposals such as galaxy mergers (major and minor), bar formation in the galaxy potential, bars within bars, turbulence in the interstellar medium, and stellar mass loss. We eliminated others – by its nature Bondi accretion cannot deal with the fundamental need to remove angular momentum from gas to allow infall from large radii, while viscous accretion via a disc extending to these radii fails for two reasons. First, the inflow timescale in a viscous disc of size ~ 1 pc is already at least 10^9 yr, even with optimistic assumptions about viscosity. This is implausibly long for an AGN timescale, and for shorter ones with more realistic viscosities we must find a reason why the gas is fed into this putative central disc at much smaller radii. Second, as we saw from (4.100), realistic accretion discs become self-gravitating at radii $R_{sg} \simeq 3 \times 10^{16}$ cm $\simeq 10^{-2}$ pc and tend to break up into stars outside this, returning us to the problem of how to remove angular momentum from the gas.

Observations find a wide variety of activity in galaxy nuclei. Bright quasars are seen out to high redshift, there are barely detectable nuclei in LINERS (low-ionization nuclear emission-line regions), and detectable but extremely weak activity in the Galactic Centre. This makes it conceivable that any one of the suggested fuelling processes may be at work in at least some galaxies. We consider first the nuclei of relatively nearby Seyfert galaxies, with SMBH masses $\sim 10^7$–$10^8 M_\odot$ accreting at rates $\sim 0.2 M_\odot \, \mathrm{yr}^{-1}$ and so luminosities $\sim 0.1 L_{\mathrm{Edd}}$. Then in a typical Seyfert episode lasting $\sim 10^6$ yr the black hole grows only by $\sim 2 \times 10^5 M_\odot$, far smaller than the gas mass potentially available in a typical spiral galaxy. Black hole feeding here cannot be through large-scale mechanisms involving spiral arms or galaxy bars, and instead must consist of small-scale events depositing a modest mass of gas with low angular momentum in the central regions of the galaxy. There are other observational constraints – we have seen already in Section 5.3 that although the radio jets and [OIII] emission in nearby Seyfert galaxies are well aligned with each other, their orientations with respect to the major axis of the galaxy are completely random.

These results prompt two important conclusions. First, the random direction of the radio jets agrees with the argument of the previous paragraph suggesting that the feeding mechanism is probably not connected with large-scale features of the host, as the directional information this carries would have to be somehow randomized. Second, the good observed correlation of radio jets and [OIII] emission shows that the inner disc and black hole spin (which fixes the radio jet axis) remain aligned with the accretion torus and outer disc. In Section 5.3 we showed that repeated small-scale fuelling ensures precisely this. The hole spin counteraligns in about one-half of all accretion events. It then spins down, and this process is more efficient than spin-up because of the larger lever arm of retrograde accretion. Once the SMBH has doubled its mass, its angular momentum always has a smaller magnitude than that of any subsequent accretion event. Then it always aligns its spin with the inner disc and so the torus. We saw in Section 7.5 that this means that successive accretion events can move the spin axis – and so probably the direction of any jets – across the sky, in a series of modest re-orientations.

In this chapter we consider whether this kind of picture can explain global features of the SMBH population of the Universe.

8.2 Chaotic Accretion and the Local AGN Luminosity Function

We showed in Section 7.5 that chaotic accretion leads at low redshift to AGN powered by a succession of randomly oriented feeding events. These are self-gravitating, and probably form stars outside the radius $R_{\mathrm{sg}} \sim 3 \times 10^{16}$ cm. These formation events probably correspond to the nuclear starbursts associated with

AGN events. In this picture the starburst does not cause the AGN episode, nor the reverse – they are different aspects of the same event. The gas within R_{sg} settles into a ring or disc in a few dynamical times t_{dyn}, where

$$t_{dyn} = \left(\frac{R_{sg}^3}{GM} \right)^{1/2} \simeq 290 \left(\frac{M}{10^8 M_\odot} \right)^{-1/2} \left(\frac{R}{0.1 pc} \right)^{3/2} yr. \tag{8.1}$$

The disc inside this radius evolves on a viscous timescale

$$t_{sg} \sim \left(\frac{H}{R} \right) \left(\frac{M}{\dot{M}} \right) = 6.6 \times 10^4 \left(\frac{L}{L_{Edd}} \right)^{-1} \eta_{0.1} yr, \tag{8.2}$$

where L, L_{Edd} are the accretion and Eddington luminosities, and $\eta = 0.1\eta_{0.1}$ the accretion efficiency. After a time $t \gtrsim t_{sg}$ the disc structure must resemble a steady disc, as given by (4.91). These show that the mass of the disc within radius R is

$$M(< R) = 2.9 \times 10^{34} \left(\frac{\alpha}{0.03} \right)^{-4/5} \eta_{0.1}^{-3/5} \left(\frac{L}{0.1 L_{Edd}} \right)^{3/5} M_8^{11/5} \left(\frac{R}{R_s} \right)^{7/5} g, \tag{8.3}$$

where R_s is the Schwarzschild radius. The disc aspect ratio is

$$\frac{H}{R} = 1.9 \times 10^{-3} \left(\frac{\alpha}{0.03} \right)^{-1/10} \eta_{0.1}^{-1/5} \left(\frac{L}{0.1 L_{Edd}} \right)^{1/5} M_8^{-1/10} \left(\frac{R}{R_s} \right)^{1/10}. \tag{8.4}$$

The mass inside R_{sg} is

$$M_{sg} = 2.8 \times 10^5 \left(\frac{\alpha}{0.03} \right)^{-2/27} \eta_{0.1}^{-5/27} \left(\frac{L}{0.1 L_{Edd}} \right)^{5/27} M_8^{23/27} M_\odot, \tag{8.5}$$

which is of course simply $(H/R)M$, where H/R is evaluated at R_{sg}. The accretion rate $\dot{M} = L/\eta c^2$ is

$$\dot{M} = 0.25 \eta_{0.1}^{-1} \left(\frac{L}{0.1 L_{Edd}} \right) M_8 M_\odot yr^{-1}, \tag{8.6}$$

so the disc evolves on the viscous timescale $t_{sg} = M_{sg}/\dot{M}$, giving

$$t_{sg} = 1.1 \times 10^6 \left(\frac{\alpha}{0.03} \right)^{-2/27} \eta_{0.1}^{22/27} \left(\frac{L}{0.1 L_{Edd}} \right)^{-22/27} M_8^{-4/27} M_\odot yr, \tag{8.7}$$

so that $t_{sg} \sim (H/R)(M/\dot{M})$.

From the disc equation (4.91) we have $\Sigma \propto \dot{M}^{7/10} R^{-3/4}$, and in a steady disc we also have the basic relation $\dot{M} \propto \nu \Sigma$, so we find that the viscosity in the disc evolves as $\nu \propto \Sigma^{3/7} R^{15/16}$. There are similarity solutions for such discs (Pringle, 1991), and these show that at late times ($t > t_{sg}$) we have

$$L \propto \dot{M} \propto t^{-19/16}. \tag{8.8}$$

Then the luminosity of an event must evolve as

$$L = L_0[1 + (t/t_{sg})]^{-19/16}, \tag{8.9}$$

where L_0 is the initial luminosity.

We can compare these simple predictions directly with observations. Heckman, Kauffmann and Brinchmann (2004) find that most of the current accretion on to black holes is on to those with masses in the range 10^7–$10^8 M_\odot$. For low-mass black holes with $M < 3 \times 10^7 M_\odot$, 50 per cent of the total mass growth is provided by only 0.2 per cent of the holes. Evidently, this shows that strong fuelling only lasts for a short time $t_{fuel} \sim 0.002 t_H$, where $t_H \sim 1.4 \times 10^{10}$ yr is the Hubble time. Then the total time when strong fuelling occurs is around $t_{fuel} \sim 3 \times 10^7$ yr. If there are N fuelling events per black hole per Hubble time this means that each event lasts around $t_{ev} \sim 3 \times 10^7/N$ yr.

This allows us to find the luminosity function that this picture predicts. We take $\alpha = 0.03$ and accretion efficiency $\eta = 0.1$ and regard black holes with mass $10^7 M_\odot$ as typical of the distribution of accreting black holes at low redshift. We assume that each of these black holes gains mass in random but identical accretion events, each having initial luminosity L_{Edd}. Then the initial disc mass is $M_{sg} \sim 6 \times 10^4 M_\odot$. The initial evolution timescale is $t_{sg} \sim 2.4 \times 10^5$ yr, and this must correspond to the length of each event, t_{ev}. So the total number of events is $N \sim 0.002 t_H/t_{sg} \sim 116$, giving the average time between events as

$$t_{rep} \sim \frac{t_H}{N} \sim \frac{t_{sg}}{0.002} \sim 1.2 \times 10^8 \text{ yr.} \tag{8.10}$$

Writing $f = L/L_{Edd}$, our assumptions imply that initially $f = 1$. The average final value is

$$f_{end} = \frac{f_{in}}{1 + \lambda_{end}^{19/16}}, \tag{8.11}$$

where $\lambda_{end} = t_{rep}/t_{sg} = 500$. Then the maximum possible range of f is $(1 + \lambda_{end})^{19/16}$. Evidently, observations cannot probe this full range: figure 3 of Heckman, Kauffmann and Brinchmann (2004) shows that for black holes of mass $10^7 M_\odot$ the observed range of f is around 40 (corresponding to $\lambda > 18$ and $N < t_H/18 t_{sg} = 540$). Inverting (8.9), we find

$$\frac{t}{t_{sg}} = f^{-16/19} - 1 \tag{8.12}$$

for $0 < t < t_{sg}$.

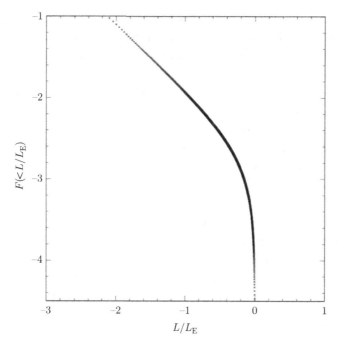

Figure 8.1 Form of the AGN luminosity function predicted by the fuelling process and subsequent disc evolution discussed here. The fraction F of sources with luminosity fraction L/L_{Edd} less than a given value is shown as a function of L/L_{Edd}. This is similar to the curves given in Figure 3 of Heckman, Kauffmann and Brinchmann (2004). Credit: King and Pringle (2007).

Then assuming that fuelling events for different black holes are independent, the fraction $F(>f)$ of black holes with luminosities $>f$ is

$$F(>f) = \frac{f^{-16/19} - 1}{f_{end}^{-16/19} - 1}, \qquad (8.13)$$

and we plot this in Figure 8.1.

This curve has a similar form to the distributions of low-mass black holes ($3 \times 10^6 M_\odot - 3 \times 10^7 M_\odot$) in figure 3 (left) of Heckman, Kauffmann and Brinchmann (2004), in the observed range $1 < f < 40$. Although the assumptions here are very simple, involving identical and independent fuelling events, the comparison is encouraging.

8.3 How Big Can a Black Hole Grow?

The central theme of this book is the close relationship between supermassive black holes and galaxy growth. The scaling relations show that the evolution of these two elements are closely related. The relationship is at least in part a consequence of the hole's ability to affect the galaxy through feedback, which we have seen can be very effective, and even potentially disruptive. Conversely, growing the galaxy seems to grow the hole's mass.

An obvious question is whether there is a limit to this coevolution, or whether a black hole can in principle reach any given mass, given a suitable host and enough time. The first attempt to answer this question was made by Natarajan and Treister (2009). They derived a limit using the idea that a self-gravitating accretion disc would blow itself away if the black hole mass reached a value $\sim 10^{10} M_\odot$, where the precise value depends on the properties of the host galaxy's dark matter halo, in agreement with observational evidence.

The main observational constraint on black hole growth is that we know from the Soltan relation that supermassive black holes grow their masses mainly by luminous accretion of gas. And because gas in galaxies must have significant angular momentum, SMBH accretion must proceed largely through a disc (as we have seen, more probably a series of discs of randomly varying orientation). We also know that an SMBH disc is likely to be self–gravitating outside the radius $R_{sg} \sim 0.01$ pc (see (4.100)) so that its outer radius cannot exceed R_{sg}, which is effectively independent of the SMBH mass. But to form a disc at all, the inner disc radius must be at least as large as the ISCO, whose size scales directly with the SMBH mass M, as

$$R_{\text{ISCO}} = f(a)\frac{GM}{c^2} = 7.7 \times 10^{13} M_8 f_5 \text{ cm}. \tag{8.14}$$

Here $f(a)$ is a dimensionless function of the SMBH spin parameter a (see Section 4.1), with $f(a) = 5f_5(a)$, so that $f_5 \simeq 1$ corresponds to prograde accretion at moderate SMBH spin rates $a \simeq 0.6$.

If $R_{\text{ISCO}} \gtrsim R_{sg}$, disc accretion is likely to be suppressed. Any gas orbiting near R_{ISCO} must be self-gravitating so must be swallowed whole, without radiating as a disc. An SMBH can in principle grow its mass in this way, effectively swallowing self-gravitating matter, but we shall see that it cannot subsequently reappear as a bright disc-accreting object, that is, a quasar or AGN.

Using (4.100) and (8.14), we see that the ISCO radius exceeds the self-gravity radius, and an accretion disc cannot form, for SMBH masses larger than

$$M_{\text{max}} = 5 \times 10^{10} M_\odot \alpha_{0.1}^{7/13} \eta_{0.1}^{4/13} (L/L_{\text{Edd}})^{-4/13} f_5^{-27/26}. \tag{8.15}$$

This is an upper limit on the mass of the SMBH in any quasar or AGN, since these systems have accretion discs.

Figure 8.2 shows the curve $M = M_{\text{max}}(a)$, for $\alpha = 0.1$ and $L = L_{\text{Edd}}$, while $\eta, f_5(a)$ are specified parametrically as functions of a (see the relations 9 and 11 in King & Pringle, 2006). The masses measured for almost all accreting SMBH

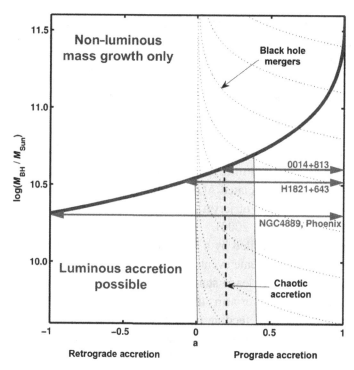

Figure 8.2 The mass limit M_{max} for luminously accreting supermassive black holes, compared with the largest observed masses. The curve shows M_{max} as a function of black hole spin parameter a, where values $a < 0$ denote retrograde accretion. Accreting SMBH must lie below the curve. In order of decreasing mass, the systems shown are $0014 + 813$, with $M = 4 \times 10^{10} M_\odot$ (Ghisellini et al., 2010), the central quasar of the H1821+643 cluster ($M = 3 \times 10^{10} M_\odot$, Walker et al., 2014), NGC 4889 (McConnell et al., 2011), and the central galaxy of the Phoenix cluster (McDonald, Bayliss & Benson, 2012), both of the latter pair having $M = 2 \times 10^{10} M_\odot$. Only the first two systems place any restrictions on the black hole spin. For 0014+813, accretion must be prograde, with $a \gtrsim 0.2$. For H1821+643, if accretion is prograde $|a|$ can have any value, but if accretion is retrograde $|a|$ must be less than about 0.1. All the other systems are compatible with any spin value. The dotted curves show the statistical effect $|a| \propto M^{-2.4}$ (Hughes & Blandford, 2003) of black hole mergers on mass growth. The attractor $|a| \to \bar{a} \simeq 0.20 M_{10}^{-0.048}$ for chaotic gas accretion is shown (dotted track), with the grey surround indicating the typical spread in $|a|$ (King, Pringle & Hofmann, 2008; see Figure 5.10). Credit: King (2016).

lie below the curve for all possible spin values $-1 \le a \le 1$. The exceptions are $0014 + 813$ (Ghisellini et al., 2010: $M \simeq 4 \times 10^{10} M_\odot$) and H1821+643 (Walker et al., 2014: $M \simeq 3 \times 10^{10} M_\odot$). For the first system the limit requires that accretion is prograde and $a \gtrsim 0.2$. For H1821+643 the limiting mass requires that $a \gtrsim -0.1$. Then prograde accretion is possible for any spin parameter $a > 0$, but retrograde accretion on to this hole with $|a| > 0.1$ is ruled out.

With maximal spin $a = 1$ prograde with respect to the sense of accretion, the normalization in (8.21) becomes $2.7 \times 10^{11} M_{\odot}$, defining the absolute maximum mass for a luminously accreting SMBH.

Values of M close to this maximum must be rather rare, because disc accretion must be almost permanently prograde as the hole grows (as we have seen, spin-down by retrograde accretion is more effective than spin-up by prograde accretion, because of the larger ISCO and therefore its lever arm). The only likely way of achieving this would be to have the hole spin permanently correlated with a fixed direction of the potential controlling the gas flow within the galaxy. This is likely to require a spin axis and associated AGN jet direction also aligned with the galaxy. As we have seen, there is no observational support for such a correlation (Nagar & Wilson, 1999; Kinney et al., 2000).

If accretion is not controlled by a large-scale potential in this way, it probably consists of multiple small-scale events, essentially random in time and orientation. Chaotic accretion like this leads statistically to spin-down, again because retrograde events have larger lever arms than prograde, and occur almost as often. We saw earlier that this type of feeding predicts an attractor $|a| \to \bar{a} \simeq 0.20 M_{10}^{-0.048}$ for large SMBH masses (shown in Figure 5.10). Other interactions also tend to decrease the magnitude $|a|$ of the spin. Hughes and Blandford (2003) show that mergers with other black holes statistically decrease the spin as $|a| \propto M^{-2.4}$. These arguments suggest that SMBH generally cross the critical $M = M_{\max}$ curve (8.15) at modest values of $|a|$, and so $M_{\max} \simeq 5 \times 10^{10} M_{\odot}$ in all but rare cases.

If we further assume that the maximum observable luminosity of an AGN respects the Eddington limit, the mass limit (8.21) translates into a luminosity limit

$$L_{\max} = 6.5 \times 10^{48} \alpha_{0.1}^{7/13} \eta_{0.1}^{4/13} f_5^{-27/26} \text{ erg s}^{-1}. \tag{8.16}$$

This predicted value is in good agreement with the highest observed QSO bolometric luminosities, for example those in the WISE survey of hot, dust-obscured galaxies ('Hot DOGs': Assef et al., 2015, Figure 4).

It is worth stressing that (8.15) and (8.16) are the limits for objects undergoing luminous accretion, not absolute limits on every black hole mass (as remarked already, non-luminous mass growth beyond $M_{\max}$ is perfectly possible).

A possible way around the limits (8.15) and (8.16) is fairly obvious from the relation $M_{sg} \sim (H/R)M$ (Equation (4.97)). Luminous disc accretion with a significantly larger scaleheight H than the very small values ($H \sim 10^{-3}R$) for standard thin-disc AGN accretion could increase M_{sg} and so R_{sg}, and thus $M_{\max}$ significantly. This could potentially occur if radiation pressure dominates gas pressure in the outer parts of the disc, that is, if much of the outer disc lies in region 'a' of Shakura and Sunyaev (1973), rather than region 'b' (gas-pressure dominant) as assumed above. The discussion by Kawaguchi, Aoki, and Collin (2004) suggests that this could happen near the self-gravity radius if the accretion rate obeys

$$\dot{M} > \dot{M}_{ba} = 4\alpha_{0.1}^{0.4} \, M_\odot \, yr^{-1}, \tag{8.17}$$

which is sub-Eddington for SMBH masses $M \gtrsim 2 \times 10^8 \alpha_{0.1}^{0.4} M_\odot$. But we know that disc region 'a' is strongly unstable on a thermal timescale, both in the context of the α-prescription (Lightman & Eardley, 1974) and as found in shearing-box simulations (Jiang, Stone & Davis, 2013). Theory and simulation do not yet tell us the likely results of this instability, so we should ask what observational constraints exist. A radiation-pressure-dominant disc region fed with gas from a part of the disc with gas pressure dominant and lying outside it presumably finds a way to supply matter to the black hole. Eddington-limited systems are observed at a wide range of black hole masses, supplied by a stable and long-lasting reservoir such as a companion star in a stellar-mass binary, so an inner region dominated by radiation pressure is perfectly compatible with AGN feeding. But to overcome the limit (8.21) an SMBH disc would need to *form* with radiation pressure dominant at its outer, self-gravitating radius, and stably feed the hole, which seems much less likely. An AGN feeding event probably involves the infall of a mass of gas, dust and possibly stars, which becomes bound to the SMBH through internal dissipation as tides act on it (see King & Pringle, 2006; King & Nixon, 2015). If radiation pressure is already dominant at the self-gravity radius this dissipation is likely to drive mass off rather than feed the AGN efficiently. Then if we assume that the outer parts of discs feeding AGN must instead have gas pressure dominant, we are again left with the limits (8.15) and (8.16).

We noted after (8.16) the SMBH can grow its mass above M_{max} provided that this does not involve luminous disc accretion. But this is unlikely to allow the black hole to reappear later as an accreting quasar by forming a disc, since $R_{ISCO} > R_{sg}$. An increase in $|a|$ that might reduce R_{ISCO} below R_{sg} and allow luminous accretion again is unlikely, as we have seen that prolonged accretion tends to spin the hole down, and disc accretion is in any case impossible for $M > 2.7 \times 10^{11} M_\odot$ for *any* value of a.

It appears, then, that an SMBH's first contact with the thick curve in Figure 8.2 probably signals its last appearance as an observable AGN. One might nevertheless detect SMBH with masses above M_{max} in other ways, perhaps through gravitational lensing or the direct effect on orbiting stars.

At masses close to but below M_{max}, luminous disc accretion in a field galaxy at the Eddington luminosity must be rare, since it is difficult to feed the SMBH at a rate high enough to produce this. For example, the dynamical infall rate $f_g \sigma^3 / G$ (where f_g is the gas fraction) is below the Eddington value except in galaxy bulges with very high velocity dispersions $\sigma \gtrsim 400 \, km \, s^{-1}$. Even for brightest cluster galaxies (BCGs) in the centres of clusters, strongly super-Eddington accretion rates appear unlikely. This probably means that SMBH in this state will not provide the strong feedback giving the M–σ relation, so SMBH close to M_{max} may be able to evolve above M–σ, and not make their hosts red and dead. The observational data (e.g. McConnell et al., 2011) support these conclusions. We see that M_{max} lies well above the M–σ relation, as host bulge velocity dispersions do not reach the required values $\sigma \gtrsim 700\,\text{km s}^{-1}$. We examine questions like this further in the next section.

8.4 Evading M–σ

We saw in Chapter 6 that there is now a fairly coherent picture of supermassive black hole growth in the local Universe. SMBH at low redshift, with a sufficient supply of surrounding gas, grow their masses M to the limiting M–σ value

$$M = M_\sigma \simeq \frac{f_g \kappa}{\pi G^2}\sigma^4 \simeq 3 \times 10^8 \sigma_{200}^4 M_\odot \qquad (8.18)$$

fixed by feedback from their accretion disc winds. At the M_σ mass, SMBH feedback changes character. The shocks of the accretion disc wind against the host galaxy's interstellar gas no longer cool, as the inverse Compton cooling of the AGN radiation field is too weak. Momentum driving changes to energy driving, and this expels the remaining bulge gas, with typical velocities

$$v_{\text{out}} \simeq 1230 \sigma_{200}^{2/3} l^{1/3}\,\text{km s}^{-1} \qquad (8.19)$$

(from (6.88)), where l is the ratio of the SMBH accretion luminosity to the Eddington value. This stops significant further growth of the black hole. This can only restart if mergers or dissipation re-supply the bulge with gas, restarting evolution towards the relation (6.2) for larger values of M and σ.

This simple picture asserts that SMBH can only grow above M_σ through mergers with other holes, or by sub-Eddington accretion exerting little feedback. Neither of these possibilities suggests that SMBH masses M could significantly exceed M_σ – this is presumably why the M–σ relation is observable at all. (We noted in Section 6.1 that SMBH below the value (6.2) are likely to be fairly hard to observe.)

But there are some clear cases where M is significantly larger than M_σ, particularly when the black hole mass M is large itself. Examples are NGC 4889 (McConnell et al., 2011) and NGC 1600 (McConnell et al., 2011; Thomas et al., 2016) in the centres of the Coma and Leo clusters, respectively. The SMBH masses here are $M = 2 \times 10^{10} M_\odot$ and $1.7 \times 10^{10} M_\odot$. From (8.18), these masses require $\sigma = 571\,\text{km s}^{-1}$ and $549\,\text{km s}^{-1}$, while their observed velocity dispersions are $\sigma = 350\,\text{km s}^{-1}$ and $293\,\text{km s}^{-1}$. Matsuoka et al. (2018) (Figure 4, and references cited therein) also give several very large measured SMBH masses. These have no measured σ, but the claimed masses would require extremely large values $\sigma \sim 500$–$1,000 \, \text{km s}^{-1}$ to be compatible with (6.2). Kormendy and Ho (2013) note that these 'monster' objects diverge even further from the other scaling relation, $M \simeq 10^{-3} M_b$, where M_b is the bulge stellar mass. Although the majority of low-redshift SMBH do conform to the simple picture discussed in Chapter 6, there are evidently some more complex cases, and we look at this question more closely here.

We have seen that SMBH mass growth is characterized by three critical masses. These are M_σ, fixed as in (8.18) by feedback, the gas mass

$$M_g \simeq \frac{2f_g \sigma^2}{G} R \qquad (8.20)$$

in the host galaxy bulge, and the maximum SMBH mass M_{max} allowed if the SMBH is to have a luminous accretion disc and so be able to exert the feedback specifying M_σ, given by

$$M_{\text{max}} = 5 \times 10^{10} \text{M}_\odot \alpha_{0.1}^{7/13} \eta_{0.1}^{4/13} (L/L_{\text{Edd}})^{-4/13} f_5^{-27/26} \qquad (8.21)$$

(see Section 8.3). Here $\alpha_{0.1}$ is the standard viscosity parameter in units of 0.1, $\eta_{0.1}$ is the black hole accretion efficiency in units of 0.1; L and L_{Edd} are, respectively, the accretion luminosity and Eddington luminosity of the black hole; and $f_5(a)$ is the spin dependence of the innermost stable circular orbit (ISCO) radius in units of a typical value $5GM/c^2$, corresponding to a prograde spin $a \sim 0.6$. Over the full range $-1 < a < 1$ the maximum mass M_{max} varies from $2 \times 10^{10} \text{M}_\odot$ up to $3 \times 10^{11} \text{M}_\odot$ (see Figure 8.2). Although further SMBH mass growth beyond M_{max} is still possible (as we discuss in the remainder of the chapter), the important point is that SMBH with $M > M_{\text{max}}$ cannot have accretion discs and so exert mechanical or other feedback on their host galaxies.

All extant observations appear compatible with the limit (8.21). The SMBH in NGC 1600 clearly lies below this limit for any spin rate, and this probably holds for NGC 4889 (see King, 2016). The most extreme system is currently the very dust-obscured WISE J104222.11+164115.3, at redshift $z = 2.52$ (Matsuoka et al., 2018) with SMBH mass $M = 10^{11} \text{M}_\odot$ and a stellar mass $\simeq 3.5 \times 10^{13} \text{M}_\odot$. The strongly blueshifted oxygen lines of this source give an outflow velocity $\sim 1,100 \, \text{km s}^{-1}$, close to that expected from (6.23). Launching a wind like this requires a luminous accretion disc. Then from (8.21) the measured mass $M = 10^{11} \text{M}_\odot$ requires the SMBH to have had a spin $a \gtrsim 0.7$ and accrete prograde at the time of launch. (We cannot say if the SMBH in WISE J104222.11+164115.3 was actually accreting at the epoch corresponding to current observations: it could conceivably have switched off after launching the observed outflow, but without of course changing either its SMBH mass or spin significantly.)

To discuss these effects we consider for simplicity a fixed value

$$M_{max} = 3 \times 10^{10} M_{\odot}, \tag{8.22}$$

which from (8.21) corresponds to a spin parameter a close to zero. This choice does not qualitatively affect the results.

For a given gas fraction f_g the ratios of the three critical masses controlling black hole growth take the simple forms

$$\frac{M_\sigma}{M_g} = \left(\frac{\sigma}{\sigma_0}\right)^2 \frac{R_0}{R} \tag{8.23}$$

$$\frac{M_g}{M_{max}} = \left(\frac{\sigma}{\sigma_0}\right)^2 \frac{R}{R_0} \tag{8.24}$$

$$\frac{M_\sigma}{M_{max}} = \left(\frac{\sigma}{\sigma_0}\right)^4, \tag{8.25}$$

where σ_0 and R_0 depend on f_g. We take $f_g = 0.5$ for 'gas-rich' cases, and $f_g = f_c \simeq 0.16$ for typical 'cosmological' conditions. Then $(\sigma_0, R_0) = (564\,\text{km s}^{-1}, 0.42\,\text{kpc})$ and $(\sigma_0, R_0) = (658\,\text{km s}^{-1}, 0.98\,\text{kpc})$, respectively, in the gas-rich and cosmological cases.

The numbers here show that we can only expect deviations from the usual low-redshift condition $M \lesssim M_\sigma$ in galaxies with very high velocity dispersions $\sigma \gtrsim 550\,\text{km s}^{-1}$. Values of σ of this order are in any case needed for galaxy bulges to acquire large gas masses $\sim 10^{12} M_{\odot}$ by redshift 6, where large SMBH masses are observed, as the gas cannot assemble faster than the dynamical rate

$$\dot{M}_{dyn} \sim \frac{f_g \sigma^3}{G}. \tag{8.26}$$

Infall only appears at the turnround time when cosmological outflow has slowed enough to allow it, so typically for a fraction ~ 0.2 of the lookback time $t_{look} \simeq 10^9$ yr at this redshift. Then requiring

$$\frac{f_g \sigma^3}{G} \times 0.2 t_{look} \gtrsim 10^{12} M_{12} M_{\odot} \tag{8.27}$$

for galaxies of mass $10^{12} M_{12} M_{\odot}$ gives an absolute lower limit, assuming continuous accretion, of

$$\sigma \gtrsim 350 - 460 \times M_{12}^{1/3}\,\text{km s}^{-1} \tag{8.28}$$

for $f_g = 0.5 - 0.16$. In agreement, simulations (e.g. van der Vlugt & Costa, 2019) find large dispersions for massive galaxies at this epoch.

The three relations (8.23), (8.24), and (8.25) divide the σ–R plane into the six regions 1–6 of Figure 8.3. This shows how SMBH grow in various cases. We first

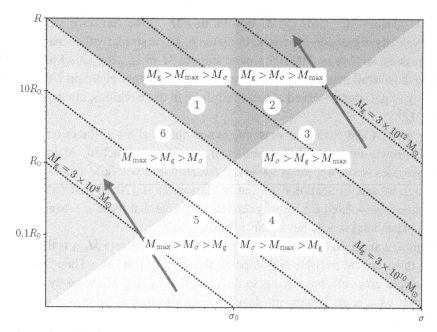

Figure 8.3 The (σ, R) plane for galaxy bulges and their central black holes. The orderings between the three critical masses M_σ, M_g and $M_{\max}$ (see (8.23), (8.24), and (8.25)) divide this into regions 1–6 as shown. The effect of minor mergers is to move a galaxy in the direction of the thick arrows. At low redshift most galaxies are observed with SMBH masses $M \lesssim M_\sigma$ and modest velocity dispersions σ in the 'normal' regions 1 and 6, depending on their total bulge gas masses M_g. Small but very dense galaxies ('blue nuggets') are observed at high redshift and necessarily with high σ (see (8.28)) in regions 2 and 3. These can grow large SMBH masses M that are below M_σ in those regions, but exceed the lower values of M_σ prevailing in region 1 if minor mergers or other processes such as dissipation move them there. This offers possible evolutionary routes to forming low-redshift galaxies whose SMBH have $M > M_\sigma$, such as NGC 1600 and NGC 4889, probably via systems like WISE J104222.11+164115.3, whose vigorous feedback despite its very high SMBH mass suggests that $M_{\max} > M > M_\sigma$ here also. Galaxies with masses $\gtrsim 10^{13} M_\odot$ must live in region 3, and their SMBH reach $M_{\max}$. This shuts off all SMBH emission and feedback, and in principle allows these holes to grow their masses still more, at the expense of their gas. The figure is drawn for a single value of $M_{\max} = 3 \times 10^{10} M_\odot$. Relaxing this to include the full range $2 \times 10^{10} M_\odot$–$3 \times 10^{11} M_\odot$ leaves the global behaviour of galaxies qualitatively unchanged, but blurs the boundaries between the regions 1–6. Credit: King and Nealon (2019).

ask how black holes grow in galaxy bulges in each region of Figure 8.3, and then consider the effects of evolution across the (σ, R) plane. In region 1 we have the familiar ordering $M_g > M_{\max} > M_\sigma$, so SMBH growth in this region terminates at $M = M_\sigma$, as expected in the simple picture sketched at the start of this section. SMBH feedback removes the excess gas ($M_g > M_\sigma$) by energy-driven outflow once M reaches M_σ. Growing M beyond this M_σ value can only happen if the bulge velocity dispersion increases. In region 6, M_σ is similarly the smallest of the three masses, so here too SMBH growth stops at $M = M_\sigma$. Regions 1 and 6 differ only in the overall mass of the galaxy, which must be larger in region 1 than 6. The majority of galaxies observed at low redshift lie in these regions, almost all having $M \le M_\sigma$.

In regions 4 and 5, the gas mass M_g is now the smallest of the critical masses. SMBH growth must stop before either M_σ or $M_{\max}$ is reached. At low redshift these galaxies would appear as low-mass dwarfs with black hole masses $M < M_\sigma$. Although we saw in Section 6.10 that Baldassare et al. (2020) found that the M–σ relation is populated for dwarf galaxies, they note that there are dwarf galaxies whose black holes scatter below this relation.

Regions 2 and 3 on Figure 8.3 are the most interesting. Here $M_{\max}$ is the smallest of the three critical masses, and in particular smaller than M_σ. Then SMBH can in principle grow all the way up to $M_{\max}$, without being hindered by feedback. To decide what actually happens depends on what physical processes control gas accretion on to SMBH, and we have emphasized several times that this is currently not understood.

Reaching $M_{\max}$ means in one sense that the barriers to further SMBH growth are reduced, as accretion does not produce radiation or feedback. Gas near the hole forms stars which can be swallowed whole, but significant mass growth requires a very large mass of gas to have lost enough angular momentum to reach the close vicinity of the hole.

So far we have considered SMBH growth in galaxy bulges that do not evolve enough to cause movement between the regions of Figure 8.3. But galaxy evolution can affect SMBH growth in several ways, for example through transformation and dissipation (Bezanson, van Dokkum & Franx, 2012), and mergers with other galaxies. In transforming bulges, star formation is quenched and σ probably remains fixed. The correlation of SMBH accretion with star formation rate means that SMBH also stop growing. This agrees with the idea that σ^4 controls both the SMBH mass and the bulge stellar mass M_b (Sections 6.5 and 6.9). Then these quiescent galaxies are likely to stay simultaneously on the Faber–Jackson (Faber & Jackson, 1976) and M–σ relations, and then necessarily on the SMBH–bulge mass relation $M \simeq 10^{-3} M_b$.

In contrast, dissipation removes energy from the gas, and makes it sink to the centre of the bulge. This increases the gravitating mass there, and therefore σ also, simultaneously reducing the effective radius R. In a similar way, if a galaxy gains mass through minor mergers with small galaxies this must reduce the velocity dispersion as $\sigma \sim M_g^{-1/2}$ (Bezanson et al., 2009). This effect is similar to growing a hot body by adding cooler (lower σ) ones – the end-product is more massive but cooler. The dispersion is likely to remain unchanged in major mergers. From (8.20) we see now that galaxies move along tracks $\sigma \propto R^{-1/4}$ (see Figure 8.3). Dissipation or mergers can reduce an initially large dispersion σ below σ_0, particularly in the central regions near the SMBH, and move galaxies from regions 2 and 3 into region 1. Here the large SMBH mass grown in regions 2 and/or 3 may exceed the new, lower value of M_σ in the centre. Then powerful SMBH feedback can restart if M has not yet reached M_{max}. Once σ in the outer regions drops below the critical value the remaining gas will be expelled and leave M above the $M–\sigma$ value defined by the smaller central velocity dispersion.

The result of this evolution may be what is seen in the system WISE J104222.11+164115.3 (Matsuoka et al., 2018) discussed earlier in this section. It appears to be blowing gas out, and this would normally require its SMBH mass $M \simeq 10^{11}M_\odot$ to be close to M_σ. But the value $\sigma \sim 1{,}208 \, \text{km s}^{-1}$ this would require is extremely high, even for a gas rich galaxy. It seems more likely instead that M is significantly larger greater than M_σ. This is what we would expect if the galaxy evolved from regions 2 or 3 into region 1. Evolution of this kind can produce galaxies like NGC 4889 and NGC 1600. The bulge has a high gas content, and the strong SMBH feedback drives much of it away before it can form stars, leaving the black hole at a mass M with $M_{max} > M > M_\sigma$, along with a rather low stellar mass for the whole galaxy. After formation in region 2 or 3 it is reasonable for a galaxy to acquire a very high SMBH mass and enter a blowout phase, as seen currently in WISE J104222.11+164115.3, before ending as a galaxy like NGC 4889 or NGC 1600 This is shown schematically in Figure 8.4. We see from Figure 8.3 that very massive galaxies (baryon masses $\gtrsim 3 \times 10^{11}M_\odot$) grown at high redshift (so with large σ; see (8.28)) are very likely to follow this kind of evolution, and so end with low stellar masses (see Figure 8.5). This may be what ultimately limits the masses of galaxies.

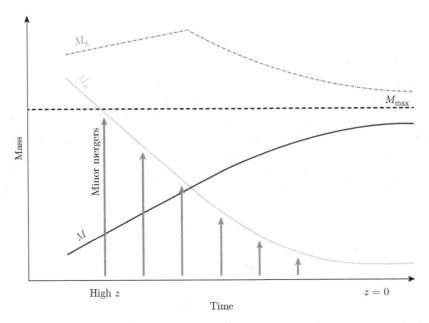

Figure 8.4 Schematic view of the evolution producing galaxies such as NGC 4889 and NGC 1600. Minor mergers gradually reduce the central velocity dispersion, reducing the critical mass M_σ and leaving the SMBH mass M above the M–σ relation, while feedback from SMBH accretion removes much of the gas mass M_g and prevents the growth of a large stellar population. Credit: King and Nealon (2019).

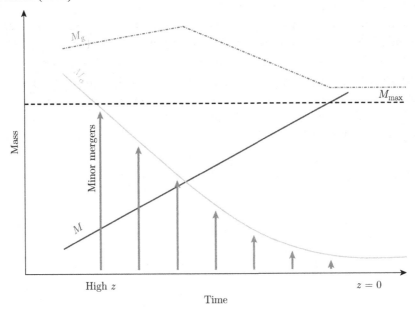

Figure 8.5 Schematic view of the evolution producing galaxies with very massive SMBH and a minimal stellar component. This is similar to that of Figure 8.4, but here the SMBH evolves beyond M_{max}. These objects would be almost undetectable by electromagnetic observations, and the low orbital frequencies of EMRIs (extreme mass-ratio inspirals) are discouraging for potential detection by LISA or eLISA. Credit: King and Nealon (2019).

Problems

Chapter 1

1.1. Consider the statement that the luminosity of a gamma-ray burst is comparable to the total luminosity of the stars in the observable Universe. Relate both luminosities to $L = c^5/G$. See Wijers (2006).

Chapter 2

2.1. The contravariant (matrix inverse $g^{\mu\nu}$) form of the Kerr metric in Boyer–Lindquist coordinates is

$$\left(\frac{\partial}{\partial s}\right)^2 = -\frac{A}{\Sigma\Delta}\left(\frac{\partial}{\partial t}\right)^2 - \frac{4Mar}{\Sigma\Delta}\frac{\partial}{\partial t}\frac{\partial}{\partial\phi} + \frac{\Delta}{\Sigma}\left(\frac{\partial}{\partial r}\right)^2 + \qquad (2.29)$$
$$\frac{1}{\Sigma}\left(\frac{\partial}{\partial\theta}\right)^2 \frac{\Delta - a^2\sin^2\theta}{\Sigma\Delta\sin^2\theta}\left(\frac{\partial}{\partial\phi}\right)^2,$$

where as before, $A = (r^2 + a^2)^2 - a^2\Delta\sin^2\theta$, $\Delta = r^2 - 2Mr + a^2$, and $\Sigma = r^2 + a^2\cos^2\theta$.

The normal vector n_μ to a surface $f(x) = 0$ is parallel to $\partial f/\partial x_\mu$, and the surface is null if

$$g^{\mu\nu}\frac{\partial f}{\partial x_\mu}\frac{\partial f}{\partial x_\nu} = 0.$$

(a) In the Kerr metric, find the two values $r_\pm$ of r for which $\Delta = 0$. Show that these surfaces are null.

(b) Find the metrics of the pair of two-surfaces $\Delta = 0, dt = 0$, and show that their areas are $A_\pm = 4\pi(r_\pm^2 + a^2)$.

2.2. Consider the motion of a light ray along the ϕ-direction, so that only u_t and u_ϕ are non-zero. What is the equation of its worldline? Show that these rays can in general be prograde or retrograde with respect to the black hole spin. If $g_{tt} = 0$, show that only the prograde sense is allowed, and interpret this result physically.

289

Chapter 3

3.1. A monatomic gas flows at speed $v \gg c_s$ into a cylinder closed at one end, where c_s is its sound speed. Show that a shock moves into the gas, and find its speed if (a) the shock is adiabatic, and (b) the shock is isothermal. Interpret these results physically in terms of the initial evolution of the flow.

3.2. A supernova explosion can be modelled as the instantaneous injection of a very large energy E into a very small volume in an ambient gas of uniform density ρ. This drives a spherical blast wave into the gas, and in the early part of the evolution the pressure within the shock is far larger than the ambient pressure, and very little of the explosion energy is radiated.

Show that the two quantities E, ρ do not define a characteristic time of lengthscale. Deduce that the properties of the blast wave must be scale-free, and determined by a dimensionless variable

$$\xi = at^\alpha \rho^\beta E^\gamma,$$

where a, α, β, γ are constants, and find the values of α, β, and γ. Find the radius and speed of the blast wave at time t as functions of E, ρ, and a value ξ_0 of the similarity variable of order unity.

Show that the pressure inside this self-similar blast wave, and its expansion speed, both fall off rapidly with radius.

3.3. The energy-conserving blast wave of Problem 3.2 eventually sweeps up enough ambient gas that it changes to a 'snowplough' state where momentum μ rather than energy E is conserved. Find the new form of the similarity variable, and show that the speed drops more rapidly than before.

3.4. When does the self-similarity assumed in Problems 3.2 and 3.3 finally break down? For a fuller discussion see AF.

3.5. In an ionization front, the state of a gas changes between neutral and ionized states over a short lengthscale. In particular, for pure hydrogen, the equation of state changes between

$$\frac{P_1}{\rho_1} = \frac{kT_1}{m_H}$$

for the neutral gas, and

$$\frac{P_2}{\rho_2} = \frac{2kT_2}{m_H}$$

in the ionized gas. The temperatures are typically $T_1 \sim 100\,\mathrm{K}$ and $T_2 \sim 10^4\,\mathrm{K}$. We can define isothermal sound speeds $c_1 = (P_1/\rho_1)^{1/2}$, $c_2 = (P_2/\rho_2)^{1/2}$ for the two gases. Assuming that, as for a shock front, we can regard a small section of the ionization front as a plane, write down the equations of conservation of mass, momentum and energy, and show that the mass densities obey

$$\frac{\rho_2}{\rho_1} = \frac{1}{2c_2^2}[c_1^2 + v_1^2 \pm \sqrt{f(v_1)}],$$

where v_1, v_2 are the velocities on each side of the front, and

$$f(v_1) = (v_1^2 - v_R^2)(v_1^2 - v_D^2)$$

where

$$v_R = c_2 + \sqrt{c_2^2 - c_1^2}$$

and

$$v_D = c_2 - \sqrt{c_2^2 - c_1^2}.$$

Deduce that the velocity of the ionization front with respect to the neutral gas must either obey $v_1 \geq v_R$ or $v_1 \leq v_D$. Here the designations 'R' and 'D' stand for 'rarified' and 'dense'. Justify these descriptions.

3.6. An accretion disc is partially supported by an external magnetic field threading it, so its local azimuthal velocity v is smaller than the Kepler value $(GM/R)^{1/2}$. Use the stress tensor (3.95) to show that the magnetic field components external to the disc satisfy

$$B_x B_z \sim \left[\frac{\mu_o}{4\pi}\right] 2\pi \Sigma \left(\frac{GM}{R^2} - \frac{v^2}{R}\right),$$

where Σ is the local disc surface density and (x, z) are Cartesian coordinates centred on the accretor, with the z axis normal to the disc plane.

3.7. A bundle of straight fieldlines of strength $B(t)$ lies parallel to the x-axis, and is surrounded by a field-free medium of constant gas pressure P. A length l of the bundle is stretched uniformly in the x-direction to a new length $L(t) = \lambda t$, with $\lambda = $ constant. The bundle remains in pressure equilibrium with its surroundings during the stretching process, so that $P_i + [4\pi/\mu_0]B^2/8\pi = P$, where P_i is the interior gas pressure. The interior temperature is kept constant during this process. Use mass conservation to find $B(t)$, the interior density $\rho_i(t)$ and the distance across the bundle, assuming that initially $[4\pi/\mu_0]B^2/8\pi \ll P$. How do $B(t)$ and ρ behave as $t \to 0, \infty$?

3.8. An ideal MHD plasma is subject to a time-independent locally plane shock. Show that the six quantities

$$\rho v_n,$$

$$B_n,$$

$$\rho v^2 + P + \left[\frac{4\pi}{\mu_0}\right]\frac{B_t^2}{8\pi},$$

$$\rho v_n v_t - \left[\frac{4\pi}{\mu_0}\right]\left[\frac{B_t B_n}{4\pi}\right],$$

$$\left(\frac{\gamma}{\gamma-1}\frac{P}{\rho} + \frac{v^2}{2}\right)\rho v + \left[\frac{4\pi}{\mu_0}\right]\left[\frac{v B_t^2}{4\pi} - \frac{B_n B_t \cdot \mathbf{v}}{4\pi}\right],$$

$$(\mathbf{v} \times \mathbf{B})_t$$

are all continuous across the shock, where the suffixes n and t denote vector components normal and tangential to the shock.

Chapter 4

4.1. For an accretion disc irradiated by a strong source at an inner radius R_*, one can show that the viscosity behaves as

$$\nu(R) = \nu_* \left(\frac{\dot{M}_c}{\dot{M}_*}\right)^{1/4} (R/R_*),$$

where $\nu_*, \dot{M}_*$ are the viscosity and accretion rate at $R = R_*$ at time $t = 0$, and $\dot{M}_c(t)$ is the accretion rate at R_* at later times t.

Show that the viscous diffusion equation (4.31) can be rewritten as

$$\frac{\partial}{\partial t}(R^{3/2}\Sigma) = \frac{3\nu_*}{R_*}\left(\frac{\dot{M}_c}{\dot{M}_*}\right)^{1/4}\left(R^{1/2}\frac{\partial}{\partial R}\right)^2 (R^{3/2}\Sigma),$$

and change variables to write this in a simpler form.

If the disc has a fixed outer edge at $R = R_0$, show that its general solution is

$$\Sigma(R, t) = \sum_{n=1}^{\infty}\left\{\frac{A_n \cos(2n-1)x}{[1 + (2n-1)^2 t/t_{\text{visc}}]^3} + \frac{B_n \sin 2nx}{1 + (2n)^2 t/t_{\text{visc}}]}\right\},$$

where

$$x = \frac{\pi}{2}\left(\frac{R}{R_0}\right)^{1/2},$$

$$t_{\text{visc}} = \frac{16R_0^2}{\pi^2 \nu_1},$$

$$v_1 = v(R_0, 0) = v_* \frac{R_0}{R_*}$$

and A_n, B_n are constants.

Show that the mass associated with the nth term of the general solution is $M_n \sim n^{-3}(1 + t/t_n)^{-3}$, with $t_n \sim t_{\mathrm{visc}}/n^2$.

Why do all these terms decay over time?

Show that

$$\dot{M}_c \sim \dot{M}_* \left[1 + \frac{t}{t_{\mathrm{visc}}} \right]^{-4}.$$

(See King, 1998.)

4.2. A certain black hole source emits a substantial fraction of its luminosity in a blackbody-like spectral component. Assuming that this is produced in a region comparable in size to the Schwarzschild radius of the hole, use the emitting area deduced from observation to provide an estimate of the black hole mass M in terms of the measured temperature. If the blackbody emission is isotropic, and sub-Eddington, show that its luminosity must satisfy

$$L_{\mathrm{bb}} < 2.3 \times 10^{44} T_{100}^{-4} \text{ erg s}^{-1},$$

where $T_{100} = kT_{\mathrm{bb}}/100\,\mathrm{eV}$ is T_{bb} measured in units of 100 eV. (See King & Puchnarewicz, 2002.)

4.3. The soft blackbody-like X-ray emission of the stellar-mass ultraluminous X-ray sources (ULXs) obeys a relation

$$L_{\mathrm{soft}} = CT^{-4},$$

(where C is a universal constant for all sources) when bright. When faint the correlation is $L \propto T^4$ instead. Considering Problem 4.1, show that the $L \propto T^{-4}$ relation follows if the source's radiation is beamed into a narrow cone of solid angle $4\pi b$, where

$$b = A\dot{m}^{-2},$$

with A a constant, and $\dot{m} = \dot{M}/\dot{M}_{\mathrm{Edd}}$ is the Eddington accretion ratio.

How can one explain the fainter sources with $L \propto T^4$? (See King, 2009a.)

4.4. The observer in Problem 4.2 must lie in the emission beam to see the bright sources, and for tighter beaming must search through a larger volume of space to find the nearest source of this kind. If $C = 73$ and $\dot{m} = 10^4 \dot{m}_4$, show that an object with $\dot{m}_4 = 1$ would lie at a minimum distance

$$D_{\text{min}} \sim \left(\frac{3}{4\pi n_g N b} \right)^{1/3} \sim 600 N^{-1/3} \dot{m}_4^{2/3} \text{ Mpc}$$

(where $n_g \sim 0.02 \text{ Mpc}^{-3}$ is the density of L^* galaxies, and N the number of sources like this per galaxy) and have an *apparent* (beamed, but assumed isotropic) luminosity

$$L_{\text{sph}} \simeq 2.2 \times 10^{45} m_* \dot{m}_4^2 \text{ erg s}^{-1},$$

where $m_* = M/10 M_\odot$ is the actual mass of the accreting black hole.

How would a 'pseudoblazar' object like this be potentially distinguishable from a genuine AGN? (See King, 2009a.)

Chapter 5

5.1. Use Equation (5.39) to find explicitly the time development of the angle θ between a misaligned hole spin vector $\mathbf{J}_h$ and the disc angular momentum $\mathbf{J}_d$. Apply this to the four cases shown in Figure 5.7.

5.2. Consider a misaligned disc around one component (mass M_1) of a binary system with a circular orbit of separation a. The (retrograde) precession caused by the companion has the frequency

$$\Omega_p = \frac{3 M_0}{4 M_1} \left(\frac{R}{a} \right)^3 \Omega \cos \theta,$$

where θ is the inclination of the disc to the orbital plane, the companion has mass M_0, $R \ll a$ the disc radius, and $\Omega = (GM_1/R^3)^{1/2}$. Find the magnitude of the precession torque per unit area $|\mathbf{G}_p|$ and compare this with the magnitude of the azimuthal and vertical torques

$$|G| = |G_{v_1}| + |G_{v_2}| = \frac{\Sigma R^2 \Omega^2}{2} \frac{H}{R} [3\alpha_1 + \alpha_2 |\psi|]$$

(Papaloizou & Pringle, 1983), where $|\psi| = R|\partial l/\partial R|$ is the warp amplitude, to give an estimate of the radius

$$R_{\text{break}} \gtrsim \left[\frac{4(\alpha_1 + \alpha_2 |\psi|/3)}{\sin 2\theta} \frac{H}{R} \frac{M_1}{M_0} \right]^{1/3} a$$

where disc breaking can be expected. Use this in the limit $\alpha_2 |\psi| \ll \alpha_1$ to give expressions for the minimum disc inclination angle for the case $M_1 < M_0$.

Constrain the break radius using the tidal limit

$$\frac{a}{R_{\text{break}}} > 2.5 \left(\frac{M_0}{M_1}\right)^{1/3}$$

to show that the minimum angle for disc breaking is given by

$$\sin 2\theta \gtrsim 0.06 \left(\frac{\alpha}{0.1}\right) \left(\frac{H/R}{0.01}\right).$$

Comment on how likely disc breaking is in this case. (See Nixon et al., 2012.)

5.3. Consider an SMBH binary with a coplanar circumbinary disc that is initially Keplerian. The binary merges, leaving a merged hole of mass M, and the merger kick gives all the disc particles an instantaneous additional velocity $\mathbf{V}$, whose projection on to the disc plane defines the positive x-axis. Show that the three components of this in cylindrical coordinates are $V_R = -V \cos\theta \sin\phi$, $V_\phi = V_k - V \cos\theta \cos\phi$ and $v_z = V \sin\theta$, where $V_k = (GM/R)^{1/2}\mathbf{e}_\phi$ is the local Kepler velocity, ϕ is the angle between the positive y-axis and the particle radius R, and θ is the usual polar angle of the disc axis.

The disc extends to sufficiently large radii that there is a radius R_v within it where $V = V_k$, and $r = R/R_v = (V/V_k)^2$. Find the specific energy ϵ of a particle in terms of V_k, r, θ, ϕ.

Show that for radii such that

$$r < r_b = (-\cos\theta + \sqrt{1 + \cos^2\theta})^2,$$

all the disc particles are bound, while for radii

$$r > r_{ub} = (\cos\theta + \sqrt{1 + \cos^2\theta})$$

all the disc particles are unbound. Show further that only particles within the azimuthal range $-\phi_b < \phi < \phi_b$, with $0 \le \phi_b \le \pi$, are bound, where

$$\cos\phi_b = \frac{r - 1}{2\sqrt{r}\cos\theta}.$$

After the merger, show that bound particles move on elliptical orbits with specific angular momentum

$$\mathbf{j} = J_k[(1 - \sqrt{r}\cos\theta \cos\phi)\mathbf{e}_z - \sqrt{r}\mathbf{e}_\phi] \qquad (5.30)$$

and eccentricity given by

$$e^2 = 1 + \frac{2j^2\epsilon}{G^2M^2}. \qquad (5.31)$$

Assuming that each particle conserves its angular momentum, show that it circularizes at a dimensionless radius $r_c = R_c/R_v$ given by

$$\frac{r_c}{r} = \left(\frac{j}{J_K}\right)^2 = (1 - \sqrt{r}\cos\theta\cos\phi)^2 + r\sin^2\theta. \qquad (5.32)$$

Show that the mean circularization radius for a ring $r_c = R_c/R_v$ is

$$\frac{r_c}{r} = \frac{1}{2\phi_b}\int_{-\phi_b}^{\phi_b}\left(\frac{j}{J_K}\right)^2 d\phi =$$

$$(1 + r\sin^2\theta) + r\cos^2\theta\left(\frac{1}{2} + \frac{\sin 2\phi_b}{2\phi_b}\right) - 2\sqrt{r}\cos\theta\frac{\sin\phi_b}{\phi_b}.$$

Show that gas at $R < R_b$ circularizes close to its original position whatever the kick angle, whereas R_c depends strongly on the kick angle for $R > R_b$. (See Rossi et al. 2010.)

Chapter 6

6.1. To prevent the further growth of the SMBH beyond the M–σ mass in an elliptical galaxy, the energy-driven outflow clearing it must push its gas out to the virial radius $R_V = 200R_{200}$ kpc. Estimate the total gas mass within R_V if it is distributed isothermally with velocity dispersion $\sigma = 200\sigma_{200}$ km s^{-1} and gas fraction f_g. Use this to find the energy required to drive the gas out to R_V. Show that the energy actually supplied to the energy-driven outflow is $E_w = (\eta^2/2)\Delta M_{BH}c^2$, where ΔM_{BH} is the increase over the M_σ value, and that

$$\frac{\Delta_{BH}}{M_\sigma} \sim \frac{0.15}{f}\frac{f_g}{0.16}\frac{R_{200}}{\eta_{0.1}^2},$$

where $f < 1$ is the efficiency of absorption of the outflow energy by the ambient gas, and $\eta_{0.1}$ is the accretion efficiency of the SMBH in units of 0.1. Use this to discuss the relative efficiency of slow- and rapidly spinning SMBH in expelling gas. Show that the fractional growth of the SMBH mass above the canonical M_σ value is significantly affected by the spin rate. (See Zubovas & King, 2019.) What effect could this have on the fitted power of σ in the M–σ relation?

6.2. Consider the effect of a continuous wind from a source in a uniform medium of density ρ. If the wind shock (at spherical radius R) exerts pressure P on the medium, show that

$$4\pi R^2 P = \frac{d}{dt}\left(\frac{4\pi}{3}R^3\rho\dot{R}\right)$$

so that

$$\frac{P}{\rho} = \frac{1}{3}R\ddot{R} + \dot{R}^2.$$

Assuming there are no radiative energy losses from the wind shock, show that

$$\frac{d}{dt}\left[2\pi R^3 P\right] = L_{Edd} - P\frac{d}{dt}\left[\frac{4\pi R^3}{3}\right].$$

Find the equation of motion for the shock and show that it has the attractor solution

$$R = \left(\frac{125L_{Edd}}{101\pi\rho}\right)^{1/5} t^{3/5},$$

or numerically

$$R \simeq 50\left(\frac{L_{39}}{\rho_{-25}}\right)^{1/5} t_5^{3/5} \text{ pc},$$

where $L_{39} = L_{Edd}/10^{39}$ erg s^{-1}, $t_5 = t/10^5$ yr, and $\rho_{-25} = \rho/10^{-25}$ g cm^{-3}.

These conditions correspond to a ULX with a lifetime of order the thermal timescale of a massive companion star, driving a wind into the interstellar medium of its host galaxy, with number density ~ 0.1 cm^{-3} H atoms per cm^3. For longer lifetimes or more powerful wind mechanical luminosities, $L_{39} \sim 10$, the predicted nebulae can have sizes ~ 500 pc. Observations (see the spherical part of the W50 nebula surrounding the extreme Galactic X-ray binary SS433 in Figure 7.6) show that many ULXs are associated with wind-blown nebulae of order 50–500 pc in size. These are larger than supernova remnants (~ 3 pc).

6.3. It is generally agreed that some process must prevent the diffuse gas in galaxy clusters from cooling. This is often thought to be via 'radio-mode' heating by jets. The SMBH in the central cD galaxy of a cluster may be sporadically fed gas at strongly super-Eddington rates, and so produce powerful quasi-spherical outflows. Is it possible for these outflows, rather than jets, to prevent the cluster gas cooling? (See King, 2009b.)

Chapter 7

7.1. Equation (6.71) gives a cubic for the velocity v_e of a radiation-driven outflow. Which term on the right-hand side must dominate if the SMBH mass is close to the M–σ value? Show that $v_e \sim 2300\sigma_{200}^{2/3}$ km s^{-1} for such masses.

Chapter 8

8.1. The curve defining the maximum luminously accreting SMBH mass in Figure 8.2 assumes that feeding from a radiation-pressure dominated disc is unlikely. What happens to the limiting mass if this restriction is lifted? (See Section 4.1 of Inayoshi & Haiman, 2016).

References

Aarseth, S. J., 2003, Ap&SS, 285, 367

Abramowicz, M. A., Czerny, B., Lasota, J. P., Szuszkiewicz, E., 1988, ApJ, 332, 646

Alloin, D., Pelat, D., Phillips, M., Whittle, M., 1985, ApJ, 288, 205

Aly, H., Dehnen, W., Nixon, C., King, A., 2015, MNRAS, 449, 65

Antonucci, R., 1993, ARAA, 31, 473

Antonucci, R. R. J., Miller, J. S., 1985, ApJ, 297, 621

Arav, N., Gabel, J. R., Korista, K. T. et al., 2007, ApJ, 658, 829

Arcodia, R., Merloni, A., Nandra, K. et al., 2021, Nature, 592, 704

Assef, R. J., Eisenhardt, P. R. M., Stern, D. et al., 2015, ApJ, 804, 27

Bahcall, J. N., Wolf, R. A., 1976, ApJ, 209, 214

Balbus, S. A., Hawley, J. F., 1991, ApJ, 376, 214

Balbus, S. A., Henri, P., 2008, ApJ, 674, 408

Baldwin, J. A., Netzer, H., 1978, ApJ, 226, 1

Baldassare, V., Reines, A. E., Gallo, E., Greene, J. E., 2015, ApJ, 809, 14

Baldassare, V., Dickey, C., Geha, M., Reines, A. E., 2020, ApJ, 898, L3

Bally, J., Harrison, E. R., 1978, ApJ, 220, 743

Bardeen, J. M., 1970, Nature, 226, 64

Bardeen, J. M., Petterson, J. A., 1975, ApJ, 195L, 65

Bardeen, J. M., Press, W. H., Teukolsky, S. A., 1972, ApJ, 178, 347

Batcheldor, D., 2010, ApJ, 711, L108

Begelman, M., 2010, MNRAS, 402, 673

Begelman, M. C., Volonteri, M., Rees, M. J., 2006, MNRAS, 370, 289

Bekenstein, J. D., 1973, Phys. Rev. D, 7, 2333

Bezanson, R., van Dokkum, P., Franx, M., 2012, ApJ, 760, 62

Bezanson, R., van Dokkum, P. G., Tal, T. et al., 2009, ApJ, 697, 1290

Binney, J. J., Tremaine, S., 2008, Galactic Dynamics, 2nd Edition, Princeton University Press

Belloni, T. M., 2010, Lect. Notes Phys., 794, 53

Bischetti, M., Maiolino, R., Carniani, S. et al., 2019, A&A, 630, 59

Blandford, R. D., Payne, D. G., 1982, MNRAS, 199, 883

Blandford, R. D., Znajek, R. L., 1977, MNRAS, 179, 433

Blundell, K. M., Bowler, M. G., 2004, ApJ, 616L, 159

Bondi, H., 1952, MNRAS, 112, 195

Bowers, R. L., Wilson, J. R., 1991, Numerical Modeling in Applied Physics and Astrophysics, Jones and Bartlett

Burderi, L., King, A. R., Szuszkiewicz, E., 1998, ApJ, 509, 85

Carter, B., 1968, Phys. Rev., 174 (5), 155
Carter, B., Luminet, J. P., 1982, Nature, 296, 211
Chandrasekhar, S., 1943, ApJ, 97, 255
Chandrasekhar, S., 1960, Proc. Natl. Acad. Sci., 46, 253
Chartas, G., Brandt, W. N., Gallacher, S. C., Garmire, G. P., 2002, ApJ, 579, 169
Choudhuri, A. B., 2010, Astrophysics for Physicists, Cambridge University Press
Cicone, C., Maiolino, R., Sturm, E. et al., 2014, A&A, 562, A21
Ciotti, L., Ostriker, J. P., 1997, ApJ, 487, L105
Clayton, D. D., 1968, Principles of Stellar Evolution and Nucleosynthesis, McGraw Hill
Collin-Souffrin S., Dumont, A. M., 1990, A&A, 229, 292
Costa, T., Sijacki, D., Haehnelt, M. G., 2014, MNRAS, 444, 2355
Coughlin, E. R., Nixon, C., Begelman, M., Armitage, P. J., Price, D. J., 2016, MNRAS, 455, 3612
Coughlin, E. R., Nixon, C. J., 2019, ApJ, 883L, 45
Cuadra, J., Armitage, P. J., Alexander, R. D., Begelman, M. C., 2009, MNRAS, 393, 1423
Cufari, M., Coughlin, E. R., Nixon, C. J., 2022, ApJ, 920, L20
Davis, T. A., Nguyen, D. D., Seth, A. C. et al., 2020, MNRAS, 496, 4061
Debuhr, J., Quataert, E., Ma, C. P., 2011, MNRAS, 412, 1341
Debye, P., Hückel, E., 1923, Physikalische Zeitschrift, 24, 185
Dehnen, W., King, A. R., 2013, ApJ, 777L, 28
de Mink, S. E., King, A. R., 2017, ApJ, 839, L7
Dehnen, W., McLaughlin, D. E., 2005, MNRAS, 363, 1057
Denney, K. D., Watson, L. C., Peterson, B. M., et al., 2009, ApJ, 702, 1353
Dietrich, M., Appenzeller, I., Wagner, S. J., et al., 1999, A&A, 352, L1
Dietrich, M., Hamann, F., Shields, J. C. et al., 2003a, ApJ, 589, 722
Dietrich, M., Hamann, F., Shields, J. C. et al., 2003b, A&A, 398, 891
Ding, Y., Li, R., Ho, L. C., Ricci, C., 2022, ApJ, 931, 77
Dŏgan, S., Nixon, C. J., King, A. R., Pringle, J. E., 2018, MNRAS, 476, 1519
Dosopoulou, F., Kalogera, V., 2016a, ApJ, 825, 70D
Dosopoulou, F., Kalogera, V., 2016b, ApJ, 825, 71D
Draine, B. T., Lee, H. M., 1984, ApJ, 285, 89
Dubner, G. M., Holdaway, M., Goss, W. M., Mirabel, I. F., 1998, AJ, 116, 1842
Dunhill, A. C., Alexander, R. D., Nixon, C. J., King, A. R., 2014, MNRAS, 445, 2285
Dyson, J. E., Williams, D. A., 1997, The Physics of the Interstellar Medium, 2nd Edition, Institute of Physics Publishing
Eddington, A. S., 1926, The Internal Constitution of the Stars, Cambridge University Press
Elvis, M., 2000, ApJ, 545, 63
Evans, C. R., Kochanek, C. S., 1989, ApJ, 346, L13
Event Horizon Telescope Collaboration, 2019, ApJ, 875L, 1
Faber, S. M., Jackson, R. E., 1976, ApJ, 204, 668
Fabian, A. C., 1999, MNRAS, 308, L39
Faucher-Giguère, C.-A., Quataert, E., 2012, MNRAS, 425, 605
Ferland, G., Baldwin, J. A., Korista, K. et al., 1996, ApJ, 461, 683
Ferrarese, L., Côté, P., Dalla Bontà, E. et al., 2006, ApJ, 644L, 21
Ferrarese L., Merritt, D., 2000, ApJ, 539, L9
Feruglio, C., Maiolino, R., Piconcelli, E. et al., 2010, A&A, 518, L155
Fragile, P. C., Blaes, O. M., Anninos, P., Salmonson, J. D., 2007, ApJ, 668, 417
Fragner, M. M., Nelson, R. P., 2010, A&A, 511, 77

Frank, J., King, A. R., Raine, D. J., 1985, Accretion Power in Astrophysics, 1st Edition, Cambridge University Press (APIA1)

Frank, J., King, A. R., Raine, D. J., 1992, Accretion Power in Astrophysics, 2nd Edition, Cambridge University Press (APIA2)

Frank, J., King, A. R., Raine, D. J., 2002, Accretion Power in Astrophysics, 3rd Edition, Cambridge University Press (APIA3)

Frank, J., Rees, M. J., 1976, MNRAS, 176, 633

Garratt-Smithson, L., Wynn, G. A., Power, C., Nixon, C. J., 2019, MNRAS, 489, 4278

Garrison-Kimmel, S., Rocha, M., Boylan-Kolchin, M., Bullock, J. S., Lally, J., 2013, MNRAS, 433, 3539

Gebhardt, K., Bender, R., Bower, G. et al., 2000, ApJ, 539, L13

Ghisellini, G., Delle Cecca, R., Volonteri, M. et al., 2010, MNRAS, 405, 387

Ghosh, P., Abramowicz, M. A., 1997, MNRAS, 292, 887

Gofford, J., Reeves, J., Tombesi, F. et al., 2013, MNRAS, 430, 60

Goldreich, P., Julian, W. H., 1969, ApJ, 157, 869

Golightly, E. C. A., Coughlin, E. R., Nixon, C. J., 2019a, ApJ, 872L, 163

Golightly, E. C. A., Nixon, C. J., Coughlin, E. R., 2019b, ApJ, 882L, 26

Guillochon, J., Ramirez-Ruiz, E., 2013, ApJ, 767, 25

Hamann, F., Ferland, G., 1992, ApJ, 391, L531

Hanni, R. S., Ruffini, R., 1973, Phys. Rev. D, 8, 3259

Häring N., Rix, H.-W., 2004, ApJ, 604, L89

Hawking, S. W., 1971, Phys. Rev. Lett., 26, 1344

Hawking, S. W., 1972, Comm. Math. Phys., 25, 152

Hernquist, L., 1990, ApJ, 356, 359

Heckman, T. M., Kauffmann, G., Brinchmann, J. et al., 2004, ApJ, 613, 109

Hills, J. G., 1988, Nature, 331, 687

Hughes, S. A., Blandford, R. D., 2003, ApJ, 585L, 101

Inayoshi, K., Haiman, Z., 2016, ApJ, 828, 110

Ishibashi, W., Fabian, A. C., Arakawa, N., 2021, MNRAS, 502, 3638

Jahnke, K., Macciò, A. V., 2011, ApJ, 734, 92

Jiang, Y.–F., Stone, J. M., Davis, S. W., 2013, ApJ, 778, 65

Kawaguchi, T., Aoki, K., Collin, S., 2004, A&A, 420, L23

Kerr, R.P., 1963, Phys. Rev. Lett., 11, 237

Kesden, M., 2012, Phys. Rev. D, 85, 024037

Khan, F. M., Andreas, J., Merritt, D., 2011, ApJ, 732, 89

King, A. R., 1998, MNRAS, 296, L45

King, A. R., 2003, ApJ, 596, L27

King, A. R., 2005, ApJ, 635, L121

King, A. R., 2009a, MNRAS, 393, L41

King, A. R., 2009b, ApJ, 695, L107

King, A. R., 2010, MNRAS, 402, 1516

King, A. R., 2016, MNRAS, 456, L109

King, A. R., Lasota, J. P., 1977, A&A, 58, 175

King, A. R., Lasota, J. P., Kundt, W., 1975, Phys. Rev. D, 12, 3037

King, A. R., Livio, M., Lubnow, H., Pringle, J. E., 2013, MNRAS, 431, 2655

King, A. R., Lubow, S. H., Ogilvie, G. I., Pringle, J. P., 2005, MNRAS, 363, 49

King, A. R., Nealon, R., 2019, MNRAS, 487, 4827

King, A. R., Nealon, R., 2021, MNRAS, 502, L1

King, A. R., Muldrew, S., 2016, MNRAS, 455, 1211

King, A. R., Nixon, C. J., 2015, MNRAS, 453, L46

King, A., Nixon, C., 2016, MNRAS, 462, 464

King, A. R., Nixon, C. J., 2018, ApJ, 857, L7

King, A. R., Pounds, K. A., 2003, MNRAS, 345, 657

King, A. R., Pounds, K. A., 2014, MNRAS, 437, L81

King, A. R., Pounds, K. A., 2015, ARAA, 53, 115

King, A. R., Pringle, J. E., 2006, MNRAS, 373, L90

King, A. R., Pringle, J. E., 2007, MNRAS, 377, L25

King, A. R., Pringle, J. E., Hofmann, J. A., 2008, MNRAS, 385, 1621

King, A. R., Pringle, J. E., Livio, M., 2007, MNRAS, 376, 1740

King, A. R., Pringle, J. E., West, R. G., Livio, M., 2004, MNRAS, 348, 111

King, A. R., Puchnarewicz, E. M., 2002, MNRAS, 336, 445

King, A. R., Zubovas, K., Power, C., 2011, MNRAS, 415, L6

Kinney, A. L., Schmitt, H. R., Clarke, C. J. et al., 2000, ApJ, 537, 152

Kollatschny, W., Weilbacher, P. M., Ochmann, M. W. et al., 2020, A&A 633, 79

Komossa, S., Bade, N., 1999, A&A, 343, 775

Komossa, S., Burwitz, V., Hasinger, G. et al., 2003, ApJ, 582, L15

Kormendy, J., Ho, L. C., 2013, ARAA, 51, 511

Kourkchi, E., Khosroshahi, H. G., Carter, D. et al., 2012, MNRAS, 420, 2819

Kraft, R. P., Forman, W. R., Jones, C. et al., 2002, ApJ, 569, 54

Kramer, M., Stairs, I. H., Manchester, R. N., et al., 2006, Science, 314, 97

Krolik, J. H., McKee, C. F., Tarter, C. B., 1981, ApJ, 249, 422

Larwood, J. D., Papaloizou, J. C. B., 1997, MNRAS, 285, 288

Leigh, N., Böker, T., Knigge, C., 2012, MNRAS, 424, 2130

Lightman, A. P., Eardley, D. M., 1974, ApJ, 187, L1

Liska, M., Chatterjee, K., Tchekhovskoy, A. et al., 2019, arXiv 1912.101192

Liska, M., Hesp, C., Tchekhovskoy, A. et al., 2021, MNRAS, 507, 983

Lobban, A., King, A. R., 2022, MNRAS, 511, 1992L

Lodato, G., Nayakshin, S., King, A. R., Pringle, J. E., 2009, MNRAS, 398, 1392

Lodato, G., Price, D. J., 2010, MNRAS, 405, 1212

Lodato, G., Pringle, J. E., 2006, MNRAS, 368, 1196

Lodato, G., Pringle, J. E., 2007, MNRAS, 381, 1287

Lubow, S. H., 1991, ApJ, 381, 259

Lubow, S. H., Ogilvie, G. I., Pringle, J. E., 2002, MNRAS, 337, 706

Lubow, S. H., Papaloizou, J. C. B., Pringle, J. E., 1994, MNRAS, 267, 235

Lynden-Bell, D., 1967, MNRAS, 136, 101

Lynden-Bell, D., 1969, Nature 223, 690

Lynden-Bell, D., Pringle, J. E., 1974, MNRAS, 168, 603

Lyubarskii, Y. E., 1997, MNRAS, 292, 679

Mahlmann, J. F., Aloy, M. A., Mewes, V., Cerdá-Durán, P., 2021a, A&A, 647, 57

Mahlmann, J. F., Aloy, M. A., Mewes, V., Cerdá-Durán, P., 2021b, A&A, 647, 58

Manzano-King, C. M., Canalizo, G., Sales, L. V., 2019, ApJ, 884, 54

Margon, B., Ford, H. C., Grandi, S. A., Stone, R. P. S., 1979, ApJ, 233, L63

Markwardt, C. B., Ögelman, H., 1995, Nature, 375, 40

Martí-Vidal, I, Marcaide, J. M., Alberdi, A. et al., 2011, A&A, 533, A111

Matsuoka, K., Toba, Y., Shidatsu, M. et al., 2018, A&A, 620, L3

McConnell, N. J., Ma, C.-P., 2013, ApJ, 764, 184

McConnell, N. J., Ma, C.-P., Gebhardt, K. et al., 2011, Nature, 480, 215

McDonald, M., et al., 2012, Nature, 488, 349

McLaughlin, D. E., King, A. R., Nayakshin, S., 2006, ApJ, 650, L37

McQuillin, R. C., McLaughlin, D. E., 2012, MNRAS, 423, 2162

Merritt, D., 2013, Dynamics and Evolution of Galactic Nuclei, Princeton University Press

Milosavljević, M., Merritt, D., 2003, ApJ, 596, 860

Miniutti, G., Saxton, R. D., Giustini, M. et al., 2019, Nature, 573, 381

Mitton, S., Ryle, M., 1969, MNRAS, 146, 221

Munroe, R., 2019, https://xkcd.com/2135

Murray, N., Quataert, E., Thompson, T., 2005, ApJ, 618, 569

Nagar, N. M., Wilson, A. S., 1999, ApJ, 526, 97

Narayan, R., Igumenshchev, I. V., Abramowicz, M. A., 2003, PASJ, 55, L69

Natarajan, P., Treister, E., 2009, MNRAS, 393, 838

Navarro, J. F., Frenk, C. S., White, S. D. M., 1996, ApJ, 462, 563

Navarro, J. F., Frenk, C. S., White, S. D. M., 1997, ApJ, 490, 493

Nixon, C. J., Cossins, P. J., King, A. R., Pringle, J. E., 2011, MNRAS, 412, 1591

Nixon, C. J., King, A. R., 2012, MNRAS, 421, 1201

Nixon, C. J., King, A. R., 2013, ApJ, 765, L7

Nixon, C. J., King, A. R., 2016, Lect. Notes Phys., 905, 45

Nixon, C. J., King, A. R., Price, D., 2013, MNRAS, 43, 1946

Nixon, C. J., King, A. R., Price, D., Frank, J., 2012, ApJ, 757, L24

Nixon, C. J., King, A. R., Pringle, J. E., 2011, MNRAS, 417, L66

Nixon, C. J., Pringle, J. E., 2010, MNRAS, 403, 1887

Nixon, C. J., Salvesen, G., 2014, MNRAS, 437, 3994

Ogilvie, G. I., 1999, MNRAS, 304, 557

Ogilvie, G. I., 2003, MNRAS, 340, 969

Ohsuga, K., Mineshige, S., 2011, ApJ, 736, 2

Paczyński, B., Wiita, P. J., 1982, A&A, 88, 23

Papaloizou, J. C. B., Lin, D. N. C., 1995, ApJ, 438, 841

Papaloizou, J. C. B., Pringle, J. E., 1977, MNRAS, 181, 441

Papaloizou, J. C. B., Pringle, J. E., 1983, MNRAS, 202, 1181

Papaloizou, J. C. B., Pringle, J. E., 1984, MNRAS, 204, 721

Peng, C. Y., 2007, ApJ, 671, 1098

Peters, P. C., 1964, Phys. Rev., 136, B1224

Peters, P. C., Mathews, J., 1963, Phys. Rev., 131, 435

Petterson, J. A., 1975, Phys. Rev. D, 12, 2218

Phinney, E. S., 1989, Proceedings of the 136th IAU Symposium, p.543

Pontzen, A., Governato, F., 2012, MNRAS, 421, 3464

Pounds, K. A., King, A. R., 2013, MNRAS, 433, 1369

Pounds, K. A., King, A. R., Page, K. L., O'Brien, P. T., 2003, MNRAS, 346, 1025

Pounds, K. A., Reeves, J. N., 2009, MNRAS, 397, 249

Pounds, K. A., Vaughan, S., 2011, MNRAS, 413, 1251

Power, C., Zubovas, K., Nayakshin, S., King, A. R., 2011, MNRAS, 413, L110

Pringle, J. E., 1991, MNRAS, 248, 754

Pringle, J. E., 1992, MNRAS, 258, 811

Pringle, J. E., 1997, MNRAS, 292, 136

Pringle, J. E., King, A. R., 2007, Astrophysical Flows, Cambridge University Press (AF)

Raj, A., Nixon, C. J., 2021, ApJ, 909, 82

Raj, A., Nixon, C.J., Doğan, S., 2021, ApJ, 909, 81

Rees, M. J., 1971, Nature, 229, 312

Rees, M. J., 1988, Nature, 333, 523

Reeves, J. N., Braito, V., 2019, ApJ, 884, 80

Reeves, J. N., O'Brien, P. T., Ward, M. J., 2003, ApJ, 593, L65

Ricci, C., Bauer, F. E., Treister, E. et al., 2017, MNRAS, 468, 1273

Richings, A. J., Faucher-Giguère, C. A., 2018, MNRAS, 474, 3673

Riffel, R. A., Storchi-Bergmann, T., 2011a, MNRAS, 411, 469

Riffel, R. A., Storchi-Bergmann, T., 2011b, MNRAS, 417, 2752

Risaliti, G., Elvis, M., 2004, Supermassive Black Holes in the Distant Universe, Ed. A. J. Barger, Kluwer Academic Publishers

Ritter, H., 1988, A&A, 202, 93

Rossi, E. M., Lodato, G., Armitage, P. J., Pringle, J. E., King, A. R., 2010, MNRAS, 401, 2021

Rosswog, S., Ramirez-Ruiz, E., Hix, W. R., 2009, ApJ, 695, 404

Rowther, S., Nealon, R., Farzana, M., 2021, ApJ, 2022, 925, 163

Rupke, D. S. N., Veilleux, S., 2011, ApJ, 729, L27

Sadowski, A., 2009, ApJ Supp, 183, 171

Salpeter, E. E., 1954, Aus. J. Phys., 7, 373

Salpeter, E. E., 1964, ApJ, 140, 796

Schawinski, K., et al., 2015, MNRAS, 451, 2517

Scheuer, P. A. G., Feiler, R., 1996, MNRAS, 282, 291

Schutte, Z., Reines, A., Greene, J., 2019, ApJ, 887, 245

Schwarzschild, K., 1916, Sitzungsberichte der Königlich Preussischen Akademie der Wissenschaften, 7, 189

Scott, N., Graham, A. W., 2013, ApJ, 763, 76

Sepinsky, J. F., Willems, B., Kalogera, V., 2007, ApJ, 660, 1624

Shakura, N. I., Sunyaev, R. A., 1973, A&A, 24, 337

Shappee, B. J., et al., 2013, Astronomer's Telegram, No. 5010

Sheng, Z., et al., 2021, ApJ, 920, L25

Shields, G. A., 1976, ApJ, 204, 330

Shlosman, I., Frank, J., Begelman, M., 1989, Nature, 338, 45

Shu, F., 2011, The Physics of Astrophysics (2 vols), University Science Books

Shu, X. W., Wang, S. S., Dou, L. M. et al., 2018, ApJ, 857, L16

Silk, J., Nusser, A., 2010, ApJ, 725, 556

Silk, J., Rees, M. J., 1998, A&A, 331, L1

Simon, J. B., Beckwith, K., Armitage, P. J., 2012, MNRAS, 422, 2685

Sirressi, M., Cicone, C., Severgnini, P. et al., 2019, MNRAS, 489, 1927

Soltan, A., 1982, MNRAS, 200, 115

Spaans, M., Silk, J., 2006, ApJ, 652, 902

Steiman-Cameron, T. Y., Durisen, R. H., 1984, ApJ, 276, 101

Sturm, E., González-Alfonso, E., Veilleux, S. et al., 2011, ApJ, 733, L16

Su, M., Slatyer, T. R., Finkbeiner, D. P., 2010, ApJ, 724, 1044

Tacconi, L. J., Genzel, R., Lutz, D. et al., 2002, ApJ, 580, 73

Thomas, J., Ma, C.-P, McConnell, N.J. et al., 2016, Nature, 532, 3490

Thomas, P. A., Vine, S., Pearce, F. R., 1994, MNRAS, 268, 253

Thompson, T. A., Quataert, E., Murray, N., 2005, ApJ, 630, 167

Tombesi, F., Cappi, M., Reeves, J. N. et al., 2010, A&A, 521, 57

Tombesi, F., Cappi, M., Reeves, J. N. et al., 2011, ApJ, 742, 44

Tombesi, F., Cappi, M., Reeves, J. N. et al., 2013, MNRAS, 430, 1120

Tombesi, F., Meléndez, M., Veilleux, S. et al., 2015, Nature, 519, 436

Tombesi, F., Sambruna, R. M., Reeves, J. N. et al., 2010, ApJ, 719, 700

Torkelsson, U., Ogilvie, G. I., Brandenburg, A. et al., 2000, MNRAS, 318, 47

van der Vlugt, D., Costa, T., 2019, MNRAS, 490, 4918

Veilleux, S., Rupke, D., Swaters, R., 2009, ApJ, 700, L149

Velikov, E. P., Sov. Phys. JEPT, 36, 1398
Wald, R. M., 1974, Phys. Rev. D. 10, 1680
Walker, S. A., Fabian, A. C., Russell, H. R., Sanders, J. S., 2014, MNRAS, 442, 2809
Wang, L., Berczik, P., Spurzem, R., Kouwenhoven, M. B. N., 2014, ApJ, 780, 164
Weisberg, J. M., Huang, Y., 2016, ApJ, 829, 55
Wijers, R. A. M. J., 2006, arXiv:astro-ph/0506218
Wijers, R. A. M. J., Pringle, J. E., 1999, MNRAS, 308, 207
Willott, C. J., McLure, R. J., Jarvis, M., 2003, ApJ, 587, L15
Zeldovich, Y. B., 1964, Soviet Astron., 9, 221
Zubovas, K., King, A. R., 2012a, ApJ, 745, L34
Zubovas, K., King, A. R., 2012b, MNRAS, 426, 2751
Zubovas, K., King, A. R., 2013, ApJ, 769, 51
Zubovas, K., King, A. R., 2014, MNRAS, 439, 400
Zubovas, K., King, A. R., 2019, MNRAS, 484, 1829
Zubovas, K., King, A. R., Nayakshin, S., 2011, MNRAS, 415, L21
Zubovas, K., Nayakshin, S., 2014, MNRAS, 440, 2625
Zubovas, K., Nayakshin, S., King, A. R., Wilkinson, M., 2013, MNRAS, 433, 3079

Index

Printed in the United States
by Baker & Taylor Publisher Services